Pancreas and Beta Cell Replacement

Regenerative and Transplant Medicine

Pancreas and Beta Cell Replacement

Volume 1

Volume Editor

Wayne J. Hawthorne
Department of Surgery, Westmead Hospital,
The Westmead Institute for Medical Research,
University of Sydney, Westmead,
NSW, Australia

Series Editor

Giuseppe Orlando
Department of Surgery, Section of Transplantation,
Wake Forest Institute for Regenerative Medicine,
Atrium Health Wake Forest Baptist,
Wake Forest School of Medicine,
Winston-Salem, United States

Academic Press is an imprint of Elsevier
125 London Wall, London EC2Y 5AS, United Kingdom
525 B Street, Suite 1650, San Diego, CA 92101, United States
50 Hampshire Street, 5th Floor, Cambridge, MA 02139, United States
The Boulevard, Langford Lane, Kidlington, Oxford OX5 1GB, United Kingdom

ISBN: 978-0-12-824011-3

For information on all Academic Press publications
visit our website at https://www.elsevier.com/books-and-journals

Publisher: Stacy Masucci
Acquisitions Editor: Elizabeth Brown
Editorial Project Manager: Samantha Allard
Production Project Manager: Swapna Srinivasan
Cover Designer: Greg Harris

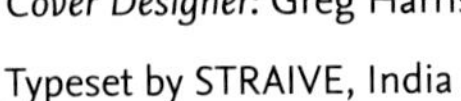

Typeset by STRAIVE, India

Contents

7. Oxygenation of the pancreas

Amy C. Kelly, Thomas M. Suszynski, and Klearchos K. Papas

8. Encapsulation devices to enhance graft survival: The latest in the development of micro and macro encapsulation devices to improve clinical, xeno, and stem cell transplantation outcomes

Thomas Loudovaris

9. Transgenic pigs for islet xenotransplantation

Peter J. Cowan

10. Xenogeneic pancreatic islet cell transplantation—Application of pig cells and techniques for clinical islet cell xenotransplantation

Jong-Min Kim, Rita Bottino, and Chung-Gyu Park

Contributors

Numbers in parentheses indicate the pages on which the authors' contributions begin.

Michael Alexander (97), Department of Surgery, University of California Irvine, Irvine, CA, United States

Amish Asthana (199), Wake Forest University School of Medicine, Winston Salem, NC, United States

Justine M. Aziz (199), Wake Forest University School of Medicine, Winston Salem, NC, United States

Appakalai N. Balamurugan (81), Center for Clinical and Translational Research, Abigail Wexner Research Institute; Department of Surgery, Nationwide Children's Hospital; Department of Pediatrics, College of Medicine, The Ohio State University, Columbus, OH, United States

Rita Bottino (167), Imagine Pharma, Devon; Allegheny Health Network, Pittsburgh, PA, United States

Lori Nicole Byers (199), Wake Forest University School of Medicine, Winston Salem, NC, United States

Antonio Citro (199), San Raffaele Diabetes Research Institute, IRCCS San Raffaele Scientific Institute, Milan, Italy

Peter J. Cowan (153), Immunology Research Centre, St Vincent's Hospital Melbourne, Fitzroy; Department of Medicine, University of Melbourne, Parkville, VIC, Australia

Kimia Damyar (97), Department of Surgery, University of California Irvine, Irvine, CA, United States

Robert A. Fisher (181), SP Global, Chantilly, VA, United States

Peter J. Friend (39), Nuffield Department of Surgical Sciences, Oxford Transplant Centre, University of Oxford, Oxford, United Kingdom

Adam Good (97), Department of Surgery, University of California Irvine, Irvine, CA, United States

Paul A. Grisales (199), Wake Forest University School of Medicine, Winston Salem, NC, United States

Wayne J. Hawthorne (1), National Pancreas and Islet Transplant Laboratories, The Westmead Institute for Medical Research; Department of Surgery, School of Medical Sciences, University of Sydney, Westmead Hospital, Westmead, NSW, Australia

Aaron Hui (63), Transplant and General Surgery, Monash Health; Transplant and General Surgery, St. Vincents Health, Melbourne, VIC Australia

David Imagawa (97), Department of Surgery, University of California Irvine, Irvine, CA, United States

Amy C. Kelly (113), University of Arizona, Department of Surgery, Institute of Cellular Transplantation, Tucson, AZ, United States

Jong-Min Kim (167), Xenotransplantation Research Center; Department of Microbiology and Immunology; Institute of Endemic Diseases; Cancer Research Institute, Seoul National University Hospital; Department of Biomedical Sciences, Seoul National University Graduate School; Biomedical Research Institute, Seoul National University Hospital, Seoul, Korea

Ahad Ahmed Kodipad (81), Center for Clinical and Translational Research, Abigail Wexner Research Institute, Nationwide Children's Hospital, Columbus, OH, United States

Jonathan R.T. Lakey (97), Department of Surgery; Department of Biomedical Engineering, University of California Irvine, Irvine, CA, United States

Thomas Loudovaris (125), Immunology and Diabetes, St Vincent's Institute for Medical Research, Fitzroy, VIC, Australia; University of Arizona, Tucson, AZ, United States

Colleen Luong (97), Department of Surgery, University of California Irvine, Irvine, CA, United States

Sri Prakash L. Mokshagundam (81), Department of Pediatrics, College of Medicine, The Ohio State University, Columbus, OH; Department of Endocrinology, University of Louisville, Louisville, KY, United States

Siddharth Narayanan (81), Center for Clinical and Translational Research, Abigail Wexner Research Institute, Nationwide Children's Hospital, Columbus, OH, United States

Jaimie D. Nathan (81), Department of Surgery, Nationwide Children's Hospital; Department of Pediatrics, College of Medicine, The Ohio State University, Columbus, OH, United States

Ann Etohan Ogbemudia (39), Nuffield Department of Surgical Sciences, Oxford Transplant Centre, University of Oxford, Oxford, United Kingdom

Gabriel Oniscu (63), University of Edinburgh; Royal Infirmary of Edinburgh, Edinburgh, Scotland

Giuseppe Orlando (209), Wake Forest University School of Medicine, Winston Salem, NC, United States

Klearchos K. Papas (113), University of Arizona, Department of Surgery, Institute of Cellular Transplantation, Tucson, AZ, United States

Chung-Gyu Park (167), Xenotransplantation Research Center; Department of Microbiology and Immunology; Institute of Endemic Diseases; Cancer Research Institute, Seoul National University Hospital; Department of Biomedical Sciences, Seoul National University Graduate School; Biomedical Research Institute, Seoul National University Hospital, Seoul, Korea

Andrea Peloso (199), Department of General and Transplantation Surgery, University of Geneva Hospitals, University of Geneva, Geneva, Switzerland

Lev T. Perelman (181), Center for Advanced Biomedical Imaging and Photonics, Division of Gastroenterology, Department of Medicine, Beth Israel Deaconess Medical Center, Harvard Medical School, Boston, MA, United States

Giuseppe Pettinato (181), Center for Advanced Biomedical Imaging and Photonics, Division of Gastroenterology, Department of Medicine, Beth Israel Deaconess Medical Center, Harvard Medical School, Boston, MA, United States

Henry Pleass (29), Department of Surgery, Westmead Hospital, Westmead; Faculty of Medicine and Health, Specialty of Surgery, University of Sydney, Sydney, NSW, Australia

Horacio Rilo (97), Department of Surgery, Zucker School of Medicine at Hofstra/Northwell, Hempstead, NY, United States

Krishna Kumar Samaga (81), Center for Clinical and Translational Research, Abigail Wexner Research Institute, Nationwide Children's Hospital, Columbus, OH, United States

John R. Savino (209), Wake Forest University School of Medicine, Winston Salem, NC, United States

Thomas M. Suszynski (113), University of Minnesota, Medical Center and M Health, Departments of Surgery (Plastic Surgery) and Orthopedic Surgery, Minneapolis, MN, United States

Andrew I. Sutherland (63), Edinburgh Transplant, Royal Infirmary of Edinburgh; University of Edinburgh, Edinburgh, Scotland

Alicia Wells (97), Department of Surgery, University of California Irvine, Irvine, CA, United States

David Whaley (97), Department of Surgery, University of California Irvine, Irvine, CA, United States

Ivana Xu (97), Department of Surgery, University of California Irvine, Irvine, CA, United States

Chapter 1

Pancreas and beta cell replacement: An overview

Wayne J. Hawthorne[a,b]

[a]National Pancreas and Islet Transplant Laboratories, The Westmead Institute for Medical Research, Westmead Hospital, Westmead, NSW, Australia,
[b]Department of Surgery, School of Medical Sciences, University of Sydney, Westmead Hospital, Westmead, NSW, Australia

Introduction

Historical perspectives of whole pancreas transplantation

The discovery of the endocrine pancreas as the cause and thus a potential treatment for diabetes began 130 years ago when von Mering and Minowski produced severe and fatal diabetes by extirpation of the canine pancreas [1]. This research provided fundamental knowledge of the role of the pancreas and provided the platform that all subsequent research into diabetes and diabetes treatments originated. As can be seen in Fig. 1, the timeline shows the development of pancreas transplantation from its initial inception to our current day successful clinical therapy, with most of the significant and impacting changes occurring over the past 130 years. Many more studies and discoveries have provided steps to improve current day outcomes; these are just some that have been landmarked over time. The remainder of this chapter describes some of the major advances seen over this time. The subsequent chapters also provide a fabulous insight into the latest technological advances that have recently occurred.

The many European researchers that followed looked to these initial experiments to base their later animal research work. Following further significant research in canine models, Minowski and Hédon were the first to develop the pancreatic autotransplantation by shifting the native pancreas to the subcutaneous space of the abdomen without disrupting the blood supply [2,3]. From this they clearly demonstrated that successful autotransplantation could prevent the development of diabetes and provide the basis for establishment of treatments, in particular pancreas transplantation as a potential option for curing patients suffering from type 1 diabetes (T1D).

Following closely the work of Minowski et al. one of the first published records of xenotransplantation of the pancreas was that of Williams, an English surgeon, who, in 1894, grafted three pieces of a pancreas obtained from a freshly slaughtered sheep into the subcutaneous tissue of a 15-year-old diabetic boy who was gravely ill with T1D. He subsequently died several days later from diabetic acidosis as a consequence of the untreated diabetes and the failure of the graft to provide any relief [4].

While there continued to be significant research into treatments for this disease, the major focus lay in the potential for transplanting the pancreas. However, in order to be able to do this, the issues of vascular anastomosis of the small vessels remained the largest obstacle to overcome. The Payr vascular anastomosis technique was a technique developed to help deal with the small vessels of the pancreas. It allowed connection of the small vessels by eversion through and over a metal cannula, allowing for easier technical anastomosis than that of Carrel's technique [5].

Realistically the first successful attempt to investigate pancreas allotransplantation began in 1913 when Hédon anastomosed the vessels of the pancreatic uncinate process into the neck vessels of a pancreatectomized dog [6]. Sadly, the graft failed to work most likely due to thrombosis and it did not provide any functional results at all. Despite this work being heralded as lifesaving, it was not followed up successfully due to the inherent technical difficulties in performing the small pancreas vascular anastomosis, and after Banting and colleagues discovered insulin in 1922 [7], research interest in pancreas transplantation slowly waned.

Ivy and Farrell reinvigorated studies into the potential of pancreas transplantation in 1926 and developed a technique for autotransplantation of the uncinate portion of the canine pancreas [8] based on the work of Minowski and Hédon [3,6]. Three to four weeks following transfer of the uncinate lobe of the pancreas to a subcutaneous site, they ligated and divided

Pancreas and Beta Cell Replacement. https://doi.org/10.1016/B978-0-12-824011-3.00011-4

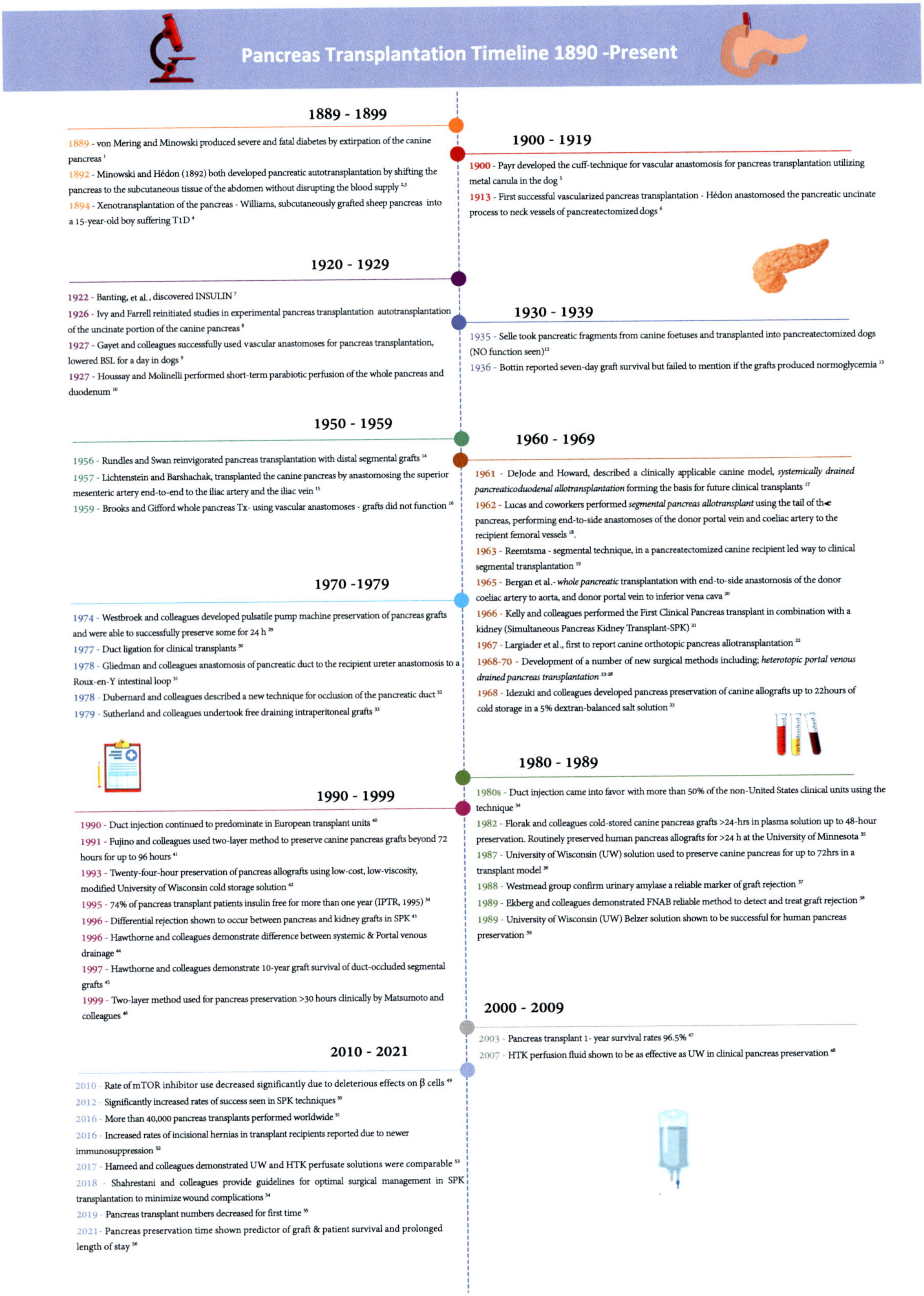

Pancreas Transplantation Timeline 1890 -Present

1889 - 1899

1889 - von Mering and Minowski produced severe and fatal diabetes by extirpation of the canine pancreas [1]

1892 - Minowski and Hédon (1892) both developed pancreatic autotransplantation by shifting the pancreas to the subcutaneous tissue of the abdomen without disrupting the blood supply [2,3]

1894 - Xenotransplantation of the pancreas - Williams, subcutaneously grafted sheep pancreas into a 15-year-old boy suffering T1D [4]

1900 - 1919

1900 - Payr developed the cuff-technique for vascular anastomosis for pancreas transplantation utilizing metal canula in the dog [5]

1913 - First successful vascularized pancreas transplantation - Hédon anastomosed the pancreatic uncinate process to neck vessels of pancreatectomized dogs [6]

1920 - 1929

1922 - Banting, et al., discovered INSULIN [7]

1926 - Ivy and Farrell reinitiated studies in experimental pancreas transplantation autotransplantation of the uncinate portion of the canine pancreas [8]

1927 - Gayet and colleagues successfully used vascular anastomoses for pancreas transplantation, lowered BSL for a day in dogs [9]

1927 - Houssay and Molinelli performed short-term parabiotic perfusion of the whole pancreas and duodenum [10]

1930 - 1939

1935 - Selle took pancreatic fragments from canine foetuses and transplanted into pancreatectomized dogs (NO function seen)[12]

1936 - Bottin reported seven-day graft survival but failed to mention if the grafts produced normoglycemia [13]

1950 - 1959

1956 - Rundles and Swan reinvigorated pancreas transplantation with distal segmental grafts [14]

1957 - Lichtenstein and Barshachak, transplanted the canine pancreas by anastomosing the superior mesenteric artery end-to-end to the iliac artery and the iliac vein [15]

1959 - Brooks and Gifford whole pancreas Tx- using vascular anastomoses - grafts did not function [16]

1960 - 1969

1961 - DeJode and Howard, described a clinically applicable canine model, *systemically drained pancreaticoduodenal allotransplantation* forming the basis for future clinical transplants [17]

1962 - Lucas and coworkers performed *segmental pancreas allotransplant* using the tail of the pancreas, performing end-to-side anastomoses of the donor portal vein and coeliac artery to the recipient femoral vessels [18].

1963 - Reemtsma - segmental technique, in a pancreatectomized canine recipient led way to clinical segmental transplantation [19]

1965 - Bergan et al.- *whole pancreatic* transplantation with end-to-side anastomosis of the donor coeliac artery to aorta, and donor portal vein to inferior vena cava [20]

1966 - Kelly and colleagues performed the First Clinical Pancreas transplant in combination with a kidney (Simultaneous Pancreas Kidney Transplant-SPK) [21]

1967 - Largiader et al., first to report canine orthotopic pancreas allotransplantation [22]

1968-70 - Development of a number of new surgical methods including; *heterotopic portal venous drained pancreas transplantation* [23-28]

1968 - Idezuki and colleagues developed pancreas preservation of canine allografts up to 22hours of cold storage in a 5% dextran-balanced salt solution [23]

1970 -1979

1974 - Westbroek and colleagues developed pulsatile pump machine preservation of pancreas grafts and were able to successfully preserve some for 24 h [29]

1977 - Duct ligation for clinical transplants [30]

1978 - Gliedman and colleagues anastomosis of pancreatic duct to the recipient ureter anastomosis to a Roux-en-Y intestinal loop [31]

1978 - Dubernard and colleagues described a new technique for occlusion of the pancreatic duct [32]

1979 - Sutherland and colleagues undertook free draining intraperitoneal grafts [33]

1980 - 1989

1980s - Duct injection came into favor with more than 50% of the non-United States clinical units using the technique [34]

1982 - Florak and colleagues cold-stored canine pancreas grafts >24-hrs in plasma solution up to 48-hour preservation. Routinely preserved human pancreas allografts for >24 h at the University of Minnesota [35]

1987 - University of Wisconsin (UW) solution used to preserve canine pancreas for up to 72hrs in a transplant model [36]

1988 - Westmead group confirm urinary amylase a reliable marker of graft rejection [37]

1989 - Ekberg and colleagues demonstrated FNAB reliable method to detect and treat graft rejection [38]

1989 - University of Wisconsin (UW) Belzer solution shown to be successful for human pancreas preservation [39]

1990 - 1999

1990 - Duct injection continued to predominate in European transplant units [40]

1991 - Fujino and colleagues used two-layer method to preserve canine pancreas grafts beyond 72 hours for up to 96 hours [41]

1993 - Twenty-four-hour preservation of pancreas allografts using low-cost, low-viscosity, modified University of Wisconsin cold storage solution [42]

1995 - 74% of pancreas transplant patients insulin free for more than one year (IPTR, 1995) [34]

1996 - Differential rejection shown to occur between pancreas and kidney grafts in SPK [43]

1996 - Hawthorne and colleagues demonstrate difference between systemic & Portal venous drainage [44]

1997 - Hawthorne and colleagues demonstrate 10-year graft survival of duct-occluded segmental grafts [45]

1999 - Two-layer method used for pancreas preservation >30 hours clinically by Matsumoto and colleagues [46]

2000 - 2009

2003 - Pancreas transplant 1- year survival rates 96.5% [47]

2007 - HTK perfusion fluid shown to be as effective as UW in clinical pancreas preservation [48]

2010 - 2021

2010 - Rate of mTOR inhibitor use decreased significantly due to deleterious effects on β cells [49]

2012 - Significantly increased rates of success seen in SPK techniques [50]

2016 - More than 40,000 pancreas transplants performed worldwide [51]

2016 - Increased rates of incisional hernias in transplant recipients reported due to newer immunosuppression [52]

2017 - Hameed and colleagues demonstrated UW and HTK perfusate solutions were comparable [53]

2018 - Shahrestani and colleagues provide guidelines for optimal surgical management in SPK transplantation to minimize wound complications [54]

2019 - Pancreas transplant numbers decreased for first time [55]

2021 - Pancreas preservation time shown predictor of graft & patient survival and prolonged length of stay [56]

FIG. 1 Timeline of the development of pancreas transplantation from its initial inception to our current day successful clinical therapy, with a significant number of the impacting changes that have occurred over the past 130 years.

the vascular pedicle. All grafts developed a collateral blood supply from the subcutaneous tissue and survived. Following closely this work, in 1927 Gayet and colleagues were the first to successfully use vascular anastomoses over cannulas for pancreas transplantation, which lowered blood sugar levels for periods of up to a day in dogs [9]. One of the first reports of pancreas perfusion was undertaken by Houssay and Molinelli [10] where they also described a short-term parabiotic perfusion of the whole pancreas and duodenum similar to that of Gayet [9]. Later, in 1929 Houssay described similar studies, resulting in normoglycemia and exocrine function in canine allografts. Their surgical techniques were based on the earlier work of Payr using the cuff technique for vascular anastomosis [11]. The French continued as the major pioneers in the field, performing pancreas transplants for physiologic studies and all groups appeared to continue to use the Payr cuff technique for their vascular anastomoses rather than the Carrel patch, developed by Carrel some twenty years earlier as a new vascular patch for use in all organs, which he later used in kidneys [12].

In 1935 Selle took pancreatic fragments from canine fetuses and from duct ligated adult glands and transplanted them into the groins of pancreatectomized dogs [13]. No graft function was demonstrated, and the grafts autolyzed within a few weeks causing significant concerns regarding their techniques. Despite this set back, in 1936 Bottin reported seven-day graft survival of pancreata transplanted into the necks of nonpancreatectomized canine recipients and followed the unmodified course of the graft death from the toxicity of rejection necrosis within eight days of the transplant; however, they failed to mention if the grafts produced normoglycemia [14].

Following a significant hiatus contributed to by the poor results seen in the earlier vascularized techniques, it was not until the late micro- and macrovascular complications of diabetes became evident and were linked to poor levels of glycemic control seen in patients on insulin, that interest in pancreas transplantation was fully renewed. With almost two decades passing, until in 1956 Rundles and Swan [15] moved the splenic tail of the pancreas with the splenic vessels into the subcutaneous abdominal tissue. After a few months they performed a residual pancreatectomy of the remnant intraabdominal pancreas and almost half of the dogs in the study didn't become diabetic. This was a major step forward demonstrating clearly that the autograft was functional, keeping the dogs normoglycemic [15].

In 1957 Lichtenstein and Barshachak transplanted the pancreas by anastomosing the superior mesenteric artery end to end to the iliac artery and the portal vein end to end to the iliac vein, in a nonpancreatectomized canine recipient. No graft material was found eight weeks posttransplant and they provided little if any insight or reference to previous research being unaware of the many years of previous successful French researchers works [16]. Two years later, Brooks and Gifford [17] allotransplanted the whole pancreas using vascular anastomoses of Carrel patches to the recipient's aorta and iliac vein along with radiation treatment of the graft, trying to dry pancreatic exocrine secretions and potentially prevent pancreatitis. Unfortunately, the grafts did not function, the primary cause of graft loss was from thrombosis, pancreatitis, and rejection which resulted in autodigestion of the graft [17]. At the same time DeJode and Howard described a clinically applicable canine technique called *systemically drained pancreaticoduodenal allotransplantation*; in this work they successfully established the precursor for the modern era of pancreas transplants utilizing the entire pancreas and including the duodenum as a conduit for exocrine drainage of the graft, forming the basis for future clinical transplants [18].

Into the current era of pancreas transplantation

In the 1960s a number of changes to techniques occurred and a number of groups commenced more clinically relevant studies. One such group was that of Lucas and coworkers, they performed the first canine *segmental pancreas allotransplant* using just the tail of the pancreas, they performed an end-to-side anastomoses of the donor portal vein and celiac artery to the recipient femoral vessels [19]. Using this much simpler technical segmental technique, the same group demonstrated two-week graft function in a canine model where the animal was pancreatectomized prior to the transplant and the graft was life supporting [20]. They also demonstrated that duct ligation of the donor's pancreas, prior to grafting, resulted in better endocrine function. Despite showing initial graft function, no immunosuppression was used, and they were unable to prove long-term function. However, despite these issues, they were successful in getting the technique to clearly work, and as such this method was to become the basis for later clinical segmental pancreas transplants to be adopted clinically for some time.

In the mid-1960s, several groups attempted to improve canine pancreas graft survival by performing duct ligation several weeks prior to transplantation [20–22]. Once again Lucas and coworkers performed *segmental pancreas allotransplant* using the tail of the pancreas, performing end-to-side anastomoses of the donor portal vein and celiac artery to the recipient femoral vessels [19]. However, this was neither successful nor clinically applicable. Interestingly, the earlier work of DeJode and Howard described a more clinically applicable canine model, *systemically drained pancreaticoduodenal allotransplantation* based on earlier work they had performed [18]. This allowed for both endocrine and exocrine function of the graft without the need for duct ligation. They transplanted the body of the pancreas and a segment of attached duodenum by anastomosing the donor portal vein and celiac artery to the recipient femoral vessels. The distal end of the donor duodenum was formed into a duodenostomy allowing exocrine drainage. Following on from this work, Bergan et al. [21] reported an intraabdominal method for *whole pancreatic* transplantation with end-to-side anastomosis of the donor celiac artery to aorta, and donor portal vein to inferior vena cava which was quite successful and provided data to demonstrate that whole pancreas grafting with vascular anastomosis was possible. Kelly and associates developed a canine model of partial pancreatic grafting. These grafts were extremely successful and supported Alloxan-induced diabetic dogs for several months. This work was to become the basis for the technique used for their first clinical transplant. *Clinical pancreas transplantation* began in December 1966 when Kelly and his associates performed a partial pancreas transplant combined with a renal transplant, this marked the commencement of the term Simultaneous Pancreas and Kidney transplant, to be abbreviated to SPK from then on. However, they did not transplant the whole pancreas; rather they removed the head of a pancreas and ligated the main duct. They then transplanted it retroperitoneally into the left iliac fossa of a 28-year-old woman who had suffered from diabetes for over 18 years [23].

The graft functioned for almost a week before exogenous insulin was once again required. Unfortunately, the patient developed pancreatitis and pancreatic fistula destroying the islet's function, necessitating the removal of the pancreas graft and approximately two months after transplantation the kidney was also removed due to vascular complications and the patient died shortly after [23]. Kelly and his colleagues continued these landmark clinical transplants, but the following ten pancreas transplants were performed as whole pancreaticoduodenal allografts with duodenostomy in four and the remainder as Roux-en-Y exocrine drainage. Nine of these patients also received a simultaneously transplanted kidney.

In the same year, Largiader and colleagues were the first to report canine orthotopic pancreas allotransplantation [24]. The pancreaticoduodenectomized dogs were transplanted with an end-to-side anastomosis between the donor pancreas celiac artery and suprarenal abdominal aorta, and another between the donor portal vein and the suprarenal inferior vena cava. Gastrointestinal drainage was restored using the donor duodenal segment in a Roux-en-Y fashion. Grafts functioned for up to 9 days [24].

With the advent of clinical transplants came several other important issues such as the need for better organ preservation techniques and the development of perfusion fluids and devices to aid perfusion. In 1968 Idezuki and colleagues developed pancreas preservation of canine allografts which they preserved for up to 22 h in hyperbaric chamber storage in a 5% dextran-balanced salt solution [25]. These experiments continued, but so did the debate over the most appropriate method for exocrine and endocrine vascular drainage of pancreas grafts. So between 1968 and 1970 a great deal of research was performed by Idezuki and other members of Lillehei's group, developing a number of canine pancreas transplant techniques which were then translated to clinical trials. They described a number of methods, including orthotopic interposition, intraabdominal heterotopic, orthotopic, and *heterotopic portal venous drained pancreas transplantation*. Portally drained transplants were performed using donor celiac artery end-to-side to recipient infrarenal aorta and an end-to-side venous anastomosis between the donor portal vein and the recipient superior mesenteric vein. Heterotopic drainage of the graft duodenum was to the jejunum of the recipient in an end-to-side fashion [25–30].

Despite the surgical techniques used for vascular anastomosis, most techniques require the use of the arterial and venous extension grafts. As can be seen in Fig. 2A, the pancreas graft has its vessels transected with liver retrieval, requiring the reconstruction of the arterial and venous blood vessels. Here you can see the use of the donor iliac vessels for extending and rejoining the donor pancreas vessels. Fig. 2A shows the donor pancreas at retrieval having had the splenic vein segment taken preferentially for the liver along with the segment of the splenic artery that has to be reconstructed. As can be seen in Fig. 2B and C, the pancreas vessels are reconstructed utilizing the donor iliac vessels to have extension grafts for both the artery and vein allowing technically easier vascular anastomosis.

Not only did the work on surgical technique continue but so did the studies into best methods of preservation, with Westbroek and colleagues performing studies into pulsatile pump machine preservation of the pancreas [31]. They were able to successfully preserve some pancreata for up to 24 h with limited success, but at that time machine preservation longer than 24 h could not be achieved.

With new techniques developed in animal models, a number of clinical transplants were performed in the next few years, including attempts at portal venous drainage. No new techniques were described in the interceding period, until in 1978 when Dubernard and colleagues described a new technique for occlusion of the pancreatic duct [32]. This new technique provided another way to transplant the graft without exocrine drainage and was later utilized in occlusion of the pancreatic duct for chronic pancreatitis [33]. During the 1970s, attempts to use pancreaticoduodenal transplantation were abandoned in favor of segmental pancreas transplantation, utilizing a number of techniques for exocrine drainage. These included duct ligation [34], anastomosis to the recipient ureter, and anastomosis to a Roux-en-Y intestinal loop [35], complete ductal obliteration with neoprene glue [32,35], and free draining intraperitoneal grafts [36]. Duct injection came into favor due to the ease of use and limited complications, and over the 1980s, it became widely used, with more than 50% of the non-US clinical units using the technique [37], a trend which dominated into the early 1990s at some European centers [38].

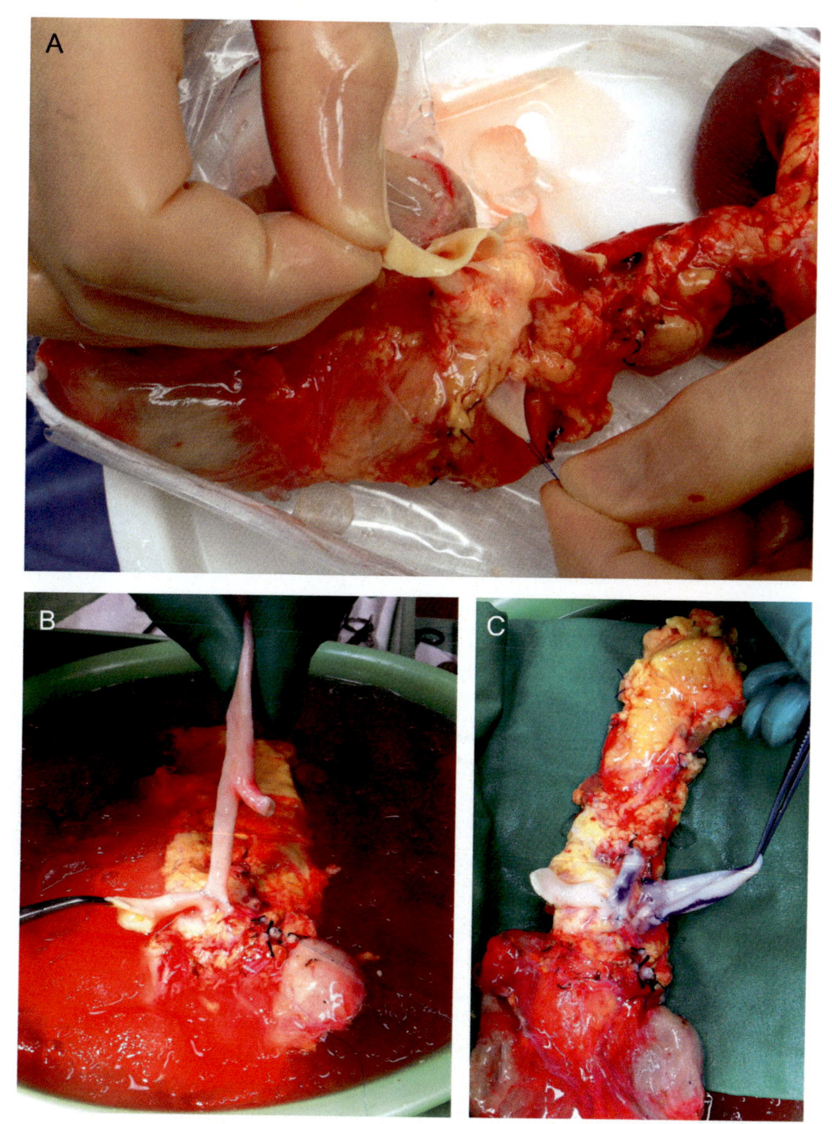

FIG. 2 (A) Donor pancreas in iced slush on the back table about to have the blood vessels reconstructed. The arterial patch held is the superior mesenteric artery which branches to the inferior pancreaticoduodenal artery to supply the head of the pancreas. The vein being held with a suture is the remnant of the portal/splenic vein that had been transected to provide the larger length of the vessel for the portal vein to the donor liver. As can be seen in (B, C), the pancreas vessels are reconstructed utilizing the donor iliac vessels. (B) Reconstruction underway utilizing the donor iliac artery anastomosed to the superior mesenteric artery bridging to the splenic artery. (C) Completed arterial reconstruction and the venous extension graft of the iliac vein to the splenic vein to replace the transected portal vein.

However, bladder drainage remained the mainstay of exocrine secretion management throughout the 1980s, especially in the majority of US units. With the increasing numbers of pancreas transplants performed, there was a subsequent increased need for enteric conversion to approximately 15%–30% of bladder-drained pancreas transplants [38]. Again, along with the surgical techniques, came changes to preservation techniques and research into the best preservation techniques and fluids. In 1982 Florak and colleagues showed that canine pancreas grafts could be cold stored for greater than 24h by using a silica gel-filtered plasma solution; they achieved up to 48-h preservation. In turn, they utilized this solution to routinely preserve human pancreas allografts for >24h at the University of Minnesota [39].

Continuous research into preservation solutions continued and the filtered plasma was soon replaced by a man-made hyperosmolar solution developed at the University of Wisconsin (UW) by Belzer and his colleagues in the late 1980s [40]. Prior to its use in the clinic, UW solution was first shown to be effective for preservation for up to 72h by Wahlberg in a canine pancreas transplant model [41]. Few if any preservation fluids have come close to the success seen with UW solution and its widespread acceptance now the mainstay of many transplant units around the world.

The various models of preservation or transplantation of the pancreas graft were not the only techniques evolving over these years, but many groups also evaluated how best to detect and treat graft rejection. The Westmead group was at the forefront of development of methods of graft rejection using both measurements of urinary amylase and also those of fine-needle aspiration under ultrasound guidance. Prieto and colleagues from Najarian's group in Minnesota first investigated the potential role of urinary amylase as a marker of graft function in a canine model of pancreas transplantation [42]. In 1986 they published a paper indicating that a decline in pancreatic amylase drained into the bladder preceded hyperglycemia as a potential marker of graft rejection [42].

These studies were confirmed by the Westmead group publishing their results of multiple bladder-drained canine pancreas transplants and the strong correlation of a rapid drop in urinary amylase as a more reliable method of detection for pancreas graft rejection than that of systemic blood sugar levels [43]. They also demonstrated that fine-needle aspiration biopsy (FNAB) was even more reliable as a means to detect and reverse rejection when the diagnosis was based on FNAB compared to a fall in urinary amylase [44]. Fasting urinary amylase levels were measured daily and percutaneous FNAB with ultrasound guidance was performed three times weekly. The diagnosis of rejection was made in alternate dogs with either a fall of urinary amylase to a level of less than 5000 IU/liter (median 7 days) or when the total corrected increment of aspirated infiltrating cells was greater than 2.6 (median 5 days). The earlier diagnosis of pancreas rejection by FNAB improved the ability of conventional antirejection therapy to reverse pancreas allograft rejection and significantly improved allograft survival [44].

In the early 1990s, a number of other methods of perfusion and perfusion media were investigated apart from the UW perfusion fluid. In particular, the group from Kobe, Japan, demonstrated impressive results by using a two-layer method with UW solution and a second layer of oxygenated perfluorochemical. They were able to preserve canine pancreas grafts beyond 72 h, the longest any other group had been able to achieve and yet with this new technique they were able to show survival for up to 96 h [45].

Due to these initial reported results the two-layer method was studied extensively in animal models by many groups and was used for pancreas preservation beyond 30 h by a number of groups, including its use clinically by Matsumoto and colleagues [46].

Around the same time as this work in Japan, the Australian group had just published data from their twenty-four-hour preservation of pancreas allografts using low-cost, low-viscosity, modified University of Wisconsin cold storage solution. [47] Using their own variation of the UW solution developed in Melbourne by Professor Vernon Marshall, the group demonstrated a cheaper option than the commercially available UW perfusion fluid that was as effective for preservation of canine pancreata for up to 24 h.

Another perfusion fluid was also investigated, initially, by the Europeans with the development of Histidine–Tryptophan–Ketoglutarate, now commonly known as HTK solution (its trade name being Custodiol). A number of studies commencing in the early 1990s investigated its specific use for multiorgan donor retrieval and for pancreas preservation. Some of the first studies were in porcine models, including studies from Leonhardt and colleagues where they directly compared its performance with other perfusion fluids such as Euro-Collins [48]. The pancreata were stored at four degrees for up to 24 h, following assessment of organ quality in a reperfusion chamber measuring physiological and biomedical parameters. They found that HTK significantly improved pancreas preservation after 24 h of ischemia, including the lactate content in the reperfusate fluid being lower, the arteriovenous flow rate was higher, and the pancreatic oxygen consumption was increased. Consequently, they concluded that pancreas preservation could be improved in vitro by protection with HTK solution [48]. This was later repeated and found to be the same in an autotransplant model in the pig [49].

Almost a decade later, Becker et al. demonstrated in retrospective analysis, in clinical data from 95 primary SPK transplants that were stratified into HTK ($n=48$) or UW ($n=47$) perfused groups. Patient/graft survival and early graft function were compared. No significant differences between 1-, 3-, and 12-month patient survival (HTK: 97.9%, 97.9%, and 95.7% vs. UW: 95.7%, 89.4%, and 89.4%, respectively) and pancreas graft survival (HTK: 87.5%, 87.5%, and 85.4% vs. UW: 87.0%, 82.6%, and 82.6%, respectively) were detected. Higher values for peak lipase were observed on day 1 in the HTK group (not reaching significance: $P=0.131$) whereas no differences were noted for amylase nor C-reactive protein [50].

More recently, Hameed and colleagues demonstrated in a systematic review and meta-analysis that HTK is comparable to UW perfusion fluid. Importantly, they found both solutions are safe for pancreas preservation, but successful pancreas transplantation depends on many factors such as donor and recipient factors, but skilled organ procurement techniques, organ preservation, and transplant experience in this field are mandatory. UW cold perfusion reduces peak serum lipase, but no quality evidence suggested UW cold perfusion improved graft survival and reduces thrombosis rates. Further research is needed to establish longer-term graft outcomes, the comparative efficacy of Celsior, and ideal perfusion volumes. The choice of organ preservation solution is only one point in this context and obviously further studies need to be undertaken to see any further potential differences in outcomes.

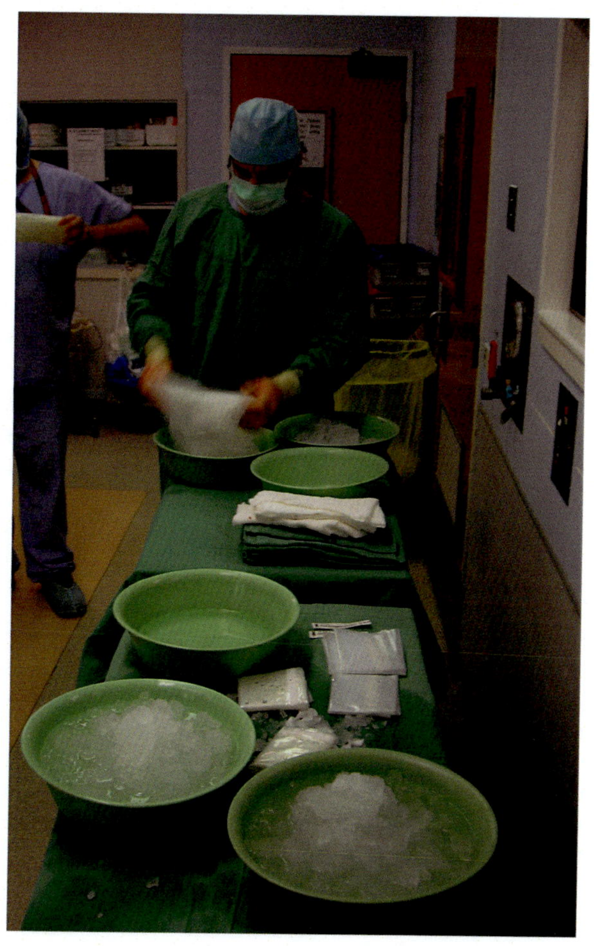

FIG. 3 Traditional back table preparation by the surgical team using iced slush as the mainstay of cooling the organ for many decades. Along with the static iced slush, UW cold perfusion is also prepared; note the multiple surgical bowls prepared for all organs including pancreas, kidneys, liver, heart, and lungs. One bowl of iced slush was traditionally used to immediately tip into the donor abdomen at cross-clamp to quench cool the organs to ~4 degrees Celsius along with the perfusion.

Fig. 3 shows the traditional back table preparation by the organ donor retrieval team using iced slush as the mainstay of cooling the organ for many decades. Along with the static iced slush, UW cold perfusion reduces peak serum lipase, but no quality evidence suggested UW cold perfusion improved graft survival and reduces thrombosis rates. Further research is needed to establish longer-term graft outcomes, the comparative efficacy of Celsior, and ideal perfusion volumes. The choice of organ preservation solution is only one point in this context and obviously further studies need to be undertaken to see any further potential differences in outcomes [51].

Most recently, Mei and colleagues have demonstrated that pancreas preservation time affects length of hospital stay after pancreas transplantation. They evaluated the results of 18,099 pancreas transplants. They found pancreas preservation time > 20 h had a significantly higher risk of graft failure than 8–12 h. Pancreas preservation time was not significantly associated with patient survival. Compared with 8 to 12 h, pancreas preservation time > 12 h had a significantly higher prolonged length of stay (PLOS) risk, increased with increased pancreas preservation time. In SPK, they also found that pancreas preservation time was positively associated with PLOS risk > 12 h. Pancreas preservation time is a sensitive predictor of PLOS. As such, we should be aiming to minimize pancreas preservation time and looking toward improvements to preservation techniques [52].

Many other issues were also investigated over the years with such things as investigation of rejection. At that time due to poor understanding of rejection and inadequate immunosuppression, the results of SPK could not be matched by pancreas transplantation alone (PTA), in part because an independent diagnosis of pancreas graft rejection being difficult. The relationship between rejection of the pancreas and rejection of the kidney was poorly understood, and it was not known whether simultaneous transplantation of both organs conferred true protection to either graft. During 1995–96, Hawthorne and colleagues performed some landmark studies that demonstrated clearly that differential rejection between simultaneously transplanted organs could occur between pancreas and kidney grafts in SPK transplants [53]. To study these questions, reliable canine allotransplant models of kidney transplantation alone (KTA), PTA, and SPK were established. Sixty-seven mongrel dogs received KTA ($n=21$), PTA ($n=23$), or SPK ($n=23$) with either no immunosuppression, low-dose cyclosporine (CsA)-based immunosuppression, or high-dose CsA-based immunosuppression. Needle core biopsy and fine-needle aspiration biopsy (FNAB) were performed at 0, 2, 4, 7, 9, 11, 14, 21, and 30 days or at the time of graft failure. Synchronous rejection was seen to occur in 73% of immunosuppressed SPK biopsies. Kidney-only rejection occurred in 23% of biopsies and pancreas-only rejection occurred in only 3% after SPK. All markers of pancreas graft rejection were poor, with the most sensitive being needle core biopsy of the simultaneously transplanted kidney. Clearly, they found a number of remarkable findings that recipients of SPK required more immunosuppression than recipients of PTA. Markers of kidney rejection were the most sensitive indicators of pancreas rejection, and independent pancreas rejection was uncommon after SPK.

As it is today, in the early 1990s, glucose homeostasis after clinical pancreas transplantation was rather complex, with the relative effect of systemic versus portal delivery of insulin remaining unresolved. In the following chapter, there is significant insight into newer findings of more recent transplant cohorts. However, the Westmead group undertook investigation of how best to demonstrate the metabolic difference between systemic and portal venous drainage [54]. In a group of thirty-two pancreatectomized dogs, they performed either systemic venous drainage (SVD) with bladder exocrine drainage ($n=16$) or portal venous drainage (PVD) with gastric exocrine drainage ($n=16$). Cyclosporine (CsA)-based immunosuppression was commenced on day -7. The effect of immunosuppression was a significant increase in fasting blood glucose level (FBGL); fasting insulin, AUC for insulin, and K values all decreased significantly. FBGL and K values remained abnormal after transplantation with no significant difference seen between SVD and PVD. However, fasting insulin became significantly lower after PVD and AUC insulin fell in both groups. CsA levels fell in both groups after transplantation, mirroring the fall in AUC insulin, and implicating CsA as a major cause of peripheral resistance to insulin. The overall outcome was that PVD did not demonstrate a significant advantage over SVD in handling an intravenous glucose challenge and as such established that both ways were effective at providing glucose homeostasis.

Many studies were also undertaken to look at the then current insulin therapies for control of glucose metabolism in patients with type I diabetes mellitus and whether they could prevent the major metabolic consequences of insulin deficiency. At that time, no therapy or regimen of insulin treatment was shown to prevent or arrest long-term complications.

After a long-term study Hawthorne and colleagues demonstrated 10-year graft survival of duct-occluded segmental grafts, and more importantly these studies demonstrated that pancreas grafts could prevent the development of the long-term microvascular complications of diabetes [55]. They demonstrated this in an experimental model of canine diabetes, using a segmental duct-occluded pancreas autograft. Long-term survival was achieved in 14 dogs for up to 5 years and in 3 dogs for 3–5 years. Glycosylated hemoglobin levels remained within normal limits over the entire study period. Fundus photography and fluorescein angiography demonstrated the absence of retinal vascular aneurisms, capillary leakage, and obliteration. Nerve conduction was normal, and histology of nerves revealed normal pathology. Renal histology also revealed no evidence of nephropathy with normal glomerular basement membranes. They clearly were able to demonstrate that pancreatic autografts are capable of providing satisfactory metabolic control for up to 5 years, thereby preventing development of the long-term microvascular complications of diabetes.

Current success rates for pancreas transplantation

Pancreas transplantation, largely in the form of SPK transplantation, has now become mainstream therapy at many transplant programs around the world, with increasing numbers being performed annually in many countries around the world; throughout the 1990s significant numbers of SPK were performed with almost a thousand transplants performed per year. With significant improvements to graft survival rates to be almost 85% at 1 year and >60% at 5 years, the initial poor results dramatically changed. The estimated half-life of a pancreas graft also increased to 7–14 years [56]. Despite such significant advances and increased survival rates, there still remained over the years significant issues with graft survival and patient morbidity.

The major issue with pancreas transplantation during this time remained in the immediate posttransplant period. Graft loss due to thrombosis remained a major issue that was and still is hard to resolve and results in take back and removal of the thrombosed organ in theaters. As can be seen in Fig. 4, this pancreas was lost due to vascular thrombosis requiring surgical removal. It remains a very uncommon event these days but nonetheless a major issue remaining for transplant surgeons. The estimated half-life of a pancreas graft at that time also increased to 7–14 years [56]. Despite such significant advances and increased survival rates, there remained over the years significant issues with graft survival and patient morbidity.

Areas that have always remained an issue are the effects of immunosuppression on wound healing and wound breakdowns, including herniation or even worse dehiscence of the wound. In 2018 Shahrestani and colleagues provided guidelines

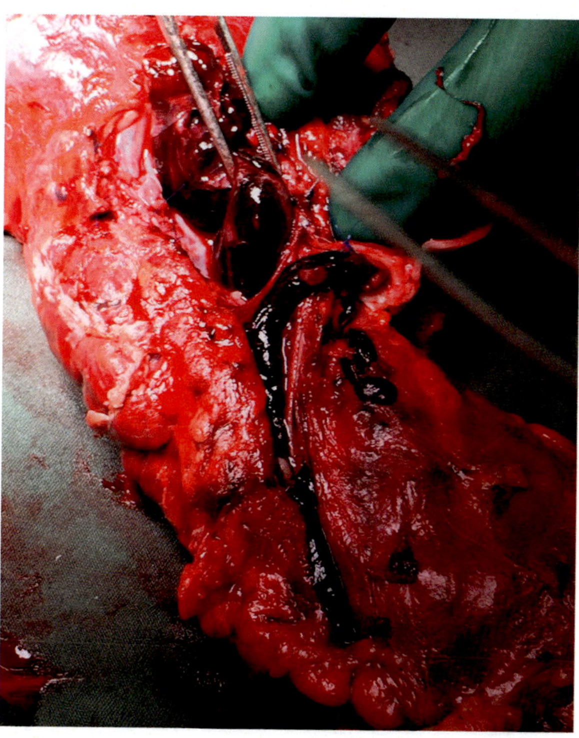

FIG. 4 Is a photo of a pancreas lost due to vascular thrombosis requiring surgical removal several days posttransplant. A very uncommon event these days but nonetheless a major issue remaining for transplant surgeons requiring reoperation and immediate graft removal.

for optimal surgical management in SPK transplantation in order to minimize wound complications [57]. The various changes over the past 50 or so years have seen a lot of change in the field of pancreas transplantation. These changes have included the development of the many different surgical techniques as described previously in this chapter, along with investigation of the immunological reactions that occur following transplantation, HLA matching, and the introduction of new and novel immunosuppressive regimen [58]. Over this period significant research also resulted in a change to how we best utilize immunosuppression, and mTOR inhibitor use was decreased significantly due to the deleterious effects seen on β cells [59]. One of the other major underlying reasons for associated issues was the immunosuppression with increased rates of incisional hernias seen in transplant recipients reported due to newer immunosuppression including the use of Sirolimus [60].

We have also seen significant advances in organ preservation, including the development of new preservation solutions and the resurgence of machine perfusion as outlined in detail in the later chapters of this book. Improvements have also come from recipients being far better managed prior to their transplant, maintaining better glycemic control, and suffering less complications than when pancreas transplants were first performed [61].

All of these changes have helped to improve the initially poor results that occurred in the very first series of pancreas transplants. Now half a century later we can see impressive results achieved worldwide with more than 40,000 pancreas transplants being performed [62]. Overall transplant survival rates have been seen to increase significantly with outcomes of graft survival also improving steadily over the past few decades, such that by the early 2000s pancreas transplant alone 1- and 4-year graft survival rates improved to be 96.5% and 85.2%, respectively [63].

However, once again we have seen some slight changes to transplantation rates and for the first time in decades, we have seen the overall number of pancreas transplants alone decrease slightly, from 1027 in 2018 to 1015 in 2019. But the number of SPK transplants increased in 2019, with a corresponding drop in pancreas-after-kidney transplants (PAKs) and pancreas transplants alone (PTAs). Both short- and long-term outcomes, including patient survival, kidney graft survival, and acute rejection-free graft survival, have shown consistent improvement over the last decades [64].

Continued research may be able to improve the results of the past three decades by developing better surgical techniques, ways of improving glucose handling, and means for detecting and preventing rejection. Ultimately, the aim of pancreas transplantation should continue to be improvement of the quality of life and amelioration of the primary and secondary complications of diabetes by establishing euglycemia [65].

Historical perspectives of islet cell transplantation

Akin to whole pancreas transplantation the original concept of pancreatic tissue or cell transplantation was proposed some 120 years ago by the Russian clinician scientist Dr. Leonid Ssobolew in Saint Petersburg around 1902 [66]. With this concept, he changed mainstream thinking that in fact there conceptually may be a way to perform cellular or tissue transplantation to treat a disease such as T1D. This propagated a number of ideas which stimulated many investigators to undertake numerous strategies which developed over time, until eventually we undertook islet transplantation for patients with T1D. The first such concept of allogeneic islet cell transplantation occurred more than a decade later in 1916 with Pybus and colleagues proposing a crude form of pancreatic fragment transplantation, a forerunner to our modern day islet cell isolation process and transplantation [67]. With the game changing discovery of exogenous insulin by Banting and Best, the next decade led to the processing of pancreatic tissue for extraction of crude insulin for direct treatment of patients rather than moving the process of cell transplantation forward. However, this in itself made significant changes in how patients suffering T1D were treated, and instead of dying they were now cured of their diabetes [67]. As can be seen in Fig. 5, the timeline shows the development of islet cell transplantation from its initial inception to our current day successful clinical therapy, with most of the significant and impacting changes occurring over the past 120 years. Many more studies and discoveries have provided steps to improve current day outcomes; these are just some that have been landmarked over time.

Some of the most basic and yet integral discoveries to issues such as the identification of the three granular cells (A, B, and D cells) of the pancreatic islet were undertaken using the then technologies and yet basic were still extraordinary achievements. Bloom described the three cells which were on the basis of granularity and differential coloration of cytoplasmic granules in 41 species [68].

Along this journey, it took another four decades before newer technologies were developed, such as the use of enzymatic digestion of the pancreatic tissue with collagenase for separation of islets from the acinar tissues, a major step in the overall process of separation of islets from the pancreas [69]. In 1965 Moskalewski undertook studies to look at how best to isolate and culture islets, this included the development of collagenase digestion for separation of islets in the Guinea Pig [70]. A few years later a variation on this theme was developed by Paul Lacy and colleagues in 1967 where they described the first use of predistension of the pancreas with collagenase for the isolation of intact islets of Langerhans from the rat pancreas [71].

There were obviously numerous other minor steps and other staged processes such as changes to density gradients for islet separation using agents such as Ficoll following initial use of agents such as sucrose. Many of these steps took years to develop and were based upon significant research, for example, demonstrating that Ficoll provided a more physiologically osmotic environment for the islets and provided a new medium for better separation of islets from the acinar tissues. Such steps provided the measures required to move isolation processes to elicit more successful outcomes [72].

Significant advances also came when a number of studies looked at the more technical challenges as to how and where islets were best transplanted, with evaluation of sites such as the intraperitoneal, subcapsular, intrasplenic, and intrahepatic sites. These investigations gave rise to a number of methods and sites of transplantation used in both research and clinical practice still currently in use [73–76].

Other integral associated aspects of islet cell transplantation, such as the cross matching of the donor and recipient, also played a role in the overall improvement to outcomes. Once originally only reliant upon the most basic ABO blood group matching, newer technologies such as the Mixed Lymphocyte Reaction (MLR) were then adopted, in addition to the use of newer immunosuppressive agents. With the advent of advances in the development of antibodies and technologies, these have further advanced to now very specific organ donor recipient cross matching and newer generation immunosuppressives [77]. These precursor studies mirrored later success in the changes to immunosuppressive protocols that lead to significantly improved clinical long-term patient outcomes, though even with small steps being undertaken there were still dramatic improvements to the overall processes. However, before comprehensive improvements were achieved, a number of other significant cofactors were required for improvement to overall outcomes. These included the role of various agents used in the processing of tissue such as the media used and simple aspects such as the evaluation and changes to pH, potassium concentration, and temperature of the isolation process [78,79]. These various changes ultimately lead to the commencement of the first clinical trials using allogeneic islets in the treatment of patients suffering from type I diabetes and severe hypoglycemic unawareness. Despite major improvements to the organ donor procedures, the processing and culture of islets, and the transplantation of the islets, the islets in this study were ultimately lost from rejection due to inadequate immunosuppression [80].

A number of groups also pursued the role of autotransplantation as a means for both performing research into refining aspects of the islet isolation process as well as also being able to provide an alternate means to treat the group of patients suffering from chronic pancreatitis [81]. The exploration of the many aspects that affect the isolation process, specifically

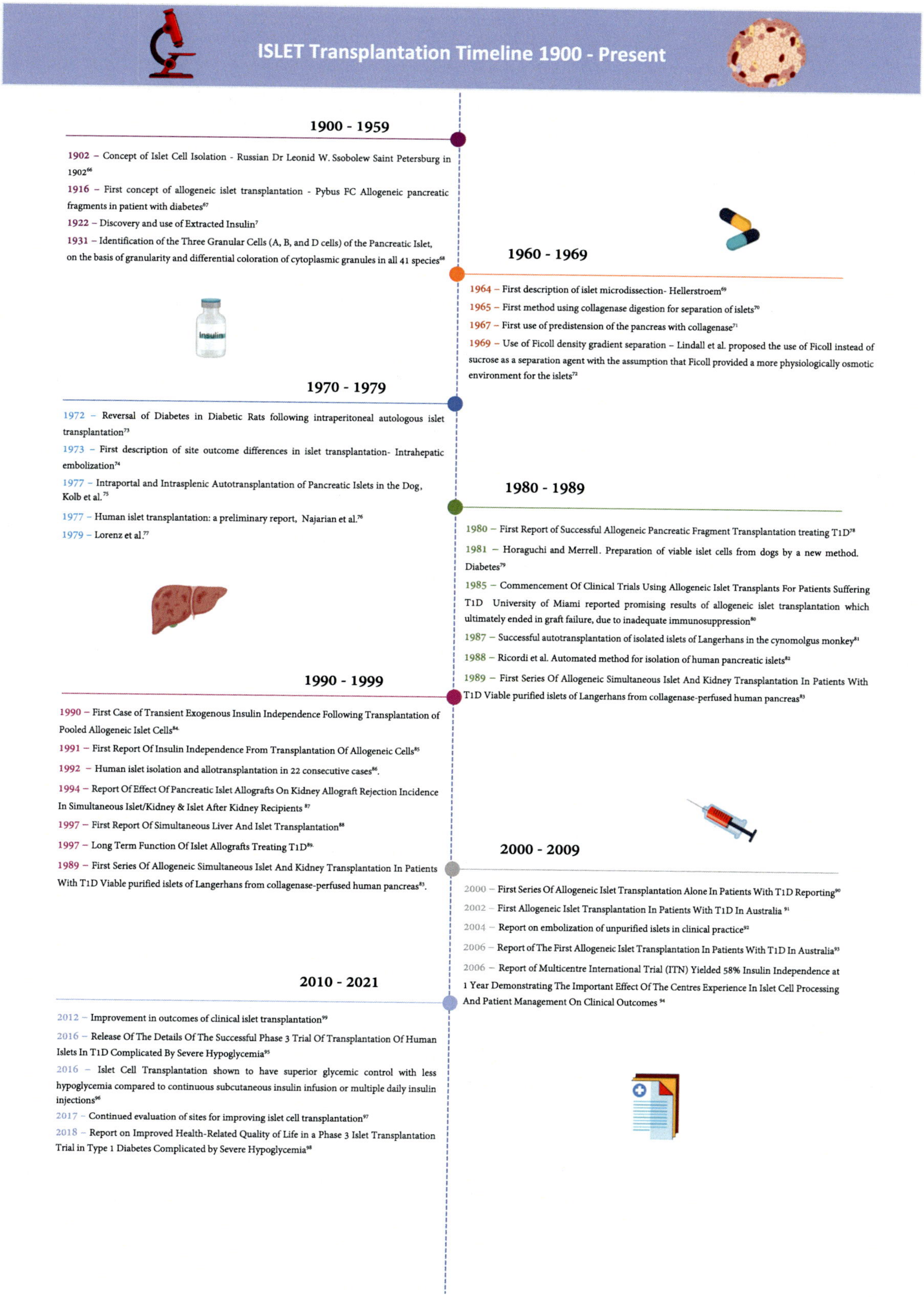

ISLET Transplantation Timeline 1900 - Present

Insulin

1900 - 1959

1902 – Concept of Islet Cell Isolation - Russian Dr Leonid W. Ssobolew Saint Petersburg in 1902[66]

1916 – First concept of allogeneic islet transplantation - Pybus FC Allogeneic pancreatic fragments in patient with diabetes[67]

1922 – Discovery and use of Extracted Insulin[7]

1931 – Identification of the Three Granular Cells (A, B, and D cells) of the Pancreatic Islet, on the basis of granularity and differential coloration of cytoplasmic granules in all 41 species[68]

1960 - 1969

1964 – First description of islet microdissection- Hellerstroem[69]

1965 – First method using collagenase digestion for separation of islets[70]

1967 – First use of predistension of the pancreas with collagenase[71]

1969 – Use of Ficoll density gradient separation – Lindall et al. proposed the use of Ficoll instead of sucrose as a separation agent with the assumption that Ficoll provided a more physiologically osmotic environment for the islets[72]

1970 - 1979

1972 – Reversal of Diabetes in Diabetic Rats following intraperitoneal autologous islet transplantation[73]

1973 – First description of site outcome differences in islet transplantation- Intrahepatic embolization[74]

1977 – Intraportal and Intrasplenic Autotransplantation of Pancreatic Islets in the Dog, Kolb et al.[75]

1977 – Human islet transplantation: a preliminary report, Najarian et al.[76]

1979 – Lorenz et al.[77]

1980 - 1989

1980 – First Report of Successful Allogeneic Pancreatic Fragment Transplantation treating T1D[78]

1981 – Horaguchi and Merrell. Preparation of viable islet cells from dogs by a new method. Diabetes[79]

1985 – Commencement Of Clinical Trials Using Allogeneic Islet Transplants For Patients Suffering T1D University of Miami reported promising results of allogeneic islet transplantation which ultimately ended in graft failure, due to inadequate immunosuppression[80]

1987 – Successful autotransplantation of isolated islets of Langerhans in the cynomolgus monkey[81]

1988 – Ricordi et al. Automated method for isolation of human pancreatic islets[82]

1989 – First Series Of Allogeneic Simultaneous Islet And Kidney Transplantation In Patients With T1D Viable purified islets of Langerhans from collagenase-perfused human pancreas[83]

1990 - 1999

1990 – First Case of Transient Exogenous Insulin Independence Following Transplantation of Pooled Allogeneic Islet Cells[84].

1991 – First Report Of Insulin Independence From Transplantation Of Allogeneic Cells[85]

1992 – Human islet isolation and allotransplantation in 22 consecutive cases[86].

1994 – Report Of Effect Of Pancreatic Islet Allografts On Kidney Allograft Rejection Incidence In Simultaneous Islet/Kidney & Islet After Kidney Recipients [87]

1997 – First Report Of Simultaneous Liver And Islet Transplantation[88]

1997 – Long Term Function Of Islet Allografts Treating T1D[89].

1989 – First Series Of Allogeneic Simultaneous Islet And Kidney Transplantation In Patients With T1D Viable purified islets of Langerhans from collagenase-perfused human pancreas[83].

2000 - 2009

2000 – First Series Of Allogeneic Islet Transplantation Alone In Patients With T1D Reporting[90]

2002 – First Allogeneic Islet Transplantation In Patients With T1D In Australia [91]

2004 – Report on embolization of unpurified islets in clinical practice[92]

2006 – Report of The First Allogeneic Islet Transplantation In Patients With T1D In Australia[93]

2006 – Report of Multicentre International Trial (ITN) Yielded 58% Insulin Independence at 1 Year Demonstrating The Important Effect Of The Centres Experience In Islet Cell Processing And Patient Management On Clinical Outcomes [94]

2010 - 2021

2012 – Improvement in outcomes of clinical islet transplantation[99]

2016 – Release Of The Details Of The Successful Phase 3 Trial Of Transplantation Of Human Islets In T1D Complicated By Severe Hypoglycemia[95]

2016 – Islet Cell Transplantation shown to have superior glycemic control with less hypoglycemia compared to continuous subcutaneous insulin infusion or multiple daily insulin injections[96]

2017 – Continued evaluation of sites for improving islet cell transplantation[97]

2018 – Report on Improved Health-Related Quality of Life in a Phase 3 Islet Transplantation Trial in Type 1 Diabetes Complicated by Severe Hypoglycemia[98]

FIG. 5 Timeline of the development of islet cell isolation and subsequent successful transplantation from the bench to the bedside showing over more than a century of research leading to its eventual clinical success and acceptance as a mainstream clinical therapy.

how best to dissociate the acinar and islet milieu, was a major focus of many groups as it was and still remains a major issue in the overall isolation process. The publication that popularized the use of more automated and systematic processes to streamline the isolation process was the same one that also has become synonymous for some of the equipment used in this process: the 1988 publication from Ricordi et al. [82] The optimization of the digestion and dissociation process was obviously a game changer and the field was able to move forward with significant improvements to the various clinical trials being undertaken. Some included combined transplantation along with other organs as simultaneous procedures where the pancreas was taken for processing while the other organs were transplanted. These initial islet cases were performed in combination with kidney and liver transplants [83,84]. The early 1990s saw major steps forward in the success of allotransplantation following the use of the newer isolation methods, use of culture and pooling of islets along with the exploration of the sites for and combinations of transplants. From these advances and adoption of others in the early 1990s we saw the first case of transient exogenous insulin independence following transplantation of pooled allogeneic islet cells being reported along with the first report of insulin independence from transplantation of allogeneic cells [84,85]. Then shortly after, in 1992 we saw the publication of results of 22 consecutive cases of islet cell transplantation at a single center, and soon thereafter further successful simultaneous transplants performed as combined transplants of kidneys and islets [86,87]. Soon after there were reports of other combinations of transplants, including the first report of simultaneous liver and islet transplantation in 1997 [88].

Ultimately these studies lead to the long-term functional (>6 years) outcomes of allogeneic islets in a series of patients transplanted by the Miami group [89], and the landmark study where 100% insulin independence was remarkably achieved in a series of alloislet transplant recipients using purified islet cell preparations [90]. Not long after this first report from the Canadian group, our Australian Islet Transplant Consortium reported similar success in a series of patients transplanted with highly purified islets from the trial commencing in 2002 at Westmead Hospital in Sydney [91]. Significant debate continued over the necessity for the actual purity of the isolation process and following this, some centers commenced not purifying their islet preparations choosing instead to transplant unpurified islets which had variable results and ultimately posed significant and increased risks for the patients such as portal vein thrombosis, segmental liver thrombosis, and exposing the patient to larger amounts of donor tissue with the effect of sensitization to higher burden of HLA and risk in multiple and future transplants [92]. Obviously subsequent to this we have learned valuable lessons with the purification of the islet preparation to demonstrate better outcomes for the long term, with purer islet isolations being performed to provide safer and less immunogenic and thrombogenic preparations, especially when these are used in multiple transplants performed into one patient. The Australian experience was testament to this with their experience in relation to more purified preparations being less immunogenic and providing better outcomes. We ourselves had a less purified preparation with significant acinar tissue contamination that we transplanted in our early series of patients and this was transplanted into a patient with shunting of the islet preparation into their left liver lobe. This patient unfortunately suffered a left partial liver lobe thrombosis induced by this less purified islet preparation. Subsequent to this early transplant, we have never since transplanted any islet preparation with less than 85% purity, choosing to selectively purify all preparations for allotransplantation [93].

Another relatively impactful factor on islet transplant outcomes is the center effect where factors that come from the center's individual experience in processing can have a dramatic effect on overall transplant outcomes. This was shown in the report of the Multicenter International Trial (ITN) trial which yielded 58% insulin independence at 1-year posttransplant, demonstrating the important effect of the center's experience in islet cell processing and patient management on clinical outcomes [94]. This can still be seen more than a decade later with the same significant differences seen between centers by the outcomes still shown to impact islet outcomes of the phase 3 trial of islet cell transplantation. The impact of the experience of the individual centers was once again seen and these were demonstrated to be as impactful as the transplant itself [95]. Ultimately the outcomes from such trials have shown that significant improvements can occur over time even with impactful effects, and that successful isolation of the islets for the clinical setting can be achieved in the majority of centers as a part of an ongoing clinical program. We have also shown that islet cell transplantation provides superior glycemic control with less hypoglycemia compared to continuous subcutaneous insulin infusion (csii) or multiple daily insulin injections and this can be achieved even for distant center transplanted patients [96]. Despite such success and dramatically improved results, the search to improve islet outcomes has remained a focus of most units, and groups such as our own continue to strive to improve by refocusing on what can be done to achieve overall improvements in the isolation process, the transplant procedure, and ultimately the long-term patient outcomes. Our own unit remains focused on the various surgical aspects, such as the site of transplantation and potential for alternate sources of donor tissue for transplantation [97].

We have also seen outcomes that appear to have improved not just in the long-term outcomes in insulin requirements for the patients but also ultimately in improving the quality of life of them. All of which are ultimately dependent upon the isolation process to provide the islets for successful outcomes to the patients [98]. Of course, after several decades of experience in the field we naturally assume that things will have improved due to advances in technology and in relation to how much we have evaluated the various factors with their influence on outcomes. Certainly, a great deal of effort has been applied to ensure this technology has advanced as a viable and safe therapy for a broader range of patients suffering from T1D. From the last few decades of experience, we can demonstrate clinical results that have progressively improved with outcomes on par with other organ transplants outcomes, in particular in terms of insulin independence and graft survival [99]. As previously outlined before, there have been significant inroads to making changes to the outcomes of the isolation procedure with the factors related to successful clinical outcomes being firstly outlined in principal and then in depth in this chapter. Particular reference has also been made to the following chapter as the influence of organ preservation has a significant effect on the overall isolation results. As such, the major influences on isolation outcomes come as part of an overall package that needs to be accounted for both prior to and during the isolation process in order to be successful.

As a result, we have seen a broader adoption of islet transplantation as a mainstream clinical therapy available to many more patients suffering T1D [99]. This chapter focuses on the process of human islet cell isolation and its role in how to optimally provide cells for successful clinical transplantation outcomes. There are a large number of processes to outline in this chapter and as such only the major ones will be the focus in this chapter. These will include donor selection, various steps in the islet isolation process, culture, and pooling pretransplantation. It is also acknowledged that there still remains the need for further ongoing improvements to islet cell isolation for transplantation; however, islet cell transplantation offers a safe and reliable therapeutic option for an increasing number of patients suffering from T1D and severe hypoglycemic unawareness [100]. In order to achieve clinical transplantation following islet isolation, we have to investigate how we can reliably and reproducibly provide islet cells that are of a sufficient number, viability, and overall quality to reach release criteria for transplantation. This process is reliant on the quality of the donor organ, which needs to be of a suitable size and quality to allow the production of enough viable islet cells. Clearly, to ensure safety of the resulting product, we must start with a donor pancreas free from any biological burden such as viral, bacterial, fungal, prion, cancer, or genetic disease from the donor. It is imperative that we ensure that we only process organs from donors that have been comprehensively screened for all commonly occurring diseases. Clinically, it is imperative to ensure the safety of our recipients and as such we should avoid accepting the pancreata for islet cell isolation and subsequent clinical transplantation until we are aware of any potential infectious risk factors. These factors depend on the history of the underlying disease of the transplanted organ, the donor, and the immunosuppressive treatment [101]. These days we can readily screen the donor patient for most common pathogens, but the frequency of pathogens does vary according to the organ and the immunosuppressive therapy given, and as such strict attention to donor details and history is paramount rather than reliance on screening alone [102].

Further to donor screening, we must also ensure that we do not cause any issues in relation to infection of the recipient from the donor or the isolation process. This commences prior to the processing of the organ, during the organ retrieval surgery where potential exists for exposure to skin, gut, respiratory, and environmental pathogens despite strict adherence to sterile technique. Over the past several years, our own group has investigated the donor retrieval transport fluid with evaluation of Bactec culture samples and found a contamination rate of as high as 60% despite the best aseptic preparation and surgical techniques [103,104]. The inherent risk of contamination does not cease there with potential risk of contamination during the isolation process. Unintentional contamination may occur during the many steps undertaken throughout the isolation and culture procedures due to the large number of manipulations of the islet cell preparation and the media that they are in prior to the transplant procedure [105].

Along with careful donor selection, screening, and optimal handling during the isolation procedure with strict adherence to sterile techniques, there remain a significant number of issues to be addressed to allow for islet cell isolation to be completed with successful outcomes. These need to be enforced in order to reach release criteria and provide safe and effective transplantation of the patient [106].

As such, all islet isolations should undergo quality control assessment to reach the release criteria decided by each individual unit as safe and effective to an appropriate standard to justify release to proceed to transplantation. This standard is determined in relation to that particular country's own government regulations and is essential to not only ensure the best possible outcome for recipients but also to prevent any potential infection from the donor. This is done by the appropriate facilities and operating procedures; our own group utilizes these as part of our overall effort to provide prevention as well as routine screening to barrier the unintentional introduction of any contamination that may occur at any point throughout the islet preparation process from commencement to completion.

Significant improvements to isolation outcomes

There has been broad acceptance of clinical alloislet transplantation as a clinical therapy for T1D and we have seen significant improvements in outcomes achieved in the past few decades. The dramatic improvement to results has helped drive the increasing use of islet cell transplantation as a clinically applicable therapy for treating patients suffering T1D and severe hypoglycemic unawareness [93,107]. These improvements to outcomes have resulted in islet cell transplantation being able to demonstrate successful outcomes as good as seen in the report from Foster last year with 92% 1-year survival rates which are extraordinary when compared to even a few years ago. These are certainly as good as or better than the currently accepted gold standard of whole pancreas transplantation that have been reported with survival rates of just over 80% at 1-year posttransplant [98]. There are numerous studies that have demonstrated dramatic improvements to clinical islet isolation techniques, contributing to significantly improved outcomes of islet transplantation. These results demonstrate that islet transplantation results in similar or better outcomes than whole pancreas transplantation, to the point where it represents a clinically viable option to achieve long-term insulin independence in selected patients with T1D and hypoglycemic unawareness [108]. Due to significant improvements to the functional and survival rates in islet cell transplantation, we have seen the widespread acceptance by governmental and health insurance providers in numerous countries including our own. Our federal government provides National Funded Centers funding for our islet transplant program where we are funded for islet isolation and transplantation as part of the ongoing clinical care of our select group of patients. As the safety of the procedure along with the immunosuppressive therapies improves and the inherent potential risks are reduced, it is clear that islet transplantation can be further utilized to treat the many patients that require islet cell transplantation for hypoglycemic unawareness.

In recent years, as many as 80% of isolations at our center result in a clinical transplant as compared to several years earlier when we processed significantly larger numbers of pancreata with lower quality donors resulting in poorer outcomes regarding isolations successfully proceeding to clinical transplant. This is shown in Fig. 6 where over the past twenty years there has been a significant improvement in outcomes. In the first two years, only a few pancreata were processed during establishment of the isolation process, followed by the next two years during which only exceptionally good donor pancreata were processed, as reflected in the high success rate. Over the next decade, we processed larger numbers of pancreata with almost 40 isolations being undertaken per year but again reliance on processing larger numbers of organs didn't necessarily result in clinical transplant outcomes. The influencing cofactors being the things that effect outcomes, rather than the reliance on numbers of pancreata processed. Over the past six years we have been highly selective in our choice of donor and pancreata to process, resulting in significantly improved percentage proceeding to clinical transplant.

The donor and donor pancreas

Significant improvements in organ donor numbers have occurred over the past decade, and in Australia we have seen almost doubling of the organ donor rate. However, despite this impressive increase, we still suffer a significant gap between supply and demand for organs to transplant the large number of recipients [109,110]. Obviously with the significant improvements to the way we now undertake and manage donor processes, improvements to organ recovery techniques, implementation of more stringent donor criteria, and improved islet cell processing techniques, there has been enhancement in organ utilization for transplantation [111]. However, the problems associated with low organ donation numbers are an ongoing global problem despite the significant investment to improve these with campaigns to educate and inform people of the benefits. In the past five years, we have seen improvements in organ donor rates and the uptake of newer methods of organ donation

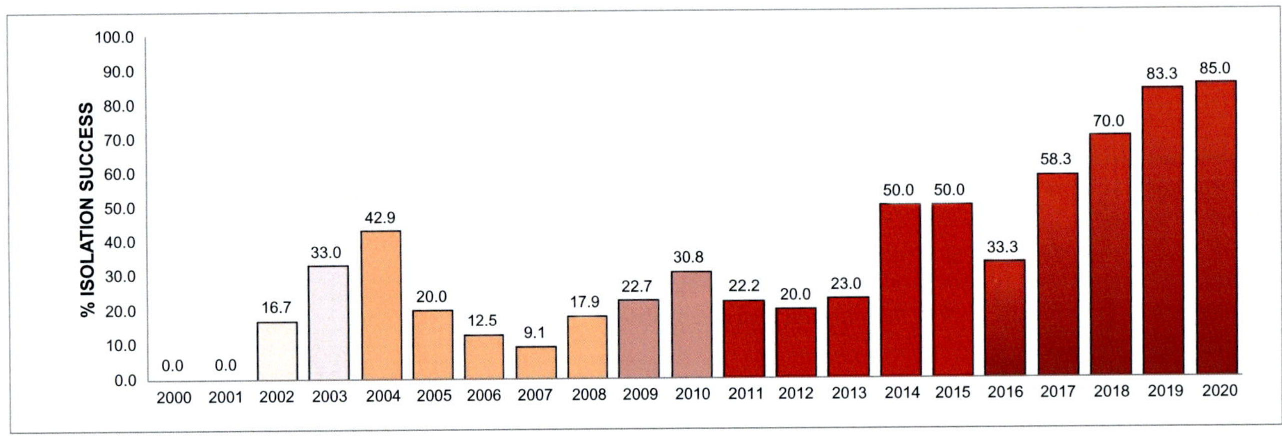

FIG. 6 Figure of our own islet outcomes over the past ten years indicating that the influencing cofactors are the things that effect outcomes not the reliance on numbers of pancreata processed.

such as the use of donation after circulatory death (DCD) and utilization of more marginal organ donation to expand the available donor pool, including the use of machine perfusion to help resuscitate these organs and generate more organs for the donor pool as will be discussed later in this chapter [103,112,113].

Despite significant improvements, there remains the fact that the pancreata for islet isolation tend to originate from less than ideal donor sources and therefore any gains to organ donor numbers may not truly reflect what can be utilized for pancreatic islet isolation and clinical transplantation [114]. However, a significant advantage of islet cell therapies is that it also provides a means upon which we can underpin any of our future advances in cellular therapies such as for xenotransplantation as a means to treat a broader range of patients that suffer from T1D [115].

In the overall process of islet cell isolation, the donor and the pancreas remain the most influential factors affecting the outcome of the isolation process and this is due to the multiple and varied cofactors that donors were subjected to over the time of their injuries, hospitalization and the organ retrieval process, ultimately affecting the outcome of the islets for release for clinical transplantation [116]. There have been a number of landmark studies that have generated algorithms that detail the most influential factors of the multiorgan cadaveric donor that play a role in islet isolation outcomes [116–118]. From both these Canadian studies, a number of influential cofactors were identified as contributing to best outcomes from islet isolations. These include donor age, body mass index (BMI), cause of death, other contributing factors such as prolonged hypotension for more than fifteen minutes, or if the patient required significant doses of vasopressors or resuscitation. These studies lead to the development of the North American Islet Donor Score (NAIDS), a scoring system utilized by a number of units as a means of donor organ selection to optimize their outcomes to result in a clinical transplant [118].

Most islet isolation units utilize a scoring system as a guide for donor selection but ultimately there are major factors that influence the outcomes as will be described later. There are major cofactors that I personally ascribe to and have successfully optimized the results of our choice in organ donor to result in a clinical transplant outcome in more than 80% of cases as can be seen in Fig. 6. These major factors include donor age, BMI, CIT, and the procuring surgical team and techniques used to retrieve the pancreas. Clearly, while these select few variables have major influences on the outcomes of islet isolations, all associated factors ultimately affect the outcomes of the isolation. We and others have shown there are many more crucial cofactors that play a role in affecting the isolation outcome which include the donor age, size (height and weight), BMI, and overall health which play significant roles as does the ultimate cause of brain death, patient resuscitation or preadmission treatment and subsequent management in hospital. The immediate and significant underpinning impactors on isolation outcome are the immediate first aid treatment, amount of hypoxic down time, whether or not resuscitation occurred at the time of first aid, the direct hospital admission and intensive care treatment, including the length of time that the patient was managed prior to organ donation. We have seen significant improvements to donor management and implementation of more stringent donor criteria that have contributed to the increased organ utilization and thus transplantation [103,111]. Obviously, all cofactors influence the overall outcomes but appropriate care and an educated decision to the optimal donor is paramount in order to select a pancreas that will provide an islet cell preparation with a higher chance of resulting in clinical transplant. These various cofactors have been intensively evaluated over decades and the most significant and influential cofactors include some of those listed as follows [116–118].

Effect of cold ischemic time (CIT)

From over 20 years of performing islet isolations I feel that by far the most impactful and as such important confounding factor affecting outcomes of islet isolation is the cold ischemic time (CIT). As a broad definition, the CIT is the time taken from the time of cessation of blood flow (cross clamp) of the donor and the commencement of cold perfusion until the time the pancreas is received into the processing laboratory and the islet isolation process commences. The cold ischemic time is acknowledged as important for all organs but its impact is particularly influential on pancreata for islet isolation. While multifactorial, it is very dependent on logistical expertise of the organ donor network and process utilized. Obviously to optimize and reduce the CIT, the pancreata should be shipped to the closest possible isolation facility. However, it requires there to be multiple facilities or centers to be able to offer this service. In Australia, this is particularly problematic as our national program covers an area of more than 7.5 million square kilometers, which is approximately twice the size of Europe or three-quarter the size of the United States. So, the importance of logistical expertise is paramount in order to minimize shipping times and ultimately CIT [107].

Over the years we have seen acceptable times for the CIT dropped from the standardly accepted 12 h as for other organs. However, our general experience and that of many clinical islet isolation and transplant centers is that a cold ischemia time beyond 8 h results in significantly reduced yields and quality of islets [119,120].

Numerous studies have previously shown that the longer the ischemic time, the worse the isolation outcome in terms of IEQ and functional capacity of the islets. Wang et al. analyzed islet isolations to identify variables for islet yield and, additionally, islet size and size distribution. They found that CIT had a significantly negative correlation with actual islet count and IEQ/g, in fact the longer the CIT the worse the outcome and that over eight hours of CIT provided a far worse outcome [106]. This has been supported by numerous other studies, and more recently Berkova et al evaluated donor factors that were associated with isolation failure and to investigate whether immunohistology could contribute to organ selection. Donor characteristics were evaluated for both successful and failed islet isolations. Samples of donor pancreatic tissue were taken for immunohistochemical examination. From numerous isolations they clearly demonstrated that a cold ischemia time of more than 8 h unfavorably affected isolation outcome [121].

Of note, several studies have sought to extend this time due to organ donor availability and logistical reasons from a limit of eight hours to as much as 12 h to be able to provide greater numbers of pancreata to process for islet isolation. However, these were not necessarily intended for transplantation which is obviously not the same as for clinical transplantation [120,122]. Lyon et al. examined the feasibility of a research-only human islet isolation and whether key criteria such as CIT and metabolic status may be relaxed and still allow successful research-focused isolations. They examined isolations over approximately 5 years and confirmed that CIT had a negative impact on isolation purity and yield and extending CIT beyond the typical clinical isolation cutoff of 12 h (to ≥ 18 h) had an impact on islet numbers and of course function [122]. Obviously the best assessment to demonstrate CIT are the findings from large clinical and collaborative trials such as the Collaborative Islet Transplant Registry (CITR) data which demonstrated that only a limited number of the clinical islet isolations used pancreata with CIT which extended past 10 h, with the mean CIT being 9.3 h for islet isolations performed between 2007 and 2010 [123].

Reinforcing this data was another in-depth analysis of transplantation outcomes in a large series of islet isolations and transplants, and this indicated donor pancreata with a mean cold ischemic time of 9.8 h provided best outcomes in regard to function [124]. In addition, another large collaborative Phase 3 Islet Transplantation Clinical Trial in Type 1 Diabetes recently demonstrated best transplant outcomes and function preferentially utilizing organ donors with less than 10 h CIT [98]. To this end in our islet transplant program we prefer not to utilize pancreata with a CIT of greater than 10 h for isolation for clinical transplantation but will process for research preparations from pancreata with a CIT of up to 12 h with the realization they will provide poorer outcomes.

Organ donor age effects

I feel that one of the most influential cofactors impacting outcomes of islet isolation is that of the donor age. Certainly, when evaluating an organ donor for isolation the first factor listed on my review sheet is donor age which has been shown to significantly influence the outcome of the islet isolation process. Age contributes on its own directly to number of islets and size of islets, but also impacts in combination with other factors that will be discussed, including their various interactions and outcomes. The age of the donor has a significant impact for many reasons both due to the size of the donor organ but also from the perspective of how to handle the pancreas and specifically how to undertake the digestion process. The younger the donor, potentially the smaller and thinner and less fat laden the organ will be. Conversely, the older the donor, the larger, fattier, and potentially more fibrotic the pancreas generally is. This is readily seen in Fig. 7 which demonstrates this point showing the sizable difference between the younger donor pancreas which is much thinner and of smaller size both when first retrieved as seen in Fig. 7A and when dissected free in Fig. 7B. This is very different as compared to the much older donor pancreas which is very fatty as seen in Fig. 7C and quite large even after cleaning off the extraneous tissues as seen in Fig. 7D.

A major influencing cofactor on the outcomes of the islet isolation is age range; as such, we try to adhere to strict acceptance criteria for utilization of donor pancreata for islet isolation. Our donor pancreas acceptance age range is from 18 years of age up to 65 years of age. However, evaluation of our own data demonstrate that we have not transplanted from any organ donor that were younger than 20 years of age or older than 60 years of age, and as such we generally transplant from organs in the donor age range of 25–55 years of age.

The role that age plays on the overall outcome is very much dependent on other external factors but as a general rule younger donor pancreata yield islets that are more resistant to hypoxic damage and more specifically any loss during the culture process. However, conversely, they are also more prone to adherence of the acinar tissue to the islets and more mantling of islets occurs during the digestion process compared to any other age group. Obviously, age can be a strong influence on the amount of fat, vascular and connective tissue, and percentage of fibrosis in the organ, but this is also impacted by the environmental factors such as lifestyle. These associated cofactors affect the speed and amount of the surrounding tissues digested and thus the number of islets and how quickly the islet cells are released from them [125].

We can evaluate the influence and role donor age plays in affecting isolation outcomes in a number of ways, but the best is to evaluate larger data sets. A number of studies including studies from our own group have investigated donor age

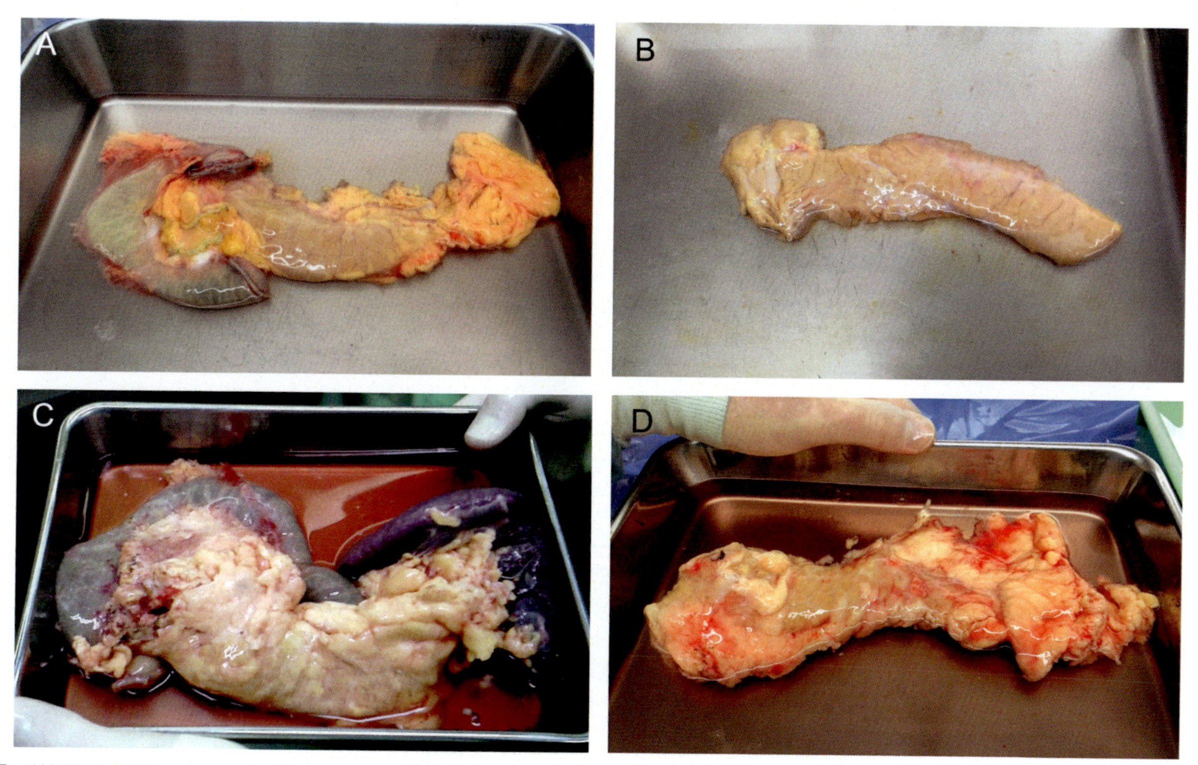

FIG. 7 (A) Photo of donor pancreas from a young slim donor prior to and (B) same pancreas post dissection as compared to (C) an older fatter donor pancreas prior to and (D) the same pancreas post dissection.

group effects on isolation outcomes. This is more readily done by firstly looking at the two major outliers to the more commonly used donor age range, these being donor pancreata from younger (<20 years of age) or from older (>65 years of age) age groups. These groups have age group-related effects with the <20 years of age group demonstrating significantly better functional outcomes than any other age groups; however, this comes at a significant cost of poorer isolation yields and lower numbers [125,126].

The younger donor and pancreas are affected by their age as the younger they are, the smaller their pancreas size and overall weight, and therefore lower numbers of smaller islet cells. This is the first issue with younger donor pancreata, but the major and most impactful is the issue of prepurification percentage of trapped or mantled islets. This has been shown in many studies to be significantly higher in younger donors, up to 44% compared to >20 years of age donor pancreata 25%. This obviously leads to a lower recovery rate in younger donors (as low as 48% compared to >20 years of age donors at 76%) and hence results in lower postpurification IEQ/g of pancreas in the younger donor (2412 ± 1789 IEQ/g) compared to >20 years of age donor (3194 ± 1892 IEQ/g, $P = 0.01$). This results in such low numbers that the younger donors do not reach release criteria numbers and as such fail to be transplanted [126].

A number of other studies have shown similar findings and have shown a strong negative correlation to IEQ/g pancreas in regard to younger donors. As discussed, this is due to the younger pancreas having more mantled and trapped islet cells and as such when they are put onto the density separation, they sustain significant losses of islets due to their density being the same as the acinar tissue entrapping them. As can be seen in Fig. 8A, the islets that are from younger donors are covered in acinar tissue that encapsulates or mantles them. The islets surrounded by the acinar tissue (mantled) are of the same density and as such centrifuges to the bottom of the density gradient and so the islets are lost into the bag tissue [127].

Conversely the donors that are from the age range older than 45 years of age provide overall better isolation results in terms of actual islet number and IEQ of pancreas with larger islets that contain higher numbers of beta cells. As can be seen in Fig. 8B, some larger islets can also be of odd shapes rather than the normal oval shaped islets and as a result can undergo significant deformation and sheer forces when being centrifuged or transplanted and as a result can be destroyed. These islets from older donors can also have lower in vivo function and as a result poorer transplantation outcomes [128]. A number of studies have indicated better outcomes from donor pancreata from donors younger than 45 years of age. One such study demonstrated significant benefits in the islet graft function parameters, which were significantly higher at 1-month follow-up in patients who had received islets from donors younger than 45 years of age. At 6-month follow-up after second or

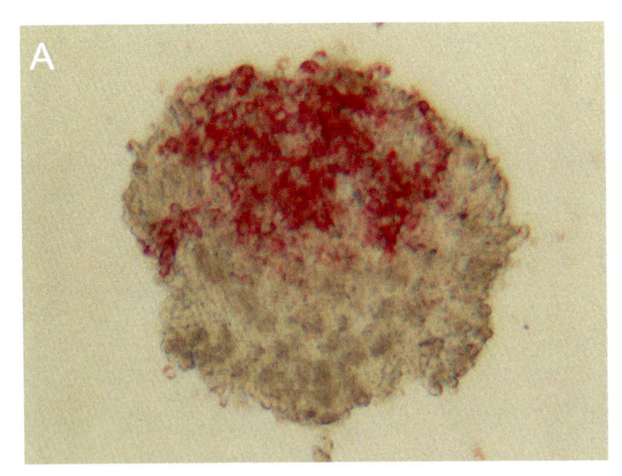

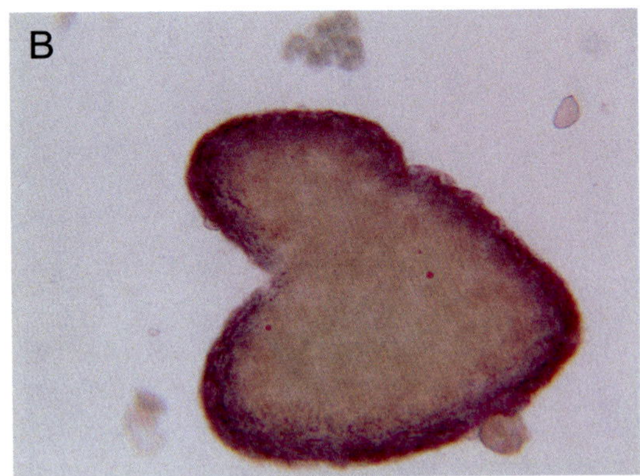

FIG. 8 (A) Photo of a mantled islet from a young donor, (B) Photo of a large oddly shaped islet from an older >45-year-old donor reflecting how I have enjoyed islet isolation over the last few decades.

third transplant and at 12-month follow-up, secretory units of islets in transplantation indices and C-peptide/glucose ratios were significantly higher in patients transplanted from donors younger than 45 years of age. This was also confirmed by the Italian group where they demonstrated a strong correlation with poorer isolation outcomes from those pancreas donors older than 50 years of age, with respect to the quality of the islet cells [129]. This was also demonstrated many years ago by the Edmonton group where they demonstrated that the insulin secretory capability of islets isolated from donors older than 50 years of age was significantly reduced as compared with their younger than 50-year-old donor age group [117]. A number of islet transplantation series reporting consistent insulin independence after single-donor islet infusion have used strict donor selection criteria, limiting the use of the donor age to less than 50 years of age to ensure better functional outcomes from a single donor [130]. However, taken in context and with the scarce and valuable donor resources we have, limiting donor age is a significant problem with regard to reducing the potential organ donor numbers available for isolation. We preferentially utilize pancreata from organ donors in the younger than 50 years age range. However, this is impractical due to the scarcity of donor organs, and as such we are forced to utilize donor pancreata for islet isolation and subsequent transplantation from organ donors into their mid-60s with reasonable results albeit with lower functional capacity [93].

This data suggests that, despite similar outcomes of the isolation procedure, islet graft function is significantly influenced by donor age. These results may have important consequences in the definition of pancreas allocation criteria. A major problem for our own program and numerous others is that organ donors within the 20–45 years age range are allocated as a group for use in whole pancreas transplantation for simultaneous pancreas and kidney transplants taking these from being utilized for islet isolation [131]. Our own transplant unit at Westmead Hospital in Sydney runs both the National Islet Cell Transplant program alongside the National Simultaneous Pancreas and Kidney (SPK) Transplant program. We prioritize donor pancreata within the 20–45-year age group to SPK transplantation. Our program works well but the majority of these pancreata are allocated to SPK. Pancreata for islet isolation generally include anything outside this age range or those within the age range exhibiting vascular disease, heavy fat infiltration, or fibrotic changes.

A number of groups such as the Italian group of Cardillo et al. analyzed the allocation protocols of all pancreas donors (2011–2012; $n=433$) in Northern Italy. Among pancreas allografts offered to vascularized organ programs, 35% were transplanted, and younger donor age was the only predictor of transplant. The most common reasons for pancreas withdrawal from the allocation process were donor-related factors. Among pancreas offered to islet programs, 48% were processed, but only 14.2% were transplanted, the most common reason for pancreas loss was due to unsuccessful islet isolation. Younger donor age and higher BMI were predictors of islet allograft transplant. As a result, they have changed the pancreas organ donor allocation strategy with equal distribution of donor pancreata between vascularized organ and islet transplant [129].

Similarly, in the United Kingdom they have adopted the same organ donor allocation strategy allocating pancreata on an alternating basis to the whole pancreas and the islet isolation teams. This decision was based upon the study where they evaluated the factors affecting outcome, together with published data on factors affecting islet transplantation, informed the development of a points-based allocation scheme for deceased donor pancreases in the UK providing equitable access for both whole organ and islet recipients through a single waiting list. Analysis of the allocation scheme 3 years after its introduction in December 2010 showed that the results were broadly as simulated, with a significant reduction in the number of long waiting patients and an increase in the number of islet transplants [132]. Sensible and fair allocation between SPK and Islet transplantation is a means by which we can potentially improve transplant rates for patients in islet programs.

Surgical retrieval

Our own group has the strong belief that organ retrieval and the team's level of training for performing the retrieval surgery can play a significant role in impacting the outcome of the islet isolation. Like all of the other cofactors there are a number of issues in the team's training and how they go about performing the retrieval that can affect islet isolation outcomes. Among these, one of the most significant to effect islet isolation outcomes in regard to islet cell yield and function is the retrieval process for a number of reasons. One of the strongest factors positively correlating with successful outcomes was when the pancreas was retrieved by a surgical team from the isolation center's hospital, and this is most likely due to an obvious interest in retrieving the pancreas for islet isolation and understanding of proper surgical retrieval techniques required to facilitate this [117].

Obviously, there have been significant improvements to organ retrieval techniques including attention to details of surgical staff training and dedication to specifically undertake pancreas retrieval as part of the multiorgan donor retrieval. Sadly, we can see these improvements negated by something as simple as the donor retrieval process, when a lack of training leads to poor handling, dissection or perfusion of the pancreas by the surgical team. In multiple discussions throughout the world, there are examples and anecdotes of such damage to pancreata due to lack of care and thus poorer outcomes. In a recent UK study, 1296 pancreata were procured from donation after brain death donors. More than 50% of recovered pancreata had at least one injury, most commonly a short portal vein (21.5%). The major contributing factors were associated with liver donation, procurement team origin, and increasing donor BMI. Of the 640 solid organ pancreas transplants performed, 238 had some form of damage [133].

Damage to the pancreas during organ recovery is more common than other organs, and meticulous surgical technique and awareness of damage risk factors are essential to reduce rates of procurement-related injuries [134]. A number of studies have demonstrated direct effects from surgical retrieval of the pancreas during procurement. By far the greatest effect of using a pancreas retrieval trained surgical team is on the pancreas retrieval outcome. Utilizing a local or dedicated team limits the potential for any damage occurring to the pancreas at retrieval by a team that is not dedicated to pancreas retrieval. More importantly, it is well known that at many centers they do not have dedicated retrieval surgeons, and this has had significant effects on the amount of damage that occurs to the pancreas [134].

Injuries significantly impact the pancreas, increasing leakage and making distension much more difficult than if a pancreas were retrieved with no damage. Obviously, this impacts on the ability to adequately distend the pancreas upon injection of enzyme into the gland for distension and dissociation. In addition to this key factor, the necessity also exists to have an adequately perfused pancreas. Inadequate perfusion impacts the organ and causes issues such as complete or partial thrombosis of the splenic vessels such as can be seen in Fig. 9A. This shows the splenic vein full of thrombus as a result of inadequate flushing due to tying off the splenic at splenectomy without adequate perfusion. Fig. 9B further demonstrates a poorly perfused pancreas with a large amount of free blood in the perfusion fluid and the gland.

Our own multiorgan donor retrieval service is also the National Pancreas and Islet transplant unit. We utilize our dedicated transplant retrieval trained surgeons who have significant experience in multiorgan donor retrieval with a dedicated focus on pancreas organ retrieval. This integrated approach to both multiorgan donation and transplantation has allowed a great focus from the time of donation to transplantation of the organ or isolation of islets. Our own unit has trained most abdominal retrieval surgeons for multiorgan retrieval for the units in Australia. We also have developed a number of "no touch" techniques to ensure no damage occurs to the pancreas at retrieval, including the recent use of the harmonic scalpel to aid in the accuracy and speed of the retrieval procedure as described by Hameed et al. As can be seen in Fig. 9B. using the no touch technique and the harmonic scalpel, a well retrieved and preserved pancreas can be easily retrieved with no damage in minimal time [135].

Romanescu et al. eloquently provided a very detailed description of the pancreatic retrieval procedure using a "no touch" technique where the spleen is used as a handle for pancreas mobilization for retrieval avoiding any direct handling of the pancreas [136].

As can be seen, we utilize and train all retrieval surgeons in the no touch technique to ensure no damage occurs to the capsule or the pancreas itself as this is critical to ensure complete distension when collagenase is infused into the pancreas to distend the organ for dissociation. Others have also demonstrated the effects of the retrieval process on islet processing outcomes. Ponte et al. demonstrated improved islet yields when the local surgical team retrieved the pancreas. Their data suggest that a sequential, integrated approach including the use of a well-trained donor surgical team can improve the success rate of islet cell processing [111].

We firmly believe that the donor team and the donor operation are very impactful and influential cofactors in the overall outcome of the islet preparation for clinical transplantation. The importance of the organ retrieval procedure cannot be underestimated as it has a huge impact on the overall outcome of the islet isolation. Therefore the same importance and value should be placed on a pancreas for islet isolation, as that of a pancreas being retrieved for whole pancreas transplantation.

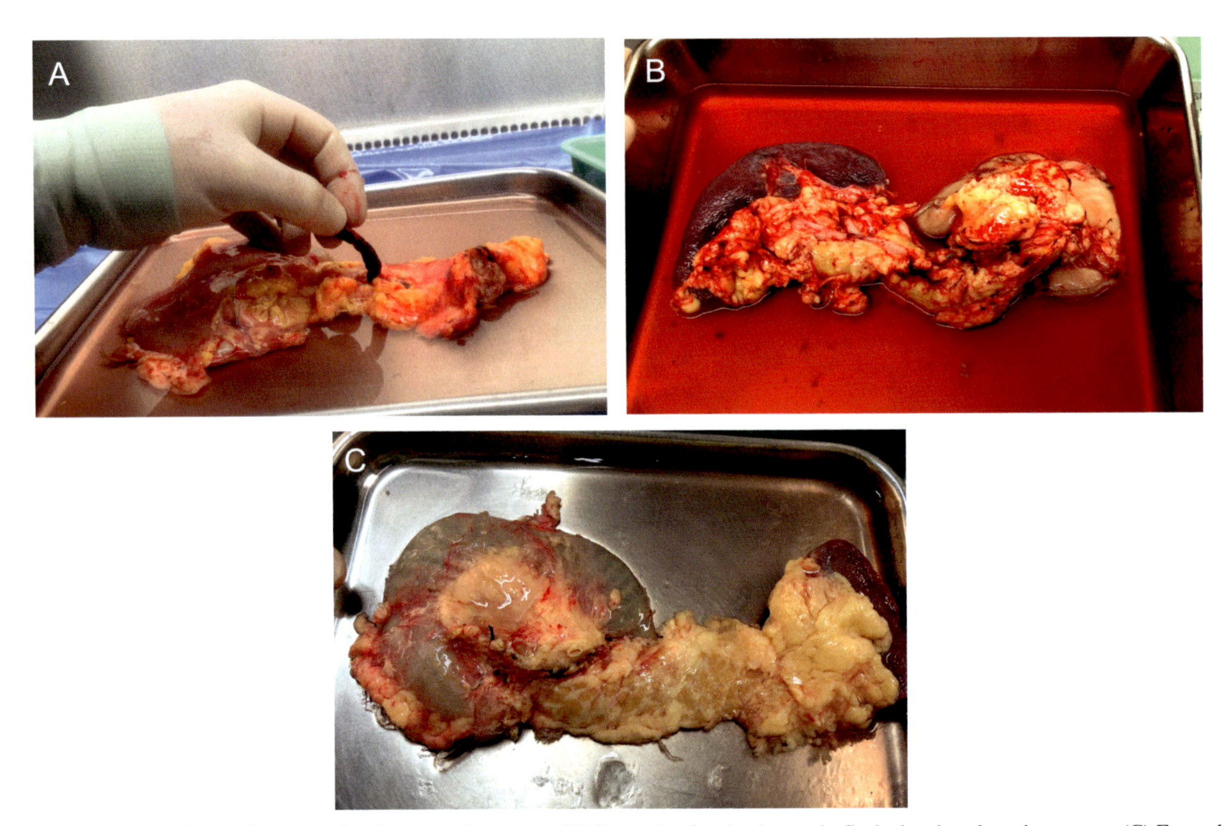

FIG. 9 (A) Example of a poorly procured and preserved pancreas. (B) Example of an inadequately flushed and packaged pancreas. (C) Example of a well retrieved and preserved pancreas for islet cell isolation with the pancreas being flushed and no remaining blood in the splenic vessels nor the transport fluid.

Effect of organ preservation and transport

Over the years significant debate has been undertaken in regard to the absolute importance of the donor operation and organ preservation solution in regard to its influence on the islet isolation outcome. The route of perfusion and the type of perfusate solution are frequent points of debate. However, the most widely used perfusion technique for donor perfusion is aortic flush of the donor utilizing cold preservation solution, predominantly University of Wisconsin (UW) solution. Despite this, a number of other perfusion solutions are utilized such as histidine–tryptophan–ketoglutarate (HTK), Celsior (CS), and Kyoto solution along with a few other regionally developed agents. Upon evaluation of CITR data from a total of 1017 islet isolations performed between 2007 and 2010, over 42% of the pancreata were perfused with UW solution, 7% with HTK, and 2.3% with CS, and unfortunately the remainder were not reported [123].

Another major factor in relation to both the organ donation procedure and the perfusion technique is the way the organ is cold stored for transportation. A number of issues remain with units not providing adequate cold flush for storage of the pancreas during transport. A volume of perfusate solution approximately five to ten times the volume of the pancreas should be utilized to ensure adequate cold preservation without danger of the organ being affected by the cold of the storage during shipping. This can be seen in Fig. 10A that shows how a pancreas should be packaged with a large volume of perfusate solution and triple bagged with iced slush inside the second bag surrounding the inner bag and pancreas. As compared to Fig. 10B which shows a pancreas that has been retrieved and inadequately packaged in a small volume of perfusate at bagging for transport.

Over the past decade, there have been a number of papers that have shown predominance in the use of UW perfusion solution during perfusion of the donor for multiorgan retrieval. However, there has also been a slight increase in the use of HTK and CS in some units despite the poorer outcomes now being obvious in regard to the use of HTK solution [103,137].

There have been studies that have comprehensively evaluated the use of UW, HTK, CS, and Kyoto solutions and despite this, there remain concerns that there is a significant need for supplementation of cold storage solutions with cytoprotective agents and oxygenation to improve pancreas and islet transplant outcomes [138,139].

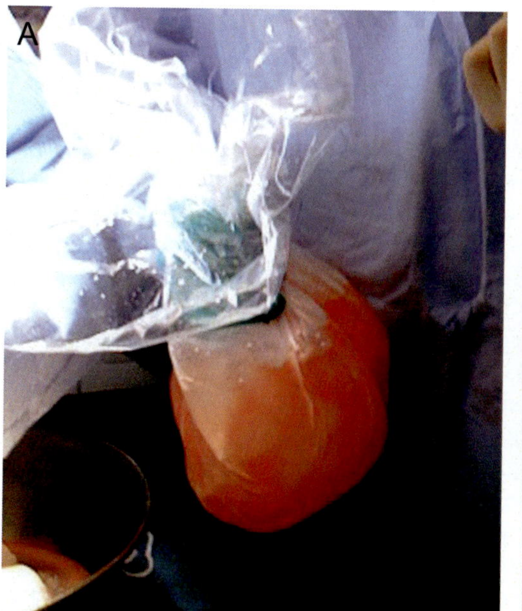

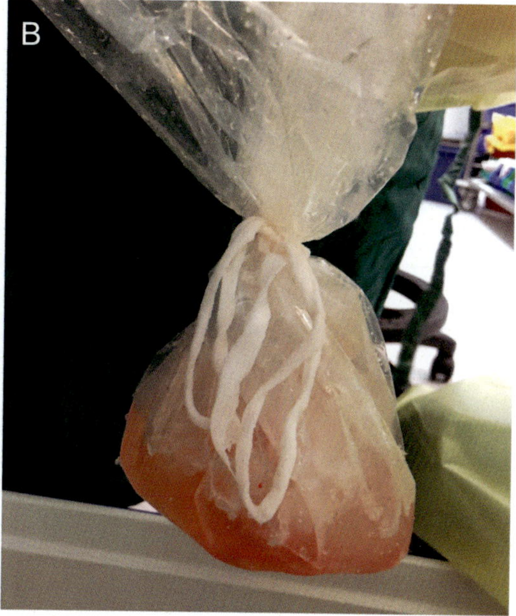

FIG. 10 (A) A photo of a pancreas postretrieval correctly packaged with a large volume of perfusate solution and triple bagged with iced slush inside the second bag surrounding the inner bag and pancreas. (B) Example of a pancreas retrieved and inadequately packaged in a small volume of perfusate barely able to cover the pancreas.

Studies that have investigated some other alternatives include that of the Institute Georges Lopez-1 (IGL-1) solution, which is a preservation solution similar to UW solution, with the ratio of Na/K reversed. The study from Niclauss et al. assessed the impact of IGL-1, UW, and CS solutions on islet isolation and clinical transplant outcomes [140].

They investigated the outcomes of islet isolations from pancreata flushed and transported with IGL-1 and compared it to the standardly used UW or CS solution. Isolation outcomes and transplant outcomes showed there was no difference in islet yields between the three groups. Interestingly, the transplant rates, β-cell secretory function, and viability were not significantly different between the three groups. They observed no difference in decreased insulin needs, C-peptide glucose ratios, β-scores, and transplant estimated function at 1- and 6-month follow-up between IGL-1, UW, and CS groups. This study demonstrated little differences between outcomes and that UW, CS, and IGL-1 were equivalent solutions with no real difference in outcomes for pancreas perfusion and cold storage before islet isolation [140].

In addition, there have been a large number of units around the world that have investigated numerous ways of infusion or treatment of the pancreas and its storage and subsequent shipping of the organ specifically for islet isolation [117,141].

In a study by Takita et al., the effects of two different solutions for pancreatic ductal perfusion (PDP) at organ procurement pancreata were evaluated by allocating them to three groups: non-PDP (control), PDP with ET-Kyoto solution, and PDP with cold storage/purification solution. Following perfusion and storage, islets were isolated by the standard method. They found no significant difference in donor characteristics, including cold ischemic time, between the three groups. All islet isolations in the PDP groups had higher IEQ and total islet yield after purification, a significant increase when compared with the control. Both ET-Kyoto solution and cold storage/purification stock solution were suitable for PDP and consistently resulted in improved isolation results. Obviously, such studies must be conservatively evaluated as there continues to be further research and newer methods to be explored [141].

Machine perfusion and persufflation

Due to the ongoing lack of suitable organ donors, the push exists to improve and increase the donors available and also to improve those donors that are in fact available. Clearly there remains a need to evaluate new methods in order to improve pancreas isolation outcomes. There have been a number of novel methods trialed in order to improve suboptimal outcomes due to hypoxia of the pancreas while undergoing perfusion, cold storage, transport, and preisolation timing. It has been postulated that enhanced oxygen delivery is a major area for improvement. A number of studies have been undertaken to investigate other options for improving oxygen delivery to the pancreas while being cold stored. Such methods being trialed include machine perfusion (both normothermic and hypothermic), perfusate oxygenation, and persufflation.

In a recent study, Leemkuil et al. evaluated the effects of hypothermic machine perfusion (HMP) on the quality of human pancreas grafts from DCD and donation after brain death (DBD) donors preserved by oxygenated HMP. Hypothermic machine perfusion was performed and results compared with those from pancreata preserved by static cold storage. They demonstrated that HMP resulted in adenosine 5′-triphosphate concentrations increasing 6,8-fold in DCD and 2,6-fold in DBD pancreata, with no signs of cellular injury, edema, or formation of reactive oxygen species. They demonstrated that islet viability and in vitro function were excellent after HMP and that oxygenated HMP is a feasible and safe preservation method for the human pancreas, increasing tissue viability and resuscitating DCD pancreata for potential use for islet isolation [142].

In an early study, Scott et al. evaluated persufflation of human pancreata obtained from brain-dead donors. They investigated the outcomes of pancreata that underwent persufflation to deliver oxygen to the organ. This was found to be beneficial as it elevated ATP levels during preservation and improved islet isolation outcomes while enabling the use of marginal donors, thus expanding the potential donor pool [143].

More recently, a large-scale trial led by Papas has demonstrated impressive results when they investigated pancreas preservation by gaseous oxygen perfusion (persufflation) compared to static cold storage (SCS). Persufflated pancreata had reduced SCS time, which resulted in islets with higher glucose-stimulated insulin secretion compared to islets from SCS only. RNA sequencing of islets from persufflated pancreata identified reduced inflammatory and greater metabolic gene expression, consistent with expectations of reducing cold ischemic exposure. Furthermore, persufflation extended the total preservation time by 50% without any detectable decline in islet function or viability. They clearly demonstrated that pancreas preservation by persufflation, rather than SCS before islet isolation, reduces inflammatory responses and promotes metabolic pathways in human islets, which results in improved β cell function and as such potentially improves isolation outcomes [144].

These encouraging studies are important in regard to increased utilization of suboptimal donors such as DCD organs for transplantation. Grafts from DCD are potentially subjected to greater ischemic insult and are at higher risk of poor outcomes. Conventional preservation techniques have been adequate to date but with the use of more marginal donors the threshold for rejecting organs decreases, and the importance of optimal preservation techniques increases. As such, the use of these newer techniques in pancreas preservation is essential to improve usage rates of potential pancreas donor organs for clinical islet isolation [145].

In review

This chapter has provided a broad overview of the major developments to the factors related to the development of both whole pancreas and islet cell isolation. With the numerous advances over the past decades from the advent of the first transplants to the latest techniques currently utilized, improvements continue to be made in the effort to provide the best possible outcomes for pancreas and islet transplantation. The major and most impacting aspects of these being organ donor selection and management, organ donor retrieval, organ perfusion, new techniques in preservation and storage, for safe and effective transplantation. Overall results from all the improvements discussed herein produce good outcomes for the whole pancreas graft, islet cells in large numbers with an even spread of islets, and various IEQ as can be seen in Fig. 11 which demonstrates islets isolated for clinical transplantation.

As a result, we have seen recipients maintaining insulin independence for almost two decades posttransplant with minimal complications [95]. Ultimately, the outcomes from such transplants have shown that significant improvements can occur over time in most centers as a part of an ongoing clinical program. We have also shown that both whole pancreas and islet cell transplantation provide superior glycemic control with less hypoglycemia compared to continuous subcutaneous insulin infusion or multiple daily insulin injections [96].

With over a century of experimental and clinical research from the never-ending striving of clinicians and researchers to achieve better outcomes, we have seen pancreas and islet transplantation grow from no more than an experimental procedure in animal models, to now be an effective clinical treatment for an ever-increasing number of patients suffering diabetes. Despite such success, these services still remain in the hands of a relatively small number of super specialized transplant units around the world offering a very viable option to treat patients suffering from diabetes. With further advances in the developments and improvements as discussed in this book, we expect ongoing improvement to clinical transplant processes and outcomes, and wider provision of such life-changing therapies to an ever-increasing number of patients suffering from diabetes.

Acknowledgments

Special thanks go to Mr. Adwin Thomas for review of the chapter and referencing, Ms. Callista Rainey for her review of the chapter, and Ms. Win Theingi for the timeline figures.

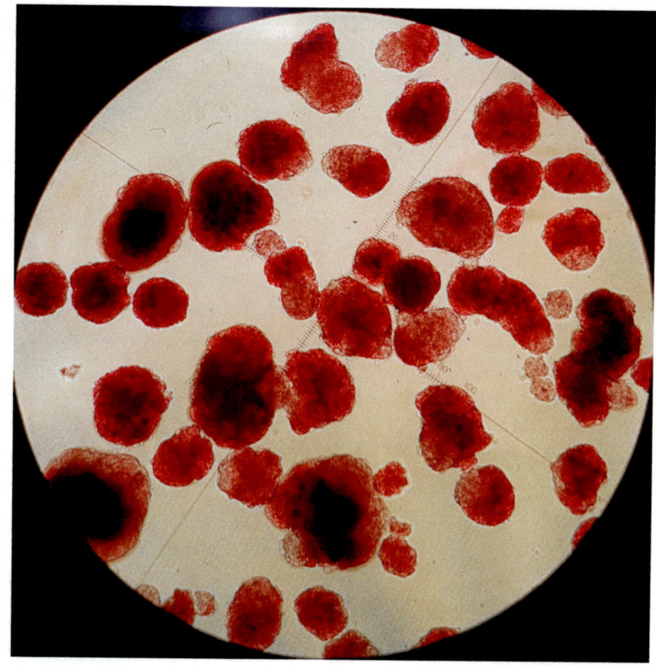

FIG. 11 A photomicrograph of islets isolated and processed for transplantation stained for counting to determine the number of islets available for transplant.

References

[1] Mering JV, Minkowski O. Diabetes mellitus nach Pankreasexstirpation. Arch Exp Pathol Pharmakol 1890;26:371–87.
[2] Hédon E. Griffe sous-cutanée du pancréas:ses resultats au pointe do one do la théorie du diabeté pancreatique. CR Soc Biol 1892;44:678.
[3] Minkowski O. Weitere Mitteilungen über den Diabetes mellitus nach extirpation des Pankreas. Arch Exp Pathol Pharmakol 1892;31(2):85–189.
[4] Williams P. Notes on diabetes treated with extract and by grafts of sheep's pancreas. Br Med J 1894;2:1303–4.
[5] Payr E. Beiträge zur Technik der Blutgefäß- und Nervennaht nebst Mittheilungen über die Verwendung eines resorbirbaren Metalles in der Chirurgie. Arch klin Chir 1900;62:67–93.
[6] Hédon E. Sur la secrétion interne du pancréas et la pathogénèse du diabète pancréatique. Arch Int Physiol 1913;13:255.
[7] Banting FG, Best CH, Collip JB, Campbell WR, Fletcher AA. Pancreatic extracts in the treatment of diabetes mellitus. Can Med Assoc J 1922;12(3):141–6.
[8] Ivy AC, Farrell JL. Contribution to the physiology of the pancreas. I. A method for subcutaneous auto-transplant of the tail of the pancreas. Am J Physiol 1926;77:474–9.
[9] RaGM G. La regulation de la secretion interne pancreatique par une processus humoral, demonstree par des transplantations de pancreas. Experience sur des animaux normaux. CR Soc Biol 1927;67:1613–20.
[10] Houssay B-A, Molinelli E-A. La greffe duodéno-pancréatique et son emploi pour déceler les décharges d'adrénaline et de sécrétine dans le sang. CR Soc Biol 1927;97:1032.
[11] Houssay B-A. Technique de la greffe pancreaticoduodenale au cou. CR Soc Biol 1929;100:138.
[12] Carrel A. La technique operatoire des anastomoses vasculaires et la transplantation des visceres. Lyon Med 1902;98:859–64.
[13] Selle WA. Studies on pancreatic grafts made with new technique. Am J Physiol 1935;113:118–9.
[14] Bottin J. Transplantation du pancréas sur la circulation carotido-jugulaire chez le chien. Survie de l'animal. Cause de la mort. CR Soc Biol 1936;121:872–6.
[15] Rundles WR, Swan H. Islet cell transplantation. Surg Forum 1957;7:502–6.
[16] Lichtenstein IL, Barschak RM. Experimental transplantation of the pancreas in dogs. J Int Coll Surg 1957;28(1 Part 1):1–6.
[17] Brooks JR, Gifford GH. Pancreatic homotransplantation. Transplant Bull 1959;6(1):100–3.
[18] Dejode LR, Howard JM. Studies in pancreaticoduodenal homotransplantation. Surg Gynecol Obstet 1962;114:553–8.
[19] Lucas Jr JF, Rogers RE, Reemtsma K. Homologous pancreatic transplantation in dogs: report of a new technique and studies of endocrine function. Surg Forum 1962;13:314–5.
[20] Reemtsma K, Lucas Jr JF, Rogers RE, Schmidt FE, Davis Jr FH. Islet cell function of the transplanted canine pancreas. Ann Surg 1963;158:645–54.
[21] Bergan JJ, Hoehn JG, Porter N, Dry L. Total pancreatic allografts in pancreatectomized dogs. Arch Surg 1965;90:521–6.
[22] Seddon JA, Howard JM. The exocrine behavior of the homotransplanted pancreas. Surgery 1966;59(2):226–34.
[23] Kelly WD, Lillehei RC, Merkel FK, Idezuki Y, Goetz FC. Allotransplantation of the pancreas and duodenum along with the kidney in diabetic nephropathy. Surgery 1967;61(6):827–37.

[24] Largiader F, Lyons GW, Hidalgo F, Dietzman RH, Lillehei RC. Orthotopic allotransplantation of the pancreas. Am J Surg 1967;113(1):70–6.
[25] Idezuki Y, Feemster JA, Dietzman RH, Lillehei RC. Experimental pancreaticoduodenal preservation and transplantation. Surg Gynecol Obstet 1968;126(5):1002–14.
[26] Idezuki Y, Goetz F, Kaufman S, Lillehei R. Functional testing and allotransplantation of the preserved canine pancreatic graft. In: Read before the first organ perfusion and preservation conference, Boston, Mass; 1968.
[27] Idezuki Y, Goetz FC, Feemster JA, Lillehei RC. In vitro assessment of pancreatic functional viability before allotransplantation. Trans Am Soc Artif Intern Org 1968;14:146–51.
[28] Idezuki Y, Goetz FC, Kaufman SE, Lillehei RC. In vitro insulin productivity of preserved pancreas: a simple test to assess the viability of pancreatic allografts. Surgery 1968;64(5):940–7.
[29] Idezuki Y, Lillehei RC, Feemster JA, Dietzman RH. Pancreaticoduodenal allotransplantation in dogs. Vasc Dis 1968;5(2):78–89.
[30] Idezuki Y, Goetz FC, Lillehei RC. Experimental allotransplantation of the preserved pancreas and duodenum. Surgery 1969;65(3):485–93.
[31] Westbroek DL, De Gruyl J, Dijkhuis CM, et al. Twenty-four-hour hypothermic preservation perfusion and storage of the duct-ligated canine pancreas with transplantation. Transplant Proc 1974;6(3):319–22.
[32] Dubernard JM, Traeger J, Neyra P, Touraine JL, Tranchant D, Blanc-Brunat N. A new method of preparation of segmental pancreatic grafts for transplantation: trials in dogs and in man. Surgery 1978;84(5):633–9.
[33] Little JM, Stephen M, Hogg J. Duct obstruction with an acrylate glue for treatment of chronic alcoholic pancreatitis. Lancet 1979;2(8142):557–9.
[34] Groth CG, Lundgren G, Arner P, et al. Rejection of isolated pancreatic allografts in patients with with diabetes. Surg Gynecol Obstet 1976;143(6):933–40.
[35] Gliedman ML, Gold M, Whittaker J, et al. Pancreatic duct to ureter anastomosis for exocrine drainage in pancreatic transplantation. Am J Surg 1973;125(2):245–52.
[36] Sutherland DE, Goetz FC, Najarian JS. Intraperitoneal transplantation of immediately vascularized segmental pancreatic grafts without duct ligation. A clinical trial. Transplantation 1979;28(6):485–91.
[37] Registry IPT. Annual report. The 5th congress of the international pancreas and islet transplant association. June 18-22; 1995. Miami Beach, Florida.
[38] Sutherland D, Gruessner R, Gores P. Pancreas and islet transplantation: an update. Transplant Rev 1994;8(4):185–206.
[39] Florack G, Sutherland DE, Heil J, Squifflet JP, Najarian JS. Preservation of canine segmental pancreatic autografts: cold storage versus pulsatile machine perfusion. J Surg Res 1983;34(5):493–504.
[40] Belzer FO. Clinical organ preservation with UW solution. Transplantation 1989;47(6):1097–8.
[41] Wahlberg JA, Love R, Landegaard L, Southard JH, Belzer FO. 72-hour preservation of the canine pancreas. Transplantation 1987;43(1):5–8.
[42] Prieto M, Sutherland DE, Fernandez-Cruz L, Heil J, Najarian JS. Urinary amylase monitoring for early diagnosis of pancreas allograft rejection in dogs. J Surg Res 1986;40(6):597–604.
[43] Ekberg H, Deane SA, Allen RD, et al. Monitoring of canine pancreas allograft function with measurements of urinary amylase. Aust N Z J Surg 1988;58(7):583–6.
[44] Ekberg H, Allen RD, Greenberg ML, et al. Percutaneous fine needle aspiration biopsy of canine pancreas allograft provides diagnosis of treatable rejection. J Surg Res 1989;47(4):348–53.
[45] Fujino Y, Kuroda Y, Suzuki Y, et al. Preservation of canine pancreas for 96 hours by a modified two-layer (UW solution/perfluorochemical) cold storage method. Transplantation 1991;51(5):1133–5.
[46] Matsumoto S, Kandaswamy R, Sutherland DE, et al. Clinical application of the two-layer (University of Wisconsin solution/perfluorochemical plus O2) method of pancreas preservation before transplantation. Transplantation 2000;70(5):771–4.
[47] Griffin AD, Hawthorne WJ, Allen RD, et al. Twenty-four-hour preservation of canine pancreas allografts using low-cost, low-viscosity, modified University of Wisconsin cold storage solution. Transplant Proc 1993;25(1 Pt 2):1595–6.
[48] Leonhardt U, Barthel M, Tytko A, et al. Preservation of the porcine pancreas with HTK and Euro-Collins solution: studies in a reperfusion system. Eur J Clin Invest 1990;20(5):536–9.
[49] Troisi R, Meester D, Regaert B, et al. Physiologic and metabolic results of pancreatic cold storage with histidine-tryptophan-ketoglutarate-HTK solution (Custodiol) in the porcine autotransplantation model. Transpl Int 2000;13(2):98–105.
[50] Becker T, Ringe B, Nyibata M, et al. Pancreas transplantation with histidine-tryptophan-ketoglutarate (HTK) solution and University of Wisconsin (UW) solution: is there a difference? JOP 2007;8(3):304–11.
[51] Hameed AM, Wong G, Laurence JM, Lam VWT, Pleass HC, Hawthorne WJ. A systematic review and meta-analysis of cold in situ perfusion and preservation for pancreas transplantation. HPB (Oxford) 2017;19(11):933–43.
[52] Mei S, Huang Z, Dong Y, et al. Pancreas preservation time as a predictor of prolonged hospital stay after pancreas transplantation. J Int Med Res 2021;49(2), 300060520987059.
[53] Hawthorne WJ, Allen RD, Greenberg ML, et al. Simultaneous pancreas and kidney transplant rejection: separate or synchronous events? Transplantation 1997;63(3):352–8.
[54] Hawthorne WJ, Griffin AD, Lau H, Ekberg H, Allen RD. The effect of venous drainage on glucose homeostasis after experimental pancreas transplantation. Transplantation 1996;62(4):435–41.
[55] Hawthorne WJ, Wilson TG, Williamson P, et al. Long-term duct-occluded segmental pancreatic autografts: absence of microvascular diabetic complications. Transplantation 1997;64(7):953–9.
[56] Gruessner RW, Gruessner AC. The current state of pancreas transplantation. Nat Rev Endocrinol 2013;9(9):555–62.
[57] Shahrestani S, Tran HM, Pleass HC, Hawthorne WJ. Optimal surgical management in kidney and pancreas transplantation to minimise wound complications: a systematic review and meta-analysis. Ann Med Surg (Lond) 2018;33:24–31.

[58] Stratta RJ, Gruessner AC, Odorico JS, Fridell JA, Gruessner RW. Pancreas transplantation: an alarming crisis in confidence. Am J Transplant 2016;16(9):2556–62.
[59] Berney T, Andres A, Toso C, Majno P, Squifflet JP. mTOR inhibition and clinical transplantation: pancreas and islet. Transplantation 2018;102(2S Suppl 1):S30–1.
[60] Shahrestani S, Tran HM, Pleass H, Hawthorne WJ. Optimal surgical management in kidney and pancreas transplantation to minimise wound complications: a systematic review and meta-analysis. Ann Med Surg (Lond) 2018;33:24–31.
[61] Larson J, Stratta R. Consequences of pancreas transplantation. J Invest Med 1994;42(4):622–31.
[62] Dholakia S, Oskrochi Y, Easton G, Papalois V. Advances in pancreas transplantation. J R Soc Med 2016;109(4):141–6.
[63] Venstrom JM, McBride MA, Rother KI, Hirshberg B, Orchard TJ, Harlan DM. Survival after pancreas transplantation in patients with diabetes and preserved kidney function. JAMA 2003;290(21):2817–23.
[64] Kandaswamy R, Stock PG, Miller J, et al. OPTN/SRTR 2019 annual data report: pancreas. Am J Transplant 2021;21(Suppl 2):138–207.
[65] Morel P, Sutherland DE, Almond PS, et al. Assessment of renal function in type I diabetic patients after kidney, pancreas, or combined kidney-pancreas transplantation. Transplantation 1991;51(6):1184–9.
[66] Ssobolew L. Zur normalen und pathologischen Morphologie der inneren Secretion der Bauchspeicheldrüse. Virchows Arch 1902;168(1):91–128.
[67] Pybus F. Notes on suprarenal and pancreatic grafting. Lancet 1924;204(5272):550–1.
[68] Bloom W. A new type of granular cell inthe islets of Langerhans of man. Anat Rec 1931;49:363–71.
[69] Hellerström C. A method for the microdissection of intact pancreatic islets of mammals. Eur J Endocrinol 1964;45(1):122–32.
[70] Moskalewski S. Isolation and culture of the islets of Langerhans of the guinea pig. Gen Comp Endocrinol 1965;5:342–53.
[71] Lacy PE, Kostianovsky M. Method for the isolation of intact islets of Langerhans from the rat pancreas. Diabetes 1967;16(1):35–9.
[72] Lindall A, Steffes M, Sorenson R. Immunoassayable insulin content of subcellular fractions of rat islets. Endocrinology 1969;85(2):218–23.
[73] Ballinger WF, Lacy PE. Transplantation of intact pancreatic islets in rats. Surgery 1972;72(2):175–86.
[74] Kemp CB, Knight MJ, Scharp DW, Ballinger WF, Lacy PE. Effect of transplantation site on the results of pancreatic islet isografts in diabetic rats. Diabetologia 1973;9(6):486–91.
[75] Kolb E, Ruckert R, Largiader F. Intraportal and intrasplenic autotransplantation of pancreatic islets in the dog. Eur Surg Res 1977;9(6):419–26.
[76] Najarian JS, Sutherland DE, Matas AJ, Steffes MW, Simmons RL, Goetz FC. Human islet transplantation: a preliminary report. Transplant Proc 1977;9(1):233–6.
[77] Lorenz D, Lippert H, Panzig E, et al. Transplantation of isolated islets of Langerhans in diabetic dogs. III. Donor selection by mixed lymphocyte reaction and immunosuppressive treatment. J Surg Res 1979;27(3):205–13.
[78] Toledo-Pereyra LH, Valjee KD, Zammit M. Important factors in islet cell transplantation: the role of pancreatic fragments' size, pH, potassium concentration and length of intraportal infusion. Eur Surg Res 1980;12(1):72–8.
[79] Horaguchi A, Merrell RC. Preparation of viable islet cells from dogs by a new method. Diabetes 1981;30(5):455–8.
[80] Piemonti L, Pileggi A. 25 Years of the Ricordi automated method for islet isolation. CellR4 Repair Replace Regen Reprogram 2013;1(1):e128.
[81] Gray D, Warnock G, Sutton R, Peters M, McShane P, Morris P. Successful autotransplantation of isolated islets of Langerhans in the cynomolgus monkey. Br J Surg 1986;73(10):850–3.
[82] Ricordi C, Lacy PE, Finke EH, Olack BJ, Scharp DW. Automated method for isolation of human pancreatic islets. Diabetes 1988;37(4):413–20.
[83] Warnock GL, Ellis DK, Cattral M, Untch D, Kneteman NM, Rajotte RV. Viable purified islets of Langerhans from collagenase-perfused human pancreas. Diabetes 1989;38(Suppl. 1):136–9.
[84] Tzakis AG, Zeng Y, Fung J, et al. Pancreatic islet transplantation after upper abdominal exeriteration and liver replacement. Lancet 1990;336(8712):402–5.
[85] Socci C, Falqui L, Davalli A, et al. Fresh human islet transplantation to replace pancreatic endocrine function in type 1 diabetic patients. Acta Diabetol 1991;28(2):151–7.
[86] Ricordi C, Tzakis AG, Carroll PB, et al. Human islet isolation and allotransplantation in 22 consecutive cases. Transplantation 1992;53(2):407.
[87] Wahoff D, Najarian J, Sutherland D, Gores P. Effect of pancreatic islet allografts on kidney allograft rejection incidence in simultaneous islet/kidney and islet after kidney recipients. In: Paper presented at: Transplantation proceedings; 1994.
[88] Smith CV, Imagawa DK, Stock PG, et al. Simultaneous islet-liver transplantation: preliminary results from the UC islet transplantation consortium. Transplant Proc 1998;30(2):295–6.
[89] Alejandro R, Lehmann R, Ricordi C, et al. Long-term function (6 years) of islet allografts in type 1 diabetes. Diabetes 1997;46(12):1983–9.
[90] Shapiro AJ, Lakey JR, Ryan EA, et al. Islet transplantation in seven patients with type 1 diabetes mellitus using a glucocorticoid-free immunosuppressive regimen. N Engl J Med 2000;343(4):230–8.
[91] Hawthorne W, Patel A, Walters S, Stokes R, Chapman J, O'Connell P. Development of an islet isolation program for clinical islet cell transplantation. Immunol Cell Biol 2004;82(2).
[92] Hyon SH, Ceballos MC, Barbich M, et al. Effect of the embolization of completely unpurified islets on portal vein pressure and hepatic biochemistry in clinical practice. Cell Transplant 2004;13(1):61–5.
[93] O'Connell PJ, Hawthorne WJ, Holmes-Walker DJ, et al. Clinical islet transplantation in type 1 diabetes mellitus: results of Australia's first trial. Med J Aust 2006;184(5):221–5.
[94] Shapiro AJ, Ricordi C, Hering BJ, et al. International trial of the Edmonton protocol for islet transplantation. N Engl J Med 2006;355(13):1318–30.
[95] Hering BJ, Clarke WR, Bridges ND, et al. Phase 3 trial of transplantation of human islets in type 1 diabetes complicated by severe hypoglycemia. Diabetes Care 2016;39(7):1230–40.
[96] Holmes-Walker DJ, Gunton JE, Hawthorne W, et al. Islet transplantation provides superior glycemic control with less hypoglycemia compared with continuous subcutaneous insulin infusion or multiple daily insulin injections. Transplantation 2017;101(6):1268–75.

[97] Stokes RA, Simond DM, Burns H, et al. Transplantation sites for porcine islets. Diabetologia 2017;60(10):1972–6.
[98] Foster ED, Bridges ND, Feurer ID, et al. Improved health-related quality of life in a phase 3 islet transplantation trial in type 1 diabetes complicated by severe hypoglycemia. Diabetes Care 2018;41(5):1001–8.
[99] Barton FB, Rickels MR, Alejandro R, et al. Improvement in outcomes of clinical islet transplantation: 1999-2010. Diabetes Care 2012;35(7):1436–45.
[100] Bertuzzi FAB, Tosca MC, Galuzzi M, Bonomo M, Marazzi M, Colussi G. Islet transplantation in pediatric patients: current indications and future perspectives. Endocr Dev 2016;30:14–22.
[101] Greenwald MA, Kuehnert MJ, Fishman JA. Infectious disease transmission during organ and tissue transplantation. Emerg Infect Dis 2012;18(8), e1.
[102] Martinez-Pourcher V. Infections in the transplant patient. Rev Prat 2015;65(8):1075–8.
[103] Shahrestani S, Webster AC, Lam VW, Yuen L, Ryan B, Pleass HC, Hawthorne WJ. Outcomes from pancreatic transplantation in donation after cardiac death: a systematic review and meta-analysis. Transplantation 2016;101:122–30.
[104] Hameed AM, Hawthorne WJ, Pleass HC. Advances in organ preservation for transplantation. ANZ J Surg 2017;87(12):976–80.
[105] Murray LMN, Fleming J, Bailey L. Use of the BacT/alert system for rapid detection of microbial contamination in a pilot study using pancreatic islet cell products. J Clin Microbiol 2014;52(10):3769–71.
[106] Wang Y, Danielson KK, Ropski A, Harvat T, et al. Systematic analysis of donor and isolation factor's impact on human islet yield and size distribution. Cell Transplant 2013;22(12):2323–33.
[107] O'Connell PJ, Holmes-Walker DJ, Goodman D, et al. Multicenter Australian trial of islet transplantation: improving accessibility and outcomes. Am J Transplant 2013;13(7):1850–8.
[108] Hedley JA, Anderson PF, et al. Australia and New Zealand islet and pancreas transplant registry annual report 2018—islet donations, islet isolations, and islet transplants. Transplant Direct 2019;5(2):e421.
[109] Hays R. Transplant community looks for ways to incentivize living organ donation. Nephrol News Issues 2016;30(1):24. 26–27.
[110] Organ and Tissue Authority PBCSA. Australian donation and transplantation 2017 activity report; 2017.
[111] Ponte GMPA, Messinger S, Alejandro A, Ichii H, Baidal DA, Khan A, Ricordi C, Goss JA, Alejandro R. Toward maximizing the success rates of human islet isolation: influence of donor and isolation factors. Cell Transplant 2007;16(6):595–607.
[112] Hameed AM, Yuen L, Pang T, Rogers N, Hawthorne WJ, Pleass HC. Techniques to ameliorate the impact of second warm ischemic time on kidney transplantation outcomes. In: Paper presented at: Transplantation proceedings; 2018.
[113] Hameed AM, Miraziz R, Lu DB, et al. Extra-corporeal normothermic machine perfusion of the porcine kidney: working towards future utilization in Australasia. ANZ J Surg 2018;88(5):E429–34.
[114] Morrissey PE, Monaco AP. Donation after circulatory death: current practices, ongoing challenges, and potential improvements. Transplantation 2014;97(3):258–64.
[115] Basta G, Montanucci P, Calafiore R. Islet transplantation versus stem cells for the cell therapy of type 1 diabetes mellitus. Minerva Endocrinol 2015;40(4):267–82.
[116] O'Gorman D, Kin T, Murdoch T, et al. The standardization of pancreatic donors for islet isolations. Transplantation 2005;80(6):801–6.
[117] Lakey JR, Warnock GL, Rajotte RV, et al. Variables in organ donors that affect the recovery of human islets of langerhans. Transplantation 1996;61(7):1047–53.
[118] Wang L-j, Kin T, O'gorman D, et al. A multicenter study: North American islet donor score in donor pancreas selection for human islet isolation for transplantation. Cell Transplant 2016;25(8):1515–23.
[119] Kaddis JS, Danobeitia JS, Niland JC, Stiller T, Fernandez LA. Multicenter analysis of novel and established variables associated with successful human islet isolation outcomes. Am J Transplant 2010;10:646–56.
[120] Kuhtreiber WM, Ho LT, Kamireddy A, Yacoub JA, Scharp DW. Islet isolation from human pancreas with extended cold ischemia time. Transplant Proc 2010;42(6):2027–31.
[121] Berkova Z, Saudek F, Girman P, et al. Combining donor characteristics with immunohistological data improves the prediction of islet isolation success. J Diabetes Res 2016;2016:4214328.
[122] Lyon J, Manning Fox JE, Spigelman AF, et al. Research-focused isolation of human islets from donors with and without diabetes at the Alberta Diabetes Institute IsletCore. Endocrinology 2016;157(2):560–9.
[123] Balamurugan A, Naziruddin B, Lockridge A, et al. Islet product characteristics and factors related to successful human islet transplantation from the Collaborative Islet Transplant Registry (CITR) 1999–2010. Am J Transplant 2014;14(11):2595–606.
[124] Andres A, Kin T, O'Gorman D, et al. Clinical islet isolation and transplantation outcomes with deceased cardiac death donors are similar to neurological determination of death donors. Transpl Int 2016;29(1):34–40.
[125] Balamurugan A, Chang Y, Bertera S, et al. Suitability of human juvenile pancreatic islets for clinical use. Diabetologia 2006;49(8):1845–54.
[126] Meier RP, Sert I, Morel P, Muller YD, et al. Islet of Langerhans isolation from pediatric and juvenile donor pancreases. Transpl Int 2014;27(9):949–55.
[127] Nagaraju SBR, Wijkstrom M, Trucco M, Cooper DK. Islet xenotransplantation: what is the optimal age of the islet-source pig? Xenotransplantation 2015;22(1):7–19.
[128] Niclauss N, Bosco D, Morel P, Demuylder-Mischler S, Brault C, et al. Influence of donor age on islet isolation and transplantation outcome. Transplantation 2011;91(3):360–6.
[129] Cardillo M, Nano R, de Fazio N, et al. The allocation of pancreas allografts on donor age and duration of intensive care unit stay: the experience of the North Italy transplant program. Transpl Int 2014;27(4):353–61.
[130] Hering BJ, Kandaswamy R, Ansite JD, et al. Single-donor, marginal-dose islet transplantation in patients with type 1 diabetes. JAMA 2005;293(7):830–5.

[131] Thwaites SE, Gurung B, Yao J, et al. Excellent outcomes of simultaneous pancreas kidney transplantation in patients from rural and urban Australia: a national service experience. Transplantation 2012;94(12):1230–5.
[132] Hudson A, Bradbury L, Johnson R, Fuggle SV, Shaw JA, Casey JJ, Friend PJ, Watson CJ. The UK pancreas allocation scheme for whole organ and islet transplantation. Am J Transplant 2015;15(9):2443–55.
[133] Ausania F, Drage M, Manas D, Callaghan C. A registry analysis of damage to the deceased donor pancreas during procurement. Am J Transplant 2015;15(11):2955–62.
[134] Andres A, Kin T, O'Gorman D, et al. Impact of adverse pancreatic injury at surgical procurement upon islet isolation outcome. Transpl Int 2014;27(11):1135–42.
[135] Hameed A, Yu T, Yuen L, et al. Use of the harmonic scalpel in cold phase recovery of the pancreas for transplantation: the westmead technique. Transpl Int 2016;29(5):636–8.
[136] Romanescu D, Gangone E, Boeti M, Zamfir R, Dima S, Popescu I. Technical aspects involved in the harvesting and preservation of the pancreas used for pancreatic islet allotransplantation. Chirurgia (Bucur) 2013;108(3):372–80.
[137] Cao Y, Shahrestani S, Chew HC, Crawford M, et al. Donation after circulatory death for liver transplantation: a meta-analysis on the location of life support withdrawal affecting outcomes. Transplantation 2016;100:1513–24.
[138] Iwanaga YS, Sutherland DE, Harmon JV, Papas KK. Pancreas preservation for pancreas and islet transplantation. Curr Opin Organ Transplant 2008;13(2):135–41.
[139] Rickels MR, Robertson RP. Pancreatic islet transplantation in humans: recent progress and future directions. Endocr Rev 2018;40(2):631–68.
[140] Niclauss N, Wojtusciszyn A, Morel P, et al. Comparative impact on islet isolation and transplant outcome of the preservation solutions Institut Georges Lopez-1, University of Wisconsin, and Celsior. Transplantation 2012;93(7):703–8.
[141] Takita M, Itoh T, Shimoda M, et al. Pancreatic ductal perfusion at organ procurement enhances islet yield in human islet isolation. Pancreas 2014;43(8):1249–55.
[142] Leemkuil M, Lier G, Engelse MA, et al. Hypothermic oxygenated machine perfusion of the human donor pancreas. Transplant Direct 2018;4(10):e388.
[143] Scott WE, Weegman BP, Ferrer-Fabrega J, et al. Pancreas oxygen persufflation increases ATP levels as shown by nuclear magnetic resonance. Transplant Proc 2010;42(6):2011–5.
[144] Kelly AC, Smith KE, Purvis WG, et al. Oxygen perfusion (persufflation) of human pancreata enhances insulin secretion and attenuates islet proinflammatory signaling. Transplantation 2019;103(1):160–7.
[145] Barlow AD, Hosgood SA, Nicholson ML. Current state of pancreas preservation and implications for DCD pancreas transplantation. Transplantation 2013;95(12):1419–24.

Chapter 2

Whole pancreas transplantation: Advantages and disadvantages, and an overview of new technologies in organ resuscitation

Henry Pleass

Department of Surgery, Westmead Hospital, Westmead, NSW, Australia; Faculty of Medicine and Health, Specialty of Surgery, University of Sydney, Sydney, NSW, Australia

Introduction

In almost exactly 100 years, Type 1 Diabetes (T1D) mellitus has been transformed from an incurable, terminal condition, to a treatable, manageable and in some instances curable condition. Banting and Best were the first to recognize and purify insulin, with the first patient being clinically treated with insulin in Toronto, Canada in 1921 [1].

Since then, Diabetes care has continued to evolve, with the advent of clinical solid organ Pancreas transplantation being reported by Kelly and Lillehei in 1967 [2]. Initially, such efforts were fraught with morbidity and mortality, and the early years of solid organ Pancreas transplantation scarred both patients and their physicians alike.

The Achilles heel of Pancreas transplantation has always been how to manage the unwanted exocrine secretions, arising from the Pancreatic duct. Initially, duct ligation and duct occlusion were employed, but it was bladder drainage of the exocrine secretions that transformed outcomes, as well as attitudes, improving both graft and patient survivals [3]. Throughout the 1980s, bladder drainage was the mainstay of exocrine secretion management, but with the advent of improved immunosuppression and the need for enteric conversion in 15% to 30% of bladder drained Pancreas transplants, techniques continued to evolve. Since 1995, primarily enteric drainage techniques have been employed in the first instance, by almost all but a few centers worldwide, as this avoids the need for subsequent exocrine diversion, as the bladder's mucosa is not exposed to pancreatic alkaline secretions.

Pancreas Transplantation, largely in the form of Simultaneous Pancreas Kidney (SPK) transplantation has now become routine in many transplant programs around the world, with increasing numbers performed annually in many countries outside the United States of America, since 2004 [4,5] (Fig. 1). The majority of patients are patients suffering from T1D, although some units have extended indications to Type 2 Diabetes Mellitus (T2D), with equivalent outcomes [4].

Advantages of pancreas transplantation

As stated, prior to the discovery of Insulin, T1D was a terminal condition, with the often-young patients slowly drifting into coma and death. However, with the development of insulin therapies and the advent of new technologies such as feedback loop systems that monitor glucose levels and adjust insulin infusions accordingly, T1D is not the same disease it once was [6].

However, the incidence of T1D and T2D are both increasing, adding financial burden to the world's health care resources [7]. Although the majority of these patients can be managed with tight and effective glucose control, some remain resistant to even the best of medical therapies and overall have higher mortality than similarly matched individuals without diabetes [8]. Not only do both T1D and T2D have an increased risk of mortality, but they also have a significantly increased risk of end-stage renal failure, in addition to retinopathy, peripheral neuropathy, autonomic neuropathy, peripheral vascular and coronary artery disease [9].

The advantages of Pancreas transplantation will be discussed in relation to each individual complication of diabetes.

Pancreas and Beta Cell Replacement. https://doi.org/10.1016/B978-0-12-824011-3.00009-6

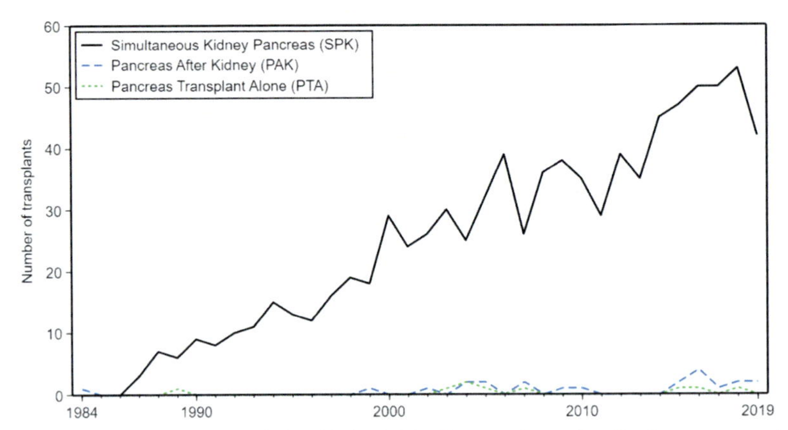

FIG. 1 Pancreas transplant over time by type, Australia and New Zealand. Islet Pancreas Transplant Registry to recognise the source of the image.

Diabetic nephropathy

Diabetic nephropathy is one of the leading causes of end-stage renal failure worldwide and affects between 10 and 30% of patients with diabetes [10]. A recent estimate from 2017, has assessed the annual cost of diabetes to the health care economy of the United States to be $327 billion, with huge implications for governments and healthcare providers around the world [11]. Although T1D can be managed with intensive medical therapy, solid organ SPK transplantation remains the widely accepted gold standard treatment for these patients who are also suffering end-stage renal failure, secondary to diabetic nephropathy [12]. The obvious advantages are that both Diabetes and renal failure can be "cured" from organs retrieved from the same deceased donor, providing theoretical protection from further diabetic nephropathy affecting the kidney allograft, but also providing the optimum pancreas graft survival, when compared with pancreas transplants alone, or pancreas transplants after kidney transplantation from two different organ donors [13].

Although only a small proportion of T1D and an even smaller proportion of T2D patients are suitable for an SPK transplantation, this surgical intervention, not only overcomes the lack of, or insulin resistance, providing a near physiological glucose control but also removes the need for dialysis for the lifespan of the Kidney transplant. This has numerous clinical and psychological benefits to the recipients, but in addition, abrogates the high mortality associated with diabetes and end-stage renal failure, when managed by either dialysis or deceased donor kidney transplant alone (KTA) [14,15].

Although T1D patients with end-stage renal failure can be managed successfully with a live kidney donor, without a reduction in long-term patient survival when compared with SPK transplantation [14]. However, almost all diabetics if given the opportunity would prefer not to be diabetic, with its an inexorable progression of their diabetic complications and limitation to their lifestyle.

Retinopathy

Diabetic retinopathy is one of the most common microvascular complications of diabetes. Currently, 450 million people are living with diabetes, with a projected increase of over 40% by 2045 [16]. As diabetic nephropathy is the major indication of pancreas transplantation in the form of SPK transplantation, patients on average have had diabetes for a quarter of a century prior to transplantation [17]. As a result of this, many have associated diabetic retinopathy and some of which are already registered blind prior to surgery.

The benefits of SPK transplantation on diabetic retinopathy are debatable, as no randomized control trial evidence exists, nor would it be ethical to perform one, simply to determine effects on eyesight. However conflicting evidence is reported on proliferative retinopathy, with one study suggesting stable disease with a functioning pancreas transplant in 86% of patients, compared with stable disease in only 43% of those managed without transplanation [18]. In another earlier article, there was no difference in the stabilization of proliferative retinopathy, whether a functioning pancreas transplant was present or not, and this was most marked in those with severe retinopthay [19]. Thus, regarding this common complication, solid organ pancreas transplantation probably has no overall benefit, although beyond four years patients do tend to have more stable disease, a proportion may well have worsened vision in the early postoperative period.

Peripheral neuropathy/autonomic neuropathy

Hyperglycaemia as seen in diabetes, causes polyneuropathy, affecting peripheral motor, sensory and autonomic nerves. In the long term, this may cause peripheral ulceration, commonly affecting the toes and feet of diabetics, leading to chronic sepsis and amputation. In addition, gross deformity of peripheral joints can arise, with Charcot joints being a further impediment to mobility, as well as an additional risk factor for ulceration.

Autonomic neuropathy can cause profound postural hypotension, as well as gastroparesis, both having a huge impact on the patient's quality of life as well as ongoing nutrition. The Minnesota group first showed in 1990, that a functioning Pancreas transplant could improve peripheral motor, sensory and autonomic nerve function at 3.5 years post transplantation when compared with a control group of patients suffering from long-term diabetes who remained on insulin [20]. They later published a series of 115 pancreas transplant recipients, who were a mixture of SPK, Pancreas After Kidney (PAK), and Pancreas Transplant Alone (PTA) procedures and compared their outcomes with a series of 92 control patients, over a 10-year period. All patients underwent clinical as well as nerve conduction studies to assess peripheral motor, sensory and autonomic function [21]. Although autonomic function and clinical assessment showed only a slight improvement in the long term Pancreas transplant recipients, the peripheral motor and sensory nerve conduction studies showed a significant improvement, compared with controls at all time points up to the end of follow up at 10 years. Similarly the Italian group from Pisa, more recently published a series of 71 PTA patients, identifying improvements in nerve conduction and autonomic function, with long-term follow-up [22].

Clinically these improvements in peripheral and autonomic neuropathy, have major a impact on quality of life, with pancreas transplant recipients having improved gastric and gastrointestinal function, provided that the neuropathy was not severe prior to transplantation. Cashion et al. studied 42 SPK transplant recipients before and up to 24 months post-transplantation to investigate gastroparesis [23]. Using electrogastrography and gastric symptom data, they demonstrated that although symptoms diminish post-transplantation, there was a shift from bradygastria to tachygastria and that gastroparesis remains a significant problem in the 2 years of follow up.

It has also been demonstrated that the intraepidermal nerve fiber density was significantly reduced in diabetic patients receiving an SPK transplant, compared with controls. With a median follow-up of 29 months, there was no significant improvement for the group as a whole, although three of the 18 patients did show some improvement in nerve fiber density on skin biopsy [24]. This would suggest that the prior chronic hyperglycemia causes irreversible nerve damage in some diabetic patients, despite the successful achievement of normoglycaemia post pancreas transplantation. Similarly, a spectral analysis of heart rate variation in a cohort of SPK recipients compared with healthy controls and diabetics managed by kidney transplant alone showed no improvement in heart rate variation during the course of follow-up [25]. Again this would suggest that autonomic neuropathy when severe enough, cannot be reversed fully by successful pancreas transplantation, at least in the follow-up period of 25 months within this study.

Macrovascular disease: Peripheral, cardiac, and neurovascular

Cardiovascular events in pancreas transplant recipients have always been the major cause of death in the post-operative period and long term [26]. Hence evidence of reversible myocardial ischemia remains a contraindication to Pancreas transplantation unless corrected either by surgery or coronary stent insertion [27]. A recent comparative study with a 10-year median follow-up showed that T1D recipients of either SPK or live donor kidney transplants had a similar degree of coronary artery disease progression on coronary angiography review [28]. However, there is no doubt that SPK transplantation provides a significant survival advantage compared with either staying on dialysis or receiving a cadaveric donor kidney. A recent ANZDATA and ANZIPTR analysis, using a probabilistic Markov model showed the average life-years saved and quality-adjusted life years, gained by an SPK transplant when compared with dialysis were 4.13 and 2.99 respectively [29]. All SPK transplant recipients were 55 years or younger, with the greatest incremental gains occurring in those under 50 years of age.

La Rocca and colleagues showed almost 20 years ago now, that patients suffering T1D and renal failure, managed with an SPK transplant had lower mortality recorded over the 7-year follow-up, compared with a similar patient cohort managed with kidney transplantation alone [30]. In addition, the same group showed using radionuclide ventriculography, that the left ventricular ejection fraction improved in the SPK cohort, with significant differences in improved cardiac function at 2 and 4 years [30].

More recently, using UK Transplant Registry data, patients with T1D survival were compared between SPK and live donor kidney transplant cohorts. On multivariate analysis, there was no association between these two groups and patient survival overall. However, if the Pancreas remained functioning, these recipients had a significantly better overall survival and kidney allograft survival, than those who received a live donor kidney alone, or those with a failed pancreas [31].

Although successful SPK transplantation appears to offer a survival advantage and improved cardiac function, it has not reduced the risk of peripheral vascular disease, as reflected by limb amputation rate [32,33]. The Dublin group showed that patients with T1D were at a greater risk of amputation following SPK transplant 9%, than if managed by a kidney transplant alone (KTA), which was only 2% in their series [33].

However, amputation rates may be influenced by both small and large vessel disease, as well as peripheral neuropathy. Richard Allen and colleagues showed from a prospective series of nerve conduction studies, that conduction velocity improved early following successful SPK transplantation, but nerve amplitude continued to slowly improve for up to 8 years [34]. Despite this improvement in neuropathy, the incidence of diabetic foot ulcers is not reduced with a functioning SPK transplant, when compared retrospectively to a cohort of kidney-only diabetic recipients with a median six-year follow up [35]. In this series, with a median follow-up of 6 years, 17% of SPK recipients and 15% of Kidney only recipients developed diabetic foot ulcers. Hence although a functioning pancreas transplant did not reduce the incidence of diabetic foot ulcers in this series, it had no impact on allograft failure. Conversely, kidney transplant only recipients had greater than a five-fold increased hazard ratio of kidney allograft failure associated with the development of a diabetic foot ulcer during follow-up.

Disadvantages of pancreas transplantation

Although solid organ Pancreas transplantation is highly successful, as discussed in the chapter above, it is not suitable or indeed available for the vast majority of Diabetic patients worldwide. It requires a major operation and in the vast majority of cases a deceased organ donor, to provide a suitable pancreas and often kidney, which limits the availability of this complex procedure.

Apart from the lack of availability of suitable donor organs, the main disadvantage to Pancreas transplantation is the risk of mortality from a major surgical procedure, followed by long-term immunosuppression [36]. Hence the need for careful recipient selection, along with donor selection, to minimize the risks to the recipient [37].

The complications of Pancreas transplantation are listed in Table 1 below and will be discussed individually, as a specific disadvantage. The complications secondary to immunosuppression are not specific to Pancreas transplantation and hence will not be individually discussed, although the heavy immunosuppression required predisposes the recipients to the standard post-operative infectious diseases, as well as post-transplant lymphoproliferative disease and malignancy [38].

Vascular thrombosis

Vascular thrombosis has remained an issue with pancreas transplantation and although the incidence has reduced over the eras, remains higher than that seen with kidney transplantation alone. A recent single center study reported an incidence of graft loss due to thrombosis as 5.6% and remains the leading cause of early graft loss [39] in pancreas transplantation. This compares for example with renal allograft loss due to vascular thrombosis in the US pediatric population of 0.8% in a comparative era cohort [40].

Although the standard use of Tacrolimus has been shown to significantly reduce Pancreas graft thrombosis, when compared with Cyslosporine [41], many centers have also tried some form of anticoagulation, although this may increase the risks of a return to theater for hemorrhage [42]. From national data, donor age, donor obesity, and smoking history, all predispose the recipient to increased pancreas graft thrombosis, although differences did not reach statistical significance [5].

TABLE 1 Complications of pancreas transplantation.

Complication
1. Vascular thrombosis
2. Pancreatic exocrine leak
3. Need for re-laparotomy
4. Peri-pancreatic collection or sepsis
5. Vascular lesion eg AV fistula or mycotic aneurysm
6. Cystic neoplasm of Pancreas

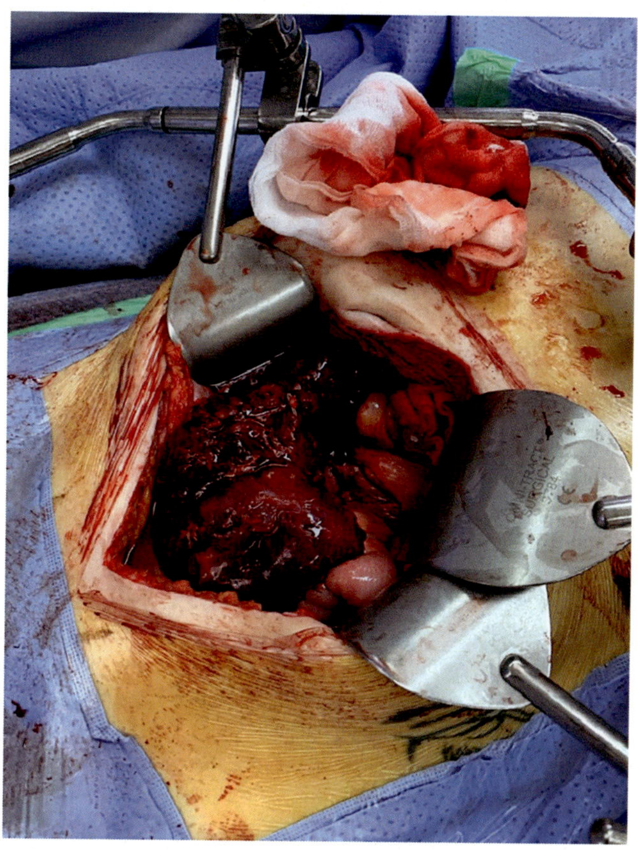

FIG. 2 Pancreatic graft thrombosis secondary to vascular rejection (Author's own image).

A recent systematic review has investigated pancreatic graft thrombosis, finding the overall incidence within the published literature of just over 7%, with the majority of these resulting in graft loss, rather than successful graft salvage [43]. It remains an unavoidable major complication of solid organ pancreas transplantation (Fig. 2).

Pancreatic exocrine leakage and re-laparotomy

Although graft thrombosis is frequently devastating if occlusive in nature, a pancreatic enteric leak is extremely morbid, with an increased length of stay and the need for drainage or relaparotomy, but rarely leads to loss of the pancreatic graft [26]. The reported incidence of an enteric leak is variable, but a recent retrospective review from the author's own unit, has found the incidence to be 3.5% in over 425 SPK transplants, with the greatest risk factor being an underlying vascular disease with a hazard ration of 4.4 fold [44].

Although an exocrine leak can be managed effectively, it is often a cause for relaparotomy, which has been reported in up to 30% of series including our own [17]. Causes are multifactorial, but ischemia, acute rejection, donor selection, and surgical technique, no doubt all play a part. From the author's experience, leaks tend to occur from the blind stapled end of the graft duodenum, rather than the surgical anastomosis, but can be successfully managed with re-exploration of the graft, surgical washout, and surgical site drainage, coupled with salvage bladder drainage if pancreas positioning allows.

Avoidance of exocrine leaks has been part of the evolution of successful pancreas transplantation, with numerous techniques employed over the last 50 years. Although the technique of bladder drainage helped to make Pancreas transplantation more routine, outcomes are significantly poorer, when compared with enteric drainage, due to the risks of cystitis and urethritis, and repeated urinary tract infections [45]. As a result, almost all units worldwide employ an enteric drainage technique, between the graft duodenum and small bowel, although some use the native stomach or recipient duodenum as the site of anastomosis [3].

Peri-pancreatic collections, mycotic aneurysms, and other complications

The need for re-laparotomy is common in many series of Pancreas transplant procedures worldwide and commonly this is due to bleeding and hematoma, with underlying sepsis increasing the need for surgical washout [17]. Although such

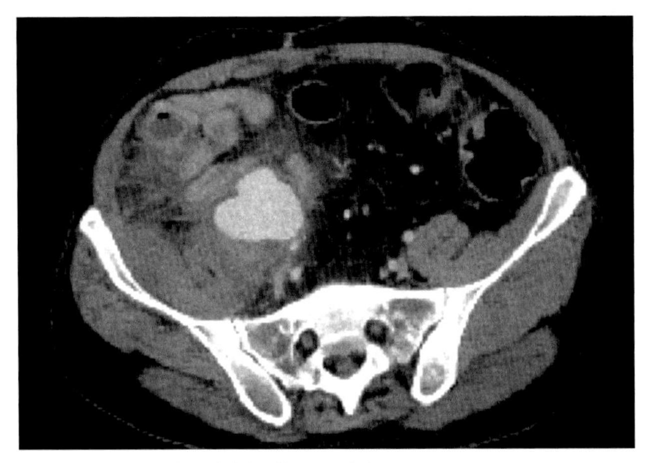

FIG. 3 CT angiogram showing pseudoaneurysm (Mycotic) of right Iliac arterial system.

complications can at times be managed with percutaneous drainage, heavy immunosuppression frequently dictates a more aggressive surgical management option for peri-pancreatic collections [46].

Mycotic aneurysm formation (Fig. 3) is a rare complication of pancreas transplantation, but a recent literature review identified that over half the published cases managed acutely with endovascular stenting, ultimately required graft excision [47]. The advantage of this interventional technique if feasible is that the acute presentation with bleed can be managed rapidly and subsequent surgical intervention can be performed in a more planned manner. In this way only, one patient death occurred in the published series, of unrelated causes.

Pancreas transplants, like the native Pancreas, are also at risk of neoplasia with graft pancreatectomy being required because of the underlying risk of neoplasia. The Wisconsin group reported cystic lesions within the pancreatic graft with an incidence of 1.8% in over 1100 transplants. In this series, the majority of lesions were pseudocysts, although two adenocarcinomas were found, in addition to four intraductal papillary mucinous neoplasms [48]. The presentation appeared within this large series at a mean of 65 months, which is similar to the author's own experience (Fig. 4) of a pancreatic graft completely replaced with main duct and branch type intraductal papillary mucinous neoplasms, 8 years after transplantation [49]. In these cases, prophylactic graft pancreatectomy is recommended to pre-empt invasive malignancy, although a local pancreatic resection may be appropriate with a localized cystic neoplasm.

New technology in pancreas resuscitation

Although the use of iced saline slush has become the standard method for organ storage for over 40 years, the concept of machine technology to store, preserve or improve organ quality is not new. Alexis Carrel, one of the early vascular surgical

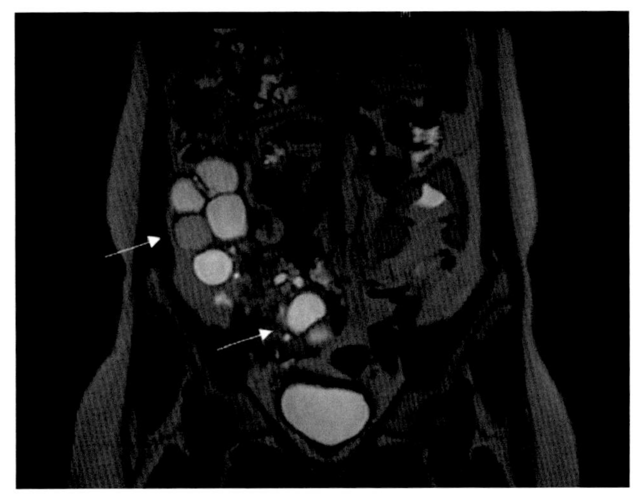

FIG. 4 MRI showing Pancreas transplant replaced with multiple large cystic neoplasms arrowed.

pioneers proposed an ex vivo pump for organs, prior to the second world war and famously made the cover of Time magazine with his co-author, Charles Lindbergh the famous aviator [50].

During the twentieth century, hypothermic machine technology was employed, particularly with Kidneys, until Opelz published evidence that iced saline slush or standard cold storage was not only significantly better in terms of one-year graft survival, but significantly simpler and cheaper [51]. Since then, iced saline slush has been the preferred organ storage and transportation method of choice and remains so to this day for the majority of transplant programs around the world.

However, with the increasing demand for suitable transplant donor organs, the donation sector has pushed the boundaries worldwide, with the increasing use of both marginal donors and donations after circulatory death donors. This change in donation practice has driven the desire and need to revisit machine perfusion technology.

Hypothermic machine perfusion

A landmark paper by Moers in 2009, proved that in a large European multi-center randomized, controlled trial, the use of hypothermic machine perfusion with deceased donor kidneys, when compared with standard cold storage, reduced the incidence of delayed graft function and reduced the risk of graft failure at 1 year [52].

Since then, many European centers have continued to use this technology with kidneys, although the benefit was not seen in the DCD cohort within the original randomized controlled trial, showing no significant difference in graft survival at 1 year [53].

The pancreas is perceived as an organ susceptible to both retrieval injury, as well as over perfusion. The technique of pancreas retrieval can significantly impact the recipient [54]. However, like the other transplantable organs, the use of DCD Pancreases for transplantation has increased, with an increased risk of graft thrombosis, but no significant difference in one-year of pancreas graft survival, when compared with brain dead donor pancreas outcomes [55,56]. This change in practice has driven recent research in the use of hypothermic machine perfusion in the human donor pancreas and the Groningen group has been the first to report dual arterial hypothermic machine perfusion system to preserve and potentially rejuvenate the human organ [57]. Five DCD human pancreata were perfused for 6h, without any evidence of cellular injury or oedema formation. In addition, ATP levels were seen to increase up to eight-fold, when compared with cold storage alone, in both DCD and DBD human pancreata [57]. Additionally, a group from Nantes, France have published a similar experience with hypothermic pulsatile perfusion, looking at 11 discarded pancreases and again have shown that they could be satisfactorily perfused for 24h, without any signs of oedema, whilst maintaining normal staining for insulin, glucagon, and somatostatin [58].

A Cochrane review recently compared hypothermic machine perfusion and sub normothermic machine perfusion with standard cold storage of kidneys [59]. In this article 14 studies reported on the incidence of delayed graft function and identified that hypothermic machine perfusion reduced the risk of this complication in both DCD and DBD kidney cohorts when compared with standard cold storage. Despite over 2000 patients included within this analysis, no reduction in primary nonfunction rate could be proven in kidney transplantation. Likewise, sub normothermic machine perfusion lacks high-level evidence as yet for kidneys, as randomized controlled trials are not present in the literature. If we extrapolate this review of technology to the pancreas, then hypothermic machine perfusion technology may have a role to play in solid organ pancreas transplantation, perhaps selectively for DCD Pancreata, which have a known increased risk of graft thrombosis [55], although no clinical studies in pancreas transplantation are as yet available.

Normothermic machine perfusion

Normothermic machine perfusion technology has not yet been utilized in clinical pancreas transplantation, but like the Liver and kidney, may have a role to play in improving islet cell function within the solid pancreas graft.

Nasralla et al. [60] have shown that this technology used within the setting of a randomized controlled trial in Liver transplantation, significantly reduced the early allograft injury, as defined by the peak transaminase enzyme rise in the first-week post perfusion. Although it did not significantly reduce the incidence of ischaemic biliopathy in the cohort of DCD Liver grafts, it did enable the overall storage time of Livers, randomized to this technology, to be extended well beyond normal clinical practice, in addition to a 50% reduction in organ discard. A trial of Kidney normothermic machine perfusion is currently underway in the UK and results are awaited with interest. Preliminary results have shown great promise in Kidney transplantation [61], which has stimulated interest in using this technology for Pancreas transplantation (Fig. 5).

Although no clinical transplant data is as yet available regarding solid organ pancreas transplantation, Barlow et al [62] have published their experience of normothermic machine perfusion of five deceased donor human pancreata, declined for transplantation. These four grafts showed stable graft physiology, with pH maintenance, ongoing insulin production, and

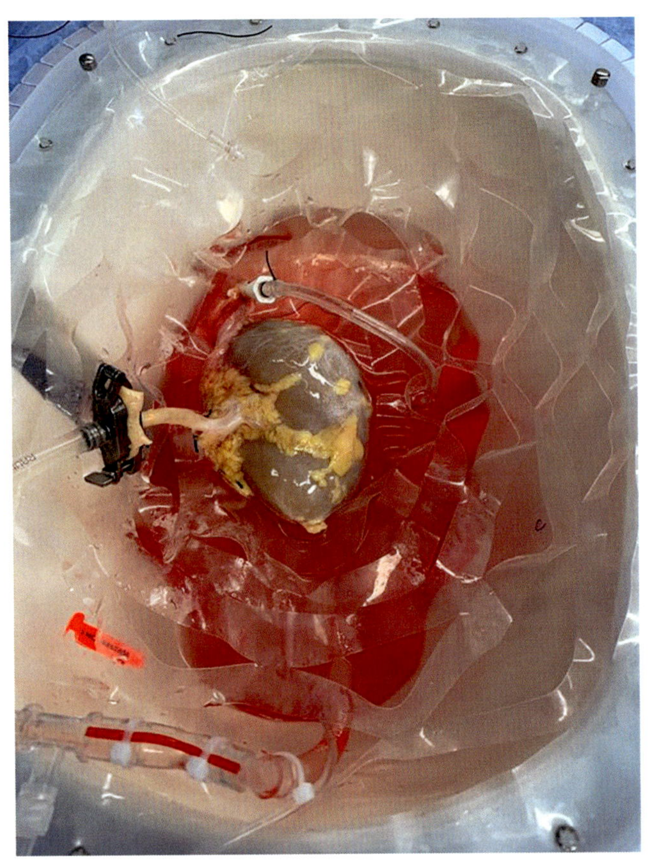

FIG. 5 Showing human kidney undergoing normothermic machine perfusion.

variable amylase and lipase levels throughout the two-hour perfusion period. Another group from the Cleveland clinic has used this NMP technology to perfuse three human pancreata, with similar findings of feasibility [63]. The main difference in this study was the short cold ischaemic time of 4h, followed by a longer duration of normothermic machine perfusion of between six and 12h. Despite this, C peptide production increased during perfusion consistent with insulin production and histological assessment showed limited pancreatic necrosis.

Conclusion

Pancreatic physiology dictates that during machine perfusion the organ requires a low flow, low-pressure environment to prevent the vascular endothelial injury occurring on graft reperfusion [64]. However, the pressure, duration, and optimal perfusate remain unproven in the clinical pancreas transplant arena. The Leicester group has compared NMP of over 10 porcine DCD pancreata at varying pressures and assessed cell death profile, exocrine and endocrine function [65]. The perfusion pressure of 20mmHg demonstrated an improved cell death profile compared with 50mmHg, although insulin release was higher in the high-pressure cohort.

Despite having advanced so far, much more work is still required to optimize this technology to increase the availability of suitable pancreata for human clinical transplantation. Localized therapies given to the pancreas whilst on machine perfusion, may be a means of reducing the risks of graft thrombosis, by targeting the graft endothelium, prior to subsequent transplantation. Like both the Liver and Kidney, the duration of machine perfusion of the pancreas is as yet unproven in the clinical setting, but will no doubt be an expanding area of research in the coming years as the burden of diabetes continues to increase worldwide.

References

[1] Vecchio I, Tornali C, Bragazzi NL, Martini M. The discovery of insulin: an important milestone in the history of medicine. Front Endocrinol (Lausanne) 2018;9:613. https://doi.org/10.3389/fendo.2018.00613. eCollection2018.

[2] Kelly WD, Lillehei RC, Merkel FK, Idezuki Y, Goetz FC. Allotransplantation of the pancreas and duodenum along with the kidney in diabetic nephropathy. Surgery 1967;61(6):827–37.

[3] El-Hennawy H, Stratta RJ, Smith F. Exocrine drainage in vascularised pancreas transplantation in the new millennium. World J Transplant 2016;6(2):255–71.

[4] Gruessner AC, Gruessner RWG. Pancreas transplantation for patients with type 1 and type 2 diabetes mellitus in the United States: a registry report. Gastroenterol Clin North Am 2018;47(2):417–41. https://doi.org/10.1016/j.gtc.2018:01.009.

[5] Webster AC, Hedley J, Robertson P, Mulley WR, Pilmore HL, Pleass H, Kelly PJ. Australia and New Zealand Ilsets and pancreas registry annula report 2018-pancreas waiting list, recipients and donors. Transplant Direct 2018;4(10). https://doi.org/10.1097/TDX.0000000000000832. eColection 2018, e390.

[6] Gawel WB, Tabor A, Kaminska H, Deja G, Jarosz-Chobot P. How modern technologies improve diabetic control. Pediatr Endocrinol Diabetes Metab 2018;2018(3):140–4. https://doi.org/10.5114/pedm.2018.80996.

[7] Mayer-Davis EJ, Lawrence JM, Dabalea D, et al. Incidence trends of type 1 and type 2 diabetes among youths, 2002-2012. N Engl J Med 2017;376(15):1419–29. https://doi.org/10.1056/NEJMoa1610187.

[8] Nelzen O, Bergqvist D, Lindhagen A. Long term prognosis for patients with chronic leg ulcers: a prospective cohort study. Eur J Vasc Endovasc Surg 1997;13(5):500–8. https://doi.org/10.1016/s1078-5884(97)80179-7.

[9] Kiire CA, Horak K, Lee KE, Klein BEK, Klein R. The period effect in the prevalence of proliferative diabetic retinopathy, gross proteinuria and peripheral neuropathy in type 1 diabetes: a longitudinal cohort study. PLoS One 2017;12(3), e174979.

[10] Guariguata L, Whiting DR, Hambleton I, et al. Global estimates of diabetes prevalence for 2013 and projections for 2035. Diabetes Res Clin Pract 2014;103:137–49.

[11] American Diabetes Association. Economic costs of diabetes in the U.S. in 2017. Diabetes Care 2018;41(5):917–28.

[12] Farney AC, Rogers J, Stratta RJ. Pancreas graft thrombosis: causes, prevention, diagnosis and intervention. Curr Opin Organ Transplant 2012;17(1):87–92.

[13] Gruessner AC, Gruessner RWG. Pancreas transplantation of US and non-US cases from 2005 to 2014 as reported to the united network for organ sharing (UNOS) and the International Pancreas Transplant Registry (IPTR). Rev Diabet Stud 2016;13(1):35–58.

[14] Ojo AO, Meier-Kriesche HU, Hanson JA, et al. The impact of simultaneous pancreas kidney transplantation on long term patient survival. Transplantation 2001;71(1):82–90.

[15] Lindahl JP, Hatmann A, Homeland R, et al. Improved patient survival with simultaneous pancreas and kidney transplantation in recipients with diabetic end stage renal disease. Diabetologia 2013;5696:1364–71.

[16] Cho NH, Shaw JE, Karuranga S, et al. IDF diabetes atlas: global estimates of diabetes prevalence for 2017 and projections for 2045. Diabetes Res Clin Pract 2018;138:271–81.

[17] Thwaites SE, Gurung B, Yao J, et al. Excellent outcomes of simultaneous pancreas kidney transplantation in patients from rural and urban Australia: a national service experience. Transplantation 2012;94(12):1230–5.

[18] Giannareli R, Coppelli A, Santini MS, et al. Pancreas transplant alone has beneficial effects on retinopthay in type 1 diabetic patients. Diabetologia 2006;49:2977–82.

[19] Scheider A, Meyer-Schwickerath E, Nusser J, Land W, Landgraft R. Diabetic retinopathy and pancreas transplantation: a 3 year follow up. Diabetologia 1991;34:S95–9.

[20] Kennedy WR, Navarro X, Goetz FC, Sutherland DER, Najarian JS. Effects of pancreas transplantation on diabetic neuropathy. N Engl J Med 1990;322:1031–7.

[21] Navarro X, Sutherland DER, Kennedy WR. Long-term effects of pancreatic transplantation with diabetics neuropathy. Ann Neurol 1997;42(5):727–36.

[22] Boggi U, Vistoli F, Amorese G, et al. Long term(5 years) efficacy and safety of pancreas transplantation alone in type 1 diabetics. Transplantation 2012;93:842–6.

[23] Cashion AK, Holmes SL, Hathaway DK, Gaber AO. Gastroparesis following kidney/pancreas transplant. Clin Transplant 2004;18(3):306–11.

[24] Boucek P, Havrdova T, Voska L, et al. Epidermal innervation in type 1 diabetic patients: a 2.5 year prospective study after simultaneous pancreas/kidney transplantation. Diabetes Care 2008;31(8):1611–2.

[25] Boucek P, Saudek F, Adamec M, Janousek L, Koznarova R, Havrdova T, Skibova J. Spectral analysis of heaert rate variation following simultaneous pancreas and kidney transplantation. Transplant Proc 2003;35(4):1494–8.

[26] Solinger HW, Odorico JS, Becker YT, et al. One thousand simultaneous pancreas-kidney transplants at a single Centre with 22 year follow up. Ann Surg 2009;250:618–30.

[27] The TSANZ. Clinical Guidelines for Organ Transplantation From Deceased Donors. Version 1.3; May 2019.

[28] Lindahl JP, Massey RJ, Hartmann A, et al. Cardiac assessment of patients with type 1 diabetes median 10 years after successful simultaneous pancreas and kidney transplantation compared with living donor kidney transplantation. Transplantation 2017 Jun;101(6):1261–7. https://doi.org/10.1097/TP0000000000001274.

[29] Shingde R, Calisa V, Craig JC, et al. Relative survival and quality of life benefits of pancreas-kidney transplantation, deceased kidney transplantation and dialysis in type 1 diabetes mellitus. Transpl Int 2020; [In press].

[30] La Rocca E, Fiorina P, di Carlo V, et al. Cardiovascular outcomes after kidney-pancreas and kidney alone transplantation. Kidney Int 2001;60(5):1964–71.

[31] Barlow A, Saeb-Parsy K, Watson CJ. An analysis of the survival outcomes of simultaneous pancreas and kidney transplantation compared to live donor kidney transplantation in patients with type 1 diabetes:a UK Transplant Registry study. Transpl Int 2017;30(9):884–92.

[32] Morrissey PE, Shaffer D, Monaco AP, Conway P, Madras PN. Peripheral vascular disease after kidney-pancreas transplantation in diabetic patients with end stage renal disease. Arch Surg 1997;132(4):358–61.

[33] MacCraith E, Davis NF, Browne C, Mohan P, Hickey D. Simultaneous pancreas and kidney transplantation: incidence and risk factors for amputation after 10 year follow up. Clin Transpl 2017;31(6) [Epub 2017 Apr 19].

[34] Allen RD, Al-Harbi IS, Morris JG, Clouston PD, O'Connell PJ, Chapman JR, Nankivell BJ. Diabetic neuropathy after pancreas transplantation: determinants of recovery. Transplantation 1997;63(6):830–8.
[35] Sharma A, Vas P, Cohen S, Patel T, Thomas S, Fountoulakis N, Karalliedde J. Clinical features and burden of new onset diabetic foot ulcers post simultaneous pancreas kidney transplantation and kidney only transplantation. J Diabetes Complications 2019;33(9):662–7.
[36] Parajuli S, Arunachalam A, Swanson KJ, et al. Outcomes after simultaneous kidney-pancreas versus pancreas after kidney transplantation in the current era. Clin Transplant 2019;33(12), e13732.
[37] Scalea JR, Redfield RR, Arpali E, Leverson G, Sollinger HW, Kaufman DB, Odorico JS. Pancreas transplantation in older patients is safe, but patient selection is paramount. Transpl Int 2016;29(7):810–8.
[38] Aref A, Zayan T, Pararajasingam R, Sharma A, Halawa A. Pancreatic transplantation:brief review of the current evidence. World J Transplant 2019;9(4):81–93.
[39] Kopp WH, Leeuwen CAT, Lam HD, Huurman VA, de Fijter JW, Schaapherder AF, Baranski AG, Braat AE. Retrospective study on detection, treatment and clinical outcome of graft thrombosis following pancreas transplantation. Transpl Int 2019;32(4):410–7.
[40] Wang C-S, Greenbaum LA, Patzer RE, Garro R, Warshaw B, George RP, Winterberg PD, Patel K, Hogan J. Renal allograft loss due to renal vascular thrombosis in the US pediatric renal transplantation. Pediatr Nephrol 2019;34(9):1545–55.
[41] Malaise J, Saudek F, Boucek P, Adamec M, Van Ophem D, Squifflet JP, EUROSPK Study Group. Tacrolimus compared with cyclosporine microemulsion in primary simultaneous pancreas kidney transplantation: the EURO-SPK3-year results. Transplant Proc 2005;37(6):2843–5.
[42] Scheffert JL, Taber DJ, Pilch NA, Chavin KD, Baliga PK, Bratton CF. Clinical outcomes associated with the early postoperative use of heparin in pancreas transplantation. Transplantation 2014;97(6):681–5.
[43] Blundell J, Shahrestani S, Lendzion R, Pleass H, Hawthorne W. Risk factors for erarly pancreatic allograft thrombosis following simultaneous pancreas-kidney transplantation: a systematic review. Clin Appl Thromb Hemost 2020;26. 107602962920942589.
[44] Hort A, Shahrestani S, Hitos K, Lee T, Allen R, Yuen L, et al. Enteric leaks from simultaneous pancreas kidney transplantation at a single centre: risk factors and management over a 20 year period. J Diabetes, Metab Complications 2021; [In Press].
[45] Siskind E, Amodu L, Pinto S, Akerman M, Jonsson J, Molmenti E, Ortiz J. Bladder versus enteric drainage of exocrine secretions in pancreas transplantation: a retrospective analysis of the united network for organ sharing database. Pancreas 2018;47(5):625–30.
[46] Daly B, O'Kelly K, Klassen D. Interventional procedures in whole organ and islet cell pancreas transplantation. Semin Interv Radiol 2004;21(4):335–43.
[47] Yao J, Vicaretti M, Lee T, Amaratunga R, Cocco N, Laurence J, et al. Endovascular management of mycotic pseudoaneurysm after pancreas transplantation: case report and literature review. Transplant Proc 2020;52(2):660–6.
[48] Al-Qaoud T, Martinez EJ, Sollinger HW, Kaufman D, Redfield R, Welch B, et al. Prevalence and outcomes of cystic lesions of the transplant pancreas: the University of Wisconsin experience. Am J Transplant 2018;18(2):467–77.
[49] Allan L, Yao J, Nahm C, Amaratunga R, Pleass H. Intraductal papillary mucinous neoplasm in a transplant pancreas: a rare occurrence. ANZ J Surg 2020. https://doi.org/10.1111/ans.16232.
[50] Carrel A, Lindbergh CA. The culture of whole organs. Science 1935;81:621.
[51] Opelz G, Terasaki PI. Advantage of cold storage over machine perfusion for preservation of cadaver kidneys. Transplantation 1982;33(1):64–8.
[52] Moers C, Smits JM, Maathuis M-HJ, et al. Machine perfusion or cold storage in deceased-donor kidney transplantation. N Engl J Med 2009;360(1):7–19.
[53] Jochmans I, Moers C, Smits JM, et al. Machine perfusion versus cold storage for the preservation of kidneys donated after cardiac death: a multicentre, randomised, controlled trial. Ann Surg 2010;252(5):756–64.
[54] Hameed A, Yu T, Yuen L, Lam V, Ryan B, Allen R, Laurence J, Hawthorne W, Pleass H. Use of the harmonic scalpel in cold phase recovery of the pancreas for transplantation: the Westmead technique. Transpl Int 2016 May;29(5):636–8.
[55] Shahrestani S, Webster AC, Lam VW, Yuen L, Ryan B, Pleass H, Hawthorne WJ. Outcomes from pancreatic transplantation in donation after cardiac death: a systematic review and meta-analysis. Transplantation 2017;101:122–30.
[56] Mittal S, Gilbert J, Friend P. Donors after circulatory death pancreas transplantation. Curr Opin Organ Transplant 2017;22(4):372–6.
[57] Leemkuil M, Lier G, Engelse MA, Ploeg R, de Koning EJ, 't Hart NA, Krikke C, Leuvenink HG. Hypothermic oxygenated machine perfusion of the human donor pancreas. Transplant Direct 2018;4(10), e388.
[58] Branchereau J, Renaudin K, Kervella D, Bernadet S, Karam G, Blancho G, Cantarovich D. Hypothermic pulsatile perfusion of human pancreas: preliminary technical feasibility study based on histology. Cryobiology 2018;85:56–62.
[59] Tingle SJ, Figueiredo RS, Moir JA, Goodfellow M, Talbot D, Wilson CW. Machine perfusion preservation versus static cold storage for deceased donor kidney transplantation. Cochrane Database Syst Rev 2019;3(3), CD011671.
[60] Nasralla D, Coussios CC, Mergental H, et al. A randomized trial of normothermic preservation in liver transplantation. Nature 2018;557(7703):50–6.
[61] Hosgood SA, Thompson E, Moore T, Wilson CH, Nicholson ML. Normothermic machine perfusion for the assessment and transplantation of declined human kidneys from donation after circulatory death donors. Br J Surg 2018;105(4):388–94.
[62] Barlow AD, Hamed M, Mallon DH, Brais RJ, Gribbl EF, Scott M, et al. Use of ex vivo normothermic perfusion for quality assessment of discarded human donor pancreases. Am J Transplant 2015;15(9):2475–82.
[63] Nassar A, Liu Q, Walsh M, Quintini C. Normothermic ex vivo perfusion of discarded human pancreas. Artif Organs 2018;42:334–5.
[64] Tersigni R, Toledo Pereyra LH, Pinkham J, Najarian JS. Pancreaticoduodenal preservation by hypothermic pulsatile perfusion for twenty four hours. Ann Surg 1975;182(6):743–8.
[65] Kumar R, Chung WY, Dennison AR, Garcea G. Ex vivo porcine perfusion models as a suitable platform for translational transplant research. Artif Organs 2017;41(9):E69–79.

Chapter 3

Pancreas resuscitation for whole pancreas transplantation

Peter J. Friend and Ann Etohan Ogbemudia
Nuffield Department of Surgical Sciences, Oxford Transplant Centre, University of Oxford, Oxford, United Kingdom

Introduction

Since the first successful pancreas transplant in 1966, significant progress has been made with respect to graft and recipient outcomes. However, despite this ostensibly being the optimal therapy to reverse all the manifestations of Type 1 diabetes mellitus (T1DM), it remains limited to the treatment of a highly selected minority of patients whose disease burden outweighs the short- and long-term morbidities still associated with Pancreas transplantation (PTx).

The first, and most obvious limitation of PTx as the treatment for what is a very common disease, is the need for a large number of suitable donor organs. Looking to a hypothetical future in which transplantation is the treatment of choice for a much wider range of patients with diabetes (including Type 2 diabetes mellitus), there is a clear disconnection between the number of deceased donors and the potential demand. This is amplified by the fact that current transplant practice requires a much more selective approach to donor pancreases than is the case for kidneys or livers. In fact, the utilization of donor organs is more similar to that seen in cardiothoracic than abdominal organ transplantation due to the significantly shorter ischemic times required, minimal handling, and selection of only ideal donor pancreases.

The demographic aspects of organ donation reflect those of the populations from which these are derived. Worldwide trends toward increasing age and body mass index (BMI) are also associated with increasing levels of other comorbidities (e.g., cardiovascular disease). These donor characteristics are associated with poorer allograft survival and graft loss [1]. In many parts of the world, the shortage of donor organs is clearly driving the increasing use of higher-risk donor organs and improving the methods of defining these. Transplant clinicians are becoming more skilled at abrogating the complications that are associated with suboptimal organs. This is illustrated by the increasing use of organs from donors declared dead by circulatory criteria (DCD), with changing practice first in liver and lung, and most recently heart transplantation. Similar trends have been seen in PTx, with increasing acceptance of older organs and organs from DCD donors, but the effect of this is more limited than in other organs because of the inherent morbidity of PTx and its vulnerability to the effects of ischemia–reperfusion injury (IRI). The severity and consequences of early postoperative complications of pancreatitis, thrombosis, and sepsis heighten the reluctance of pancreas transplant surgeons to deviate far from the "ideal pancreas donor" which constitutes a relatively small subset of today's organ donors.

The ideal pancreas donor [2] is conventionally described as one with a diagnosis of brain stem death (DBD), aged between 10 and 45 years, with body mass index (BMI) less than 30 kg/m^2 and without a history of excessive alcohol intake. Donors outwith this definition are therefore classified "extended criteria" donors (ECD) or "marginal." Donors after circulatory death (DCD), although not technically classified as ECD, are also regarded as high risk. There is increasing evidence that DCD status is a substantial risk factor, but that such donors can be used successfully, when not associated with other high-risk characteristics (especially older age), and with comparable outcomes to those of DBD organs, albeit with utilization rates that are markedly lower [3].

Currently the universal method of pretransplant pancreas preservation is static cold storage (SCS). The results of today's PTx practice internationally show that this is a viable method. However, the frequency and severity of early complications [2] including graft loss show that this is an area where innovation is much needed: early morbidity is largely related to ischemia–reperfusion injury (IRI), manifesting as graft pancreatitis. It is clear that attenuating IRI must be a key focus of research in the quest to improve the scope and benefit of PTx.

Pancreas and Beta Cell Replacement. https://doi.org/10.1016/B978-0-12-824011-3.00001-1

In this chapter we address the issue of pancreas preservation and consider the ways in which innovations in the field may reduce morbidity, improve outcomes, and thereby shift the threshold for offering this treatment to patients with more complex diabetes. We will (i) summarize the history of preservation strategies and their limitations; (ii) describe the molecular events that occur during organ ischemia and reperfusion, illustrating pathways that may provide therapeutic targets; and (iii) discuss the potential of machine perfusion as a means not only to improve pancreas allograft preservation, but also to enable viability assessment and resuscitation/repair of marginal donor organs.

History and current strategies of pancreas preservation for transplantation

The aim of organ preservation is to preserve tissue viability during the period between perfusion in the donor and reperfusion in the recipient. Static cold storage (SCS) is the universal cornerstone for preservation of all solid organs, and the pancreas is no exception. SCS comprises the flushing out of donor blood with a cold organ preservation solution and subsequent storage under sterile conditions at melting ice temperature (conventionally regarded as 4°C) in an icebox until implantation. The constituents of cold storage preservation solutions have evolved, although with little fundamental change since the introduction of University of Wisconsin (UW) solution [4,5]. The essential effects of solutions designed for SCS are (i) to slow cellular metabolism by cooling and (ii) to minimize the ill effects of fluid and electrolyte shifts as the functions of cell membranes cease due to cooling.

Cooling tissue to ice temperature reduces metabolic function and energy requirements, thereby enabling tissues to tolerate prolonged periods of ischemia before cellular energy stores (ATP) are depleted. This very simple approach to improving the tolerance to hypoxia has served the field of clinical organ transplantation for 60 years. The significant advantages are that it is logistically easy, affordable, and effective enough to be the gold standard for such a long time. In pancreas preservation, prolonged ischemia times (e.g., 17.2 ± 4.4 h [4]) can be achieved when SCS is used with UW solution.

However, SCS has fundamental limitations. It slows but does not completely halt cellular metabolism: anaerobic metabolism continues, albeit at a reduced rate, and there is, therefore, a point at which irreversible changes occur, thereby limiting the utility of this preservation strategy. Much of the injury associated with the transplantation of cold-stored organs occurs at the point of reperfusion in the recipient, the process known as Ischemia–Reperfusion Injury (IRI) during which the provision of oxygen and resumption of normal temperature lead to the production of radical oxygen species (ROS)—a primary effector of preservation injury. Much is now known about the molecular processes involved and that mitochondrial physiology is central to the mechanism. These molecular processes will be discussed later. The immediate consequences of IRI include edema, acidosis, depletion of antioxidants, accumulation of metabolites, apoptosis, and necrosis.

The clinical consequences of Ischemia–Reperfusion (IR) are particularly profound in the pancreas. Whereas in the kidney, IR can lead to acute injury to tubular epithelium and delayed graft function (a reversible injury, defined as continued recipient dialysis while the kidney graft recovers), in the pancreas it is manifested as reperfusion pancreatitis. It is likely that this is essentially the same process as sometimes seen following cardiopulmonary bypass and as well as with hypovolemic shock states [6,7].

The severe consequences of IRI increase the risk profile of PTx and largely explain more conservative pancreas acceptance practices and high organ discard rates: the clinical penalty for IRI is too great to be warranted in patients who could probably afford to wait for a lower-risk organ. NHSBT registry data (2019–20) from the United Kingdom (UK) revealed that pancreas retrieval occurred in 28% of deceased organ donors, and that 52% of these were ultimately discarded [8]. In the same time frame, in the United States of America (USA) it was reported that 7% of deceased donors resulted in pancreas retrieval with 25% of these discarded [9]. The 52% discard rate of retrieved organs in the UK is in stark difference to the discard rates reported for other retrieved organs: kidneys (14%), livers (21%), hearts (5%), and lungs (12%) [8]. Part of this difference reflects the difficulty of assessing the quality of a pancreas before it has been retrieved: many organs are discarded once shown to be steatotic, fibrotic, or nodular, features which are difficult or impossible to predict before visualization. The lower rate of pancreas retrieval and also of postretrieval discard in the USA probably reflects a more selective approach to organ acceptance, itself a reflection of the greater distances and costs of retrieval.

The PTx perioperative period is associated with increased surgical morbidity and mortality compared to kidney transplantation alone (KTA). Sung et al. [10] compared patient and graft survival for 7308 simultaneous kidney and pancreas (SPK) transplantations versus 4653 KTA patients with a diagnosis of T1DM over a 10-year period using registry data. They observed that in early posttransplantation, the death rate was initially significantly higher for SPK patients with a hazard ratio of HR of 1.5, but then rapidly decreasing to a similar rate to KTA at 3 months. This analysis is helpful in quantifying the higher risk entailed in the immediate posttransplant period following PTx. An immediate conclusion is that there is great value to be obtained from any novel approach to organ preservation that reduces the effects of IRI following PTx. The indications for PTx would be greatly expanded if the morbidity and mortality of the procedure could be reduced to levels seen in KTA.

The improvement of PTx outcomes seen over the last four decades has been achieved by incremental advances in many areas, but particularly in strategies to reduce IRI, most notably:

- Minimization of the period of preservation.
- Optimization of SCS solutions. Notably, University of Wisconsin was initially developed and tested for pancreas transplantation [5].
- Improved selection of donor organs by defining the criteria that determine outcome (e.g., the Pancreas Donor Risk Index PDRI).

It is intuitive that minimization of the period of preservation (cold ischemia) reduces ischemia-related allograft injury. Tactics to minimize ischemia times involve accepting local donors, particularly in the case of DCD organs, which have undergone exposure to a period of warm ischemia. Estimating the degree of hypoxic injury is now often based upon donor warm functional ischemia, defined as when the mean arterial pressure drops below 50 mmHg and/or oxygen saturation less than 70%, rather than the timing of donor cardiac arrest. In terms of organ allocation algorithms, distance (from donor to transplant hospitals) is a very approximate surrogate [11,12] for cold ischemia time and can be incorporated in organ allocation. Similarly, donor warm functional ischemia is a variable which is not known until after the allocation has been made and is therefore relevant as a factor in whether to accept/reject the organ at the point of retrieval.

It can be appreciated that any policy to limit cold or warm ischemia tends to have the effect of restricting both organ utilization and broader sharing, both undesirable consequences. A more reliable predictive method to assess the viability of an organ would have great value as this would impact on the very large number of currently discarded organs, many of which would probably function well if clinicians had access to better information. The current basis upon which acceptance decisions are made is extremely imprecise and therefore generates a high level of uncertainty. Clinicians, faced with the consequence of a donor organ that might result in a severe postoperative complication in a patient that could wait for a better lower-risk organ, are more likely to choose the "safer" option and decline.

In this context, the increasing use of DCD pancreases is an important precedent as well as one which is likely to inform other innovations. In the parallel context of liver transplantation, increasing use of DCD organs has driven the rapid development of novel perfusion (both cold and warm) technologies. In pancreas transplantation, good results (comparable to DBD) have been achieved with DCD organs, but only at the price of a more selective approach to organ acceptance (mostly with respect to age and body mass index) [13,14]. There is a strong case for testing the use of machine perfusion technology in DCD pancreas transplants to explore whether the advantages shown in the liver, especially for oxygenation during hypothermic machine perfusion, are replicated in the pancreas setting.

Despite the perceived (and actual) risks, there is a good justification for the use of higher-risk organs. The EXPAND trial conducted in Eurotransplant countries compared outcomes in SPK recipients for ECD (defined as age 50–60 years or BMI > 30) to those with standard criteria donors [15]. At one-year posttransplantation there were identical allograft survival rates but, notably, a reduction of 9% in waiting list times for those patients that received ECD transplants. Clearly, this result casts doubt as to the value of the current definition of ECD, but it also illustrates the gains that might result if we were able to systematically use a higher proportion of ECD organs.

Organ preservation solutions (OPS)

OPS have evolved to enable longer cold preservation times, also to target the negative side effects of cold storage observed in the early days of transplantation, particularly reperfusion edema and vascular thromboses. University of Wisconsin solution (UW) rapidly became the OPS of choice for PTx and largely remains so to the present day. Newer solutions for PTx include histidine–tryptophan–ketoglutarate (HTK), Institut Georges Lopez-1 (IGL-1), and Celsior. All these solutions generally follow a similar formula which includes impermeant agents and colloids to minimize edema, antioxidants for ROS scavenging, buffers to maintain physiological pH, energy substrates to support Adenosine Triphosphate (ATP) production, and electrolytes for ionic homeostasis.

A number of studies have investigated comparisons between different OPSs for PTx [16–18]. No study, however, identified any significant differences in outcome, although, indeed, none of these studies was powered sufficiently to provide unequivocal evidence.

Donor assessment

A number of donor characteristics are independently associated with increased risk of postoperative complications: these have formed the basis of a number of risk-scoring algorithms or donor risk indices. Pancreas donor risk index (PDRI) was developed by Axelrod and colleagues just over a decade ago using retrospective USA registry data that includes gender, age, race, BMI, height, cause of death, cold ischemia time (CIT), serum creatinine, pancreas transplant type (SPK or pancreas alone), and DCD status [19]. These factors, shown to be independently predictive of one-year pancreas graft survival, were weighted and combined to generate a risk score.

The Preprocurement Pancreas Allocation Suitability Scale (P-PASS) was developed and utilized in the Eurotransplant allocation system [20]. However, this is based on the factors that are currently used by clinicians in deciding whether to accept an organ and is therefore predictive of utilization rather than graft survival. It does not address the vital clinician's question of "what is likely to happen if I accept this organ?"

Any risk scoring algorithm requires validation if it is to be used in a population that is different from the one in which it was developed. The PDRI (developed in the USA) when tested in a UK population accurately discriminated graft survival for SPK but showed no association between graft survival and PDRI quartile in the Pancreas transplant alone (PTA) and Pancreas after kidney (PAK) groups [21]. The PDRI requires inclusion of the cold ischemia time, which is typically not available at the time of organ offering, and therefore a factor that must be estimated. While this is a limitation, on the other hand, to ignore the anticipated cold ischemia time would weaken the predictive capability, as this is a variable with one of the strongest effects on outcome.

The surgeon's assessment both at retrieval and on the back-table includes estimation of the degree of interlobular fat or fibrosis, and the presence of atherosclerosis, aberrant anatomy, or graft injury (during or prior to retrieval). These factors are taken in combination with the donor factors already considered to inform the final decision regarding the use of the graft. Assessment at this point by an experienced surgeon is a key part of pancreas transplantation.

The impact of ischemia and ischemia–reperfusion in the pancreas

As described, the perioperative period of PTx is associated with the highest morbidity and mortality and these outcomes are inextricably linked to IRI. In the current practice of transplantation, IRI is unavoidable as the pancreas allograft is exposed to ischemia during organ explant and preservation.

IRI is manifested in different ways in different transplant organs. In kidney transplantation it typically appears as delayed graft function, whereas in the liver it is marked by early allograft dysfunction and (later) by ischemic cholangiopathy. In both cases, there is an increase in the risk of primary nonfunction. Despite these risks, the inexorable demand from patients on transplant waiting lists has driven the use of marginal kidneys and livers: this has been to a much greater extent than in PTx. The difference is mainly due to the differing consequences of IRI in different organ types and partly due to the lower waiting list pressure for PTx.

IRI of the pancreas causes acinar necrosis, edema, and endothelial disruption which lead to graft pancreatitis and thrombosis, which combine as the main cause for early graft loss as well as the most important cause of patient morbidity [22–24]. Having established the central importance of IRI in PTx, the mechanisms involved in ischemia and reperfusion will now be reviewed in order to highlight potential therapeutic strategies. The various issues that contribute to both are discussed individually as well as in their combined effects in order to guide an understanding of their dynamics.

Ischemia

Although SCS provides hypothermia that reduces metabolic rate it does not halt it, therefore some energy requirements remain. Deprivation of oxygen terminates the ability of cells to carry out oxidative phosphorylation, the mechanism by which oxygen is used for the production of adenosine triphosphate (ATP). The response of cells in hypoxia is to shift to anaerobic respiration in order to maintain ATP production, although at a lower rate.

With reduced ATP production due to hypoxia and continued metabolic demands, a number of events occur [25]: ionic pump failure, acidosis, ROS production, and antioxidant depletion.

Ionic pump failure: One of the earlier events that occurs during ischemia is the failure of ATP-dependent ionic pumps, e.g., sodium–potassium pumps (Na^+/K^+ ATPase) that are responsible for maintaining electrochemical gradients across the cell membrane. As a consequence, sodium and other ions (e.g., calcium) start to move back to ionic equilibrium, resulting in higher levels of intracellular sodium and causing water to follow leading to cellular edema. Influx of cytosolic calcium leads to activation of calcium-dependent proteases such as trypsin, elastase, phospholipase A2, and calpain. Additionally, in SCS, there is no flow to flush out accumulated activated enzymes that leak into the tissues from cellular and endothelial defects, contributing to tissue degradation and necrosis.

Acidosis: In hypoxia there is a cellular switch from aerobic to anaerobic respiration to maintain ATP production. In this process pH falls due to the accumulation of lactic acid. Cellular acidosis consequently activates intracellular trypsinogen and endonucleases which both lead to cell injury and apoptosis, respectively.

ROS production: During ischemia, small amounts of ROS are produced, but a number of pathways initiated during ischemia potentiate ROS production during reperfusion. The main instigators accumulate in the mitochondria and include succinate (a universal signature of ischemic tissue [26]) which builds up due to the reverse catalysis by complex II (namely succinate dehydrogenase) of the electron transport chain [27], reduced cytochromes, and oxidized nicotinamide adenine dinucleotide. Another well studied precursor of ROS generation is via the accumulation of xanthine oxidase [25]. In normal conditions there is an abundance of xanthine dehydrogenase in the pancreas, but in ischemia it undergoes enzymatic conversion to xanthine oxidase by the increased cytosolic calcium.
Antioxidant depletion: ROS are normally neutralized by endogenously produced antioxidants. In ischemia ROS accumulation exceeds antioxidant capacity.

Reperfusion and ischemia–reperfusion injury (IRI)

IRI is a paradoxical phenomenon: reintroduction of blood flow and oxygenation is associated with an increase in the rate of cellular and tissue injury occurring. The consequence of the IRI cascade is the generation of an acute inflammatory response, the severity of which is dependent on the initial status of the organ [28].

Cell injury with ROS: At reperfusion there is reoxidization of the accumulated succinate, which drives ROS generation by reverse electron transport at complex I [29]. Increased levels of xanthine oxidase produce the aggressive oxygen superoxide anion (O_2^-) and hydrogen peroxide (H_2O_2). ROS production combined with increased cytosolic calcium causes opening of mitochondrial permeability transition pore which allows the release of cytochrome C, succinate, and mitochondrial DNA, the last of which is a type of damage-associated molecular pattern (DAMP). These elements induce apoptosis and necrosis. A significant property of DAMPs is activation of the immune response and amplification of IRI.
Endothelial dysfunction: IRI causes injury to endothelial cells and the endothelial glycocalyx. Injured endothelial cells produce vasoactive factors such as endothelin-1 which leads to reduction of microcirculatory blood flow.
The glycocalyx is a ubiquitous thin layer that covers the luminal surface of endothelial cells.
Its structure is an important determinant of vascular permeability that allows water and small molecules to pass through but repels negatively charged plasma proteins to the center of the lumen. Its degradation and shedding caused by the process of IRI leads to vascular hyperpermeability and interstitial edema [30].
Activation of polymorphonuclear leukocytes: Damaged endothelium and ROS cause activation of polymorphonuclear leukocytes. Injured endothelium upregulates intercellular adhesion molecule ICAM-1 expression, contributing to leucocytes' adherence to the capillary endothelium, endothelial breakdown, increased permeability, and transmigration of the leucocytes into the tissue. Leucocyte endothelial adherence also leads to occlusion of venules and impairment of blood flow. With increased vascular permeability red cells and debris leave the vasculature and block lymphatic drainage, further contributing to tissue edema and interstitial hemorrhage. Increased permeability also allows large molecules (e.g., activated proteases) into the pancreatic tissue exacerbating ongoing damage. IRI activates macrophages which release proinflammatory cytokines and proteolytic enzymes, driving further injury.
Coagulopathy: Damaged endothelium activates coagulation which causes intravascular microthrombosis that contributes to poor flow. Additionally, intravascular fibrin deposits contribute to thrombosis.
Microcirculatory failure: This is believed to be a key feature of pancreas IRI and has been demonstrated in experimental models of pancreatitis [31] and in clinical studies [32,33] using intravital imaging and measurement of tissue oxygen pressures. The anatomy and physiology of the pancreas might explain its susceptibility to microcirculatory failure during ischemia and IRI.
The parenchyma of pancreas is divided into individual lobules by connective tissue septa. Lobules predominately consist of exocrine acinar cells, among which are embedded the highly vascularized endocrine islets of Langerhans. A pancreatic lobule is normally supplied by a single end artery that divides into arterioles. The initial larger arterioles preferentially supply the islets first before draining from their many vasa efferentia to reach the acinar intercapillary bed in a manner called the insulo-acinar portal system [34]. Lobular capillaries are fenestrated and have greater permeability compared to other tissues [35] increasing susceptibility to extravasation and edema in disease states. Additionally, a few animal studies have described the presence of a physiological arteriovenous shunt [36] (Fig. 1) that does not contribute to the nutritive capillary supply to the lobule and drains directly from the end arteriole into the postcapillary venule.

In pathological states such as IRI, where there is microcirculatory dysfunction in the form of vasoconstriction, the presence of this shunt continues to divert blood away from pancreatic tissue, exacerbating the reperfusion insult. Some authors believe this attenuation of lobular capillary flow is a major contributor to pancreatic IRI as it potentiates acinar damage and intracellular activation of proteases due to ischemia [37].

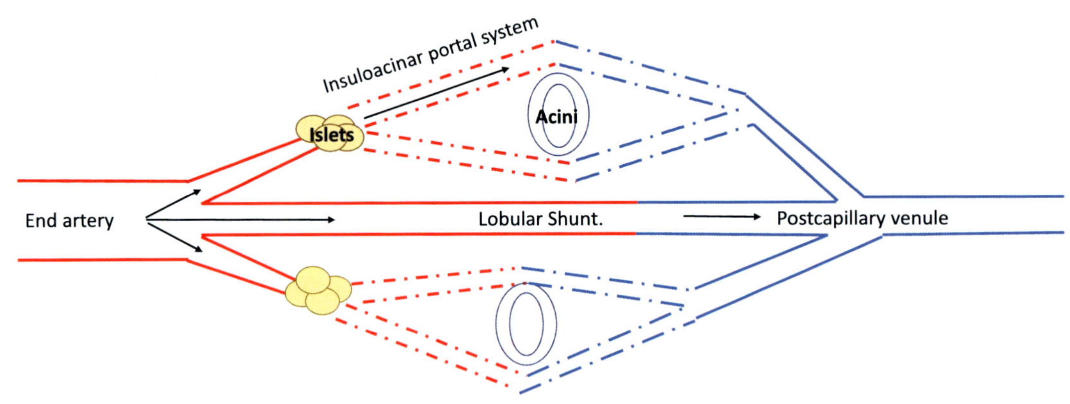

FIG. 1 Diagrammatic representation of the pancreatic lobular capillary system and lobular shunt.

Graft pancreatitis

This is informally described by transplant surgeons as the appearances of graft edema, saponification, and peri-pancreatic fluid and shares similar features observed in native pancreatitis. There is no formal consensus on the definition of graft pancreatitis.

Histological description, based on experimental models of severe pancreatitis and human samples of the disease, shows vessel necrosis, infiltration of polymorphonuclear leucocytes, increased vessel fenestration, and disruption of the endothelial lining [38].

Radiological imaging of early pancreatitis may reveal normal appearances despite clinical features and raised serum levels of C-reactive protein (CRP), amylase, and lipase. Later appearances can show loss of surface undulations on the pancreas due to gland swelling. In severe pancreatitis, contrast computed tomography (CT) may show peri-graft fluid and stranding, graft heterogenicity, hypodense regions of poor perfusion, and associated collections [39].

Acute graft pancreatitis (within the first 72h of transplantation) occurs in all allografts, as a subsequence of IRI [40]. The severity varies and can be exacerbated by other factors including donor age, steatosis, microvascular disease, and prolonged CIT. Early graft pancreatitis (within three months) occurs with an incidence of 35%–38% [24,40,41] and is the most common reason for early relaparotomy (38% of post PTx relaparotomies) [41].

The most severe complications of pancreatitis include intraabdominal abscess, fluid collections, pancreatic necrosis, vascular thrombosis, pseudocyst formation, and consequent systemic inflammatory response (SIRS) in 10%–20% of recipients diagnosed with early graft pancreatitis [40]. The management of postoperative abdominal complications is a matter of fine and appropriate medical and surgical judgment and broad clinical experience.

Early relaparotomy (occurring within the first three months posttransplantation or during the initial hospital stay) is associated with a significantly higher rate of graft loss compared to recipients without relaparotomy, mainly due to the associated rates of graft pancreatectomy (SPK 50%, PAK 71%, and PTA 61%) [41].

Resolution of severe pancreatitis involves parenchymal replacement with adipose tissue and fibrosis [42], which decreases the functional mass of the allograft.

The steatotic pancreas and IRI

The process of IRI is particularly injurious for steatotic pancreases. Pancreases with fat infiltration are less commonly used due to increased vulnerability to IRI and therefore have a significantly high rate of organ discard. In a recent 11-year analysis of UK pancreas utilization, "fatty pancreas" was the leading reason for nonutilization of retrieved organs [43].

Obesity leads to fat infiltration of several organs including the pancreas. Animal studies have demonstrated that when steatotic pancreases are placed under oxidative stress similar to preservation conditions in transplantation, fat-derived cytokines are released to promote a local inflammatory process that contributes to organ dysfunction [44]. Indeed, obesity is a proven risk factor that worsens the severity and outcome of native pancreatitis [45–48].

Using in vitro coculture study of acinar cells and adipocytes, Navina et al. [38] demonstrated that unsaturated fatty acids are proinflammatory releasing intracellular calcium, inhibiting mitochondrial complexes I and V and causing necrosis. This observation was validated in a murine model of induced pancreatitis: by inhibiting lipolysis in obese mice there was prevention of systemic inflammation, renal failure, lung injury, and reduced pancreatic necrosis and mortality. The same group reviewed histological slides from debrided necrotic pancreases from patients with severe pancreatitis and found that that the worst damage was found where there were necrosed adipocytes adjacent to necrosed pancreatic parenchyma, suggesting a link between necrosis severity and location of intraparenchymal fat.

By understanding the relationship between IRI and intraparenchymal fat, it may be possible to devise therapies targeted at managing lipotoxicity prior to transplantation in order to attenuate the effect of IRI. Our increasing understanding of the molecular processes that underlie IRI is leading to opportunities to devise effective interventions. Potential strategies include organ defatting, mitochondrial protection, and thrombolysis, all of which might be delivered prior to transplantation.

Resuscitation of the pancreas for whole organ transplantation

The definition of "resuscitation" includes "revival, renewal, restoration to health, restoration to life after death" [49]. In relation to a donor organ, the term implies not only recovery of function at a cellular and whole organ level but also the conversion from the status of nonviable (untransplantable) to that of transplantable.

As discussed previously, conventional static cold storage is designed to slow the rate of deterioration of an organ, primarily by reducing metabolic activity, stabilizing the flux of electrolytes and other molecules across cell membranes, and preventing damage due to cell swelling. This does not result in an organ that is more viable at the end of the process than it was at the beginning, and does not, therefore, constitute resuscitation.

The diversity of methods used to resuscitate donor organs is based, essentially, upon one or more of three variables:

1. Temperature
2. Oxygen delivery
3. Machine perfusion

Strategies for the active resuscitation of the pancreas have not reached clinical practice to date. This is in contrast to practice in other organs, which have seen clinical trials and commercial deployment of ex situ hypothermic (with and without oxygen) systems in kidney [50] and liver [51] transplantation and normothermic perfusion systems in liver [52], kidney [53], lung [54], and heart [55].

All the methods listed before rely upon perfusion of the organ once it has been retrieved from the donor. An exception to this is normothermic regional perfusion (NRP), which is discussed in the next chapter in greater depth. NRP is a method used in DCD retrievals whereby the circulation to the abdominal organs is restored using extracorporeal membrane oxygenation following the declaration of death. This has the effect of restoring the cellular energy stores in the abdominal organs following a period of circulatory arrest, essentially converting the status of the organs to one that is comparable to the DBD situation. Data in relation to pancreas transplantation are limited to a few reports documenting pancreas retrievals and transplantation post-NRP, comprising in the UK, two simultaneous pancreas and kidney transplants with one islet transplant [56], in Spain five PTx [57], and in USA one PTx [58].

Experimental ex situ preservation strategies that have been explored in respect to whole PTx include

- Two-layer method (TLM)
- Persufflation with oxygen
- Machine perfusion
  - Hypothermic
  - Normothermic

Two-layer method

The two-layer method (TLM) was designed to allow oxygen diffusion to the cold-stored pancreas during SCS. It uses the combination of a conventional organ preservation solution (University of Wisconsin) and a specialized oxygen carrier, a perfluorocarbon (PFC) to provide oxygenation. The unique capability of the PFC is its ability to bind reversibly to large amounts of oxygen, estimated at 20–25 times greater than in simple solution [59].

The hydrophobic property of PFC and the different densities of the solutions cause the two to separate as discrete layers, enabling the pancreas to be suspended at the interface, sandwiched between the organ-preserving UW and oxygen-rich PFC.

In the context of pancreas preservation, TLM was first described in 1988 [60] by Kuroda et al. using perfluorodecalin as the PFC and Euro-Collins as the preservation solution, later to be replaced with UW.

Initial canine studies suggested effectiveness of the TLM using UW and PFC for up to 96h [61] for dogs weighing 12–18kg in an autotransplantation model. Matsumoto et al. described the only reported application of TLM in clinical PTx [62], comparing 10 pancreases preserved by the TLM to 44 cases preserved in UW in SCS. The reported outcomes included graft condition at reperfusion, complications, and three-month graft function. After a mean CIT of 16.5 h in TLM and 18.1 h in the conventional SCS group they observed no edema at reperfusion in the TLM group compared to 23.3% in the SCS group, and largely better graft appearances (described as soft pancreas and normal appearances of a globally perfused pancreas and duodenum). They reported no significant differences in the time to insulin independence, graft loss, complications, or acute rejection. As this study was neither randomized nor powered to show superiority, the authors concluded that TLM is at least equivalent to SCS.

However, these positive outcomes have not been replicated in any further large animal or human studies. There is a clear impression that TLM is less effective in larger organs than had been hoped, and that this is due to poor oxygen diffusion to deeper tissue layers in these much larger organs. Papas et al. investigated the pO_2 profiles in sections of pancreas preserved in TLM using a fiber-optic sensor to develop a model that predicts oxygenated volume fraction [63]. The model predicted that for a pancreas 2.5 cm thick only 15% oxygenated volume fraction would be achieved, suggesting TLM would be able to only oxygenate a small volume fraction of human pancreases. These findings led the authors to conclude that an alternative route of oxygen delivery is needed and that this should be the vasculature.

Persufflation (PSF)

PSF or vascular gas perfusion delivers humidified oxygen through the organ's vasculature. This technique has been investigated in the heart [64], liver [65], kidney [66], and small intestine [67]. PSF has been most extensively evaluated in the kidney, in which early studies helped determine the optimal route of delivery (i.e., anterograde via the main arterial system or retrograde via venous system), oxygen flow/pressure, partial pressures, and temperature. Due to some reported detrimental outcomes of decreased global reperfusion in the kidney after anterograde PSF, retrograde PSF was investigated; historically this involved making small pinpricks to the surface of the organ to allow escape of gas [68], an intervention which can cause bleeding on reperfusion along with initiation of inflammation in whole pancreas transplantation and the lack of ability to adequately infuse digestive enzyme in pancreases used for islet cell isolation.

PSF of pancreases has been identified as a promising means to improve whole pancreas and islet cell transplantation with a few published studies describing its feasibility [69,70]. Scott III et al. compared the effectiveness of persufflation to the TLM using both porcine and nontransplanted human pancreases from DBD donors [70]. Pancreases were immersed in Histidine–tryptophan–ketoglutarate (HTK) solution at 4°C, and 20 mL/minute of 40% humidified oxygen was delivered to either/both the superior mesenteric and splenic arteries in the human pancreases or celiac trunk in the porcine pancreases.

The effectiveness of oxygenation delivered by the two interventions TLM versus PSF was investigated by magnetic resonance (MR), comparing the ratio of ATP to inorganic phosphate to assess bioenergetic status. Additionally, MR was used to assess the homogeneity of persufflation. MR demonstrated that persufflation effectively reached greater than 90% of the tissue, and also showed that TLM was not effective as a means of oxygen delivery, with undetectable ATP levels compared to effective restoration of ATP levels (following prior depletion due to hypoxia during SCS) in the persufflation group. The findings from this study support the hypothesis that TLM is only effective in the smaller rodent and canine models, as reported in the literature, and that persufflation may be a feasible and effective means to deliver oxygen.

Machine perfusion (MP)

Machine perfusion constitutes the delivery of a perfusate through the organ's vasculature in a continuous or pulsatile manner. This is a technology that has been used in kidney transplantation for several decades, although high-level evidence of its benefit has only been reported in more recent years [50,71]. The postulated benefits of MP over SCS are as follows:

- Flush-out of metabolic products and enzymes
- Maintenance of the microvasculature and reduction of vascular resistance
- Delivery of oxygen and other nutrients
- Delivery of therapeutics
- Viability assessment by means of functional testing

The earliest description of a perfusion apparatus is probably in 1866 by Elias Cyon in Ludwig's laboratory in Germany, where the isolated perfused heart of a frog was kept beating for 48 h [72]. For purposes of transplantation, Alexis Carrel and Charles Lindbergh focused on developing an apparatus that would take on the role of the heart and lungs and also keep an ex situ organ free of infection [73]. In 1935 they constructed the world's first functional pump oxygenator.

Over a century since the initial proof of concept, isolated perfusion for preservation of the pancreas has still not materialized. The roadblocks for the establishment of pancreas MP are several: determination of optimal perfusion composition, flow rate and pressures due to the susceptibility of the pancreas to becoming edematous during perfusion. Experience has revealed that the pancreas does not tolerate high perfusion pressures, therefore, achieving a pressure sufficient enough for nutritional needs but without causing endothelial injury and edema is still a significant fundamental challenge.

The feasibility of pancreas MP has been investigated at two temperature ranges: hypothermic (0–12°C) and normothermic (35–38°C). The reader is directed to three excellent reviews [74–76] that provide the detailed perfusion protocols of the published studies focused on pancreas machine perfusion that are described in this chapter.

Hypothermic machine perfusion (HMP)

HMP combines the benefits of hypothermia with dynamic MP. It is an intrinsically safe strategy because preservation easily reverts to SCS in the event of machine failure. An essential disadvantage is that functional assessments cannot be carried out due to the nonphysiological temperature of HMP.

Much of the evidence regarding HMP comes from kidney transplantation, and more recently liver transplantation [51]. Some of the strongest evidence on simple (nonoxygenated) HMP is from the multicenter trial carried out in the Eurotransplant region [71]. Compared to SCS, HMP is associated with a significantly reduced risk of delayed graft function in deceased donor kidneys [71], including specifically ECD kidneys [77] and DCD kidneys [78]. Conversely, and at around the same time, no advantage of HMP over SCS was shown by Watson et al. [79]. One of the possible reasons for the different conclusions of these trials was that the Eurotransplant study mandated that the perfusion devices were transported to the donor hospital so that perfusion was continuous from the point of retrieval to implantation, whereas the UK study allowed a period of SCS during transport of the organ from donor hospital to transplant center.

Mechanistically the core of IRI is mitochondria dysfunction and there is recently renewed focus in the delivery of oxygenated hypothermia to enable normal electron flow for ATP recovery. A recent (2021) multicenter, randomized controlled trial in liver transplantation—Dual Hypothermic Oxygenated machine Perfusion (DHOPE), compared this intervention to SCS controls in DCD livers [51]. DHOPE involved the delivery of UW supplemented with glutathione for two hours at 10°C via both the portal vein (5 mmHg mean pressure) and hepatic artery (25 mmHg mean pressure) with 500 mL/min of oxygen to achieve a perfusate oxygen partial pressure target of > 100 kPa. The investigators observed significant results in the primary outcome with a 68% reduction in symptomatic ischemic nonanastomotic biliary strictures both at six months posttransplant and overall, 6% versus 18%.

Similarly positive outcomes have been described in the COMPARE trial (2020): this Phase III, randomized, double-blind, multicenter trial investigated the benefit of oxygenated HMP to nonoxygenated HMP in paired DCD kidneys [50]. The HMP device was taken to the donor hospital to enable immediate connection and perfusion. UW perfusate was delivered at a temperature of 1–4°C and fixed mean pressure of 25 mmHg. In the oxygenated arm 100% oxygen was delivered at 100 mL/min, resulting in perfusate pO_2 of 600 mmHg. There were no significant differences in the primary outcome (estimated glomerular filtration rate at 12 months), but fewer severe complications, fewer biopsy proven rejection episodes, and a lower rate of graft failure in the oxygenated HMP arm.

With respect to pancreas HMP, two of the earlier reports were published in 1975 by Tersigni [80] and Brynger [81] both using a canine DBD model of pancreas transplantation. Tersigni and colleagues preserved procured canine pancreaticoduodenal segments by pulsatile, hypothermic perfusion at 6°C with a pulse rate of 60/min and gaseous flow rates of oxygen at 2 L/min and carbon dioxide at 30 mL/min. Varying HMP perfusion pressures were studied (5, 10, and 25 mmHg) and HMP perfusion was carried out for a duration of 24 h. Two perfusate compositions were investigated: the first was cryoprecipitated pooled plasma with medication additives plus mannitol to achieve a pH range of 7.4 to 7.6 and osmolarity of 260–280 mOs/L. The second perfusate comprised a similar pooled plasma base but with the addition of 200 mL/L human albumin, dextrose, and the exclusion of mannitol. The pH range was 7.4 to 7.6 with a higher osmolarity ranging between 300 and 310mOs/L. After perfusion, allogeneic transplantation was performed into the right groin of the recipient animals after an autologous pancreatectomy. In total there were 5 experimental groups defined by perfusate used and perfusion pressure. Additionally, a control group underwent transplantation without perfusion. After revascularization it was observed that with the exception of the combined higher osmolarity and 25 mmHg pressure cohort (group V), the other experimental groups showed macroscopic evidence of edema, hemorrhage, venous congestion, and absence of duodenal peristalsis. Group V displayed minimal edema and duodenal peristalsis. Furthermore, this group had the longest graft and recipient survival (18.3 ± 4.5 days), even longer than the fresh transplanted control recipients (mean 14.6 days). The remaining HMP groups had poorer survival rates ($< 4.7 \pm 1.4$ days) due to early posttransplant (1–5 days) occurrence of vascular thrombosis, hemorrhagic pancreatitis, and ischemic duodenal necrotic as evidenced on autopsy and histology. The only abnormal histologic characteristic reported for the longest survivors (control and group V) was minimal interlobular edema. In 2 dogs, one each from the control group and group V, features of duodenal rejection were seen (mononuclear cellular infiltrate, predominately in the submucosa), and a third dog (control group) features of pancreas rejection (induration, atrophy, and diffuse interstitial fibrosis).

Brynger [81] studied the posttransplant outcomes of duct-ligated canine allografts after 24 h of preservation. The experimental groups included grafts preserved with continuous hypothermic perfusion between 6°C and 8°C, allografts stored in standard cold storage alone, and a control group of pancreases that were transplanted immediately without preservation. Intravenous azathioprine was given as a single immunosuppression agent. The primary outcome was evidence of early endocrine function posttransplant (i.e., normoglycemia) and response to intravenous glucose stimulation at 4–8 days posttransplant. The HMP perfusate comprised 20% human albumin in a solution with an extracellular-type electrolyte composition. Perfusion pressures of 50/36-44 mmHg gave an average flow rate of 95 mL/min. A gas mixture of 99% oxygen and 1% carbon dioxide was delivered into the organ chamber at a rate of 40 mL/min without a membrane.

At the end of HMP, significant edema was observed with an increase in organ weight of 135%–275%. There was no edema observed in the SCS or control groups. On reperfusion at the time of transplantation, the initial edema observed in the HMP group resolved considerably after 15 min. Early (<24 h) fatal postoperative bleeding rates were similar for the HMP and SCS groups and thought to be due to coagulopathy in the pancreatectomized recipients. Blood glucose levels were normalized in all groups, correlating with insulin secretion, and responses in the glucose tolerance tests were in the same range as those prior to pancreatectomy. These findings led to the author's conclusion that HMP and SCS can both preserve viable allografts for up to 24 h.

Florack et al. (1983) [82] investigated the efficacy of pulsatile machine perfusion compared to cold storage in the preservation of canine open-duct segmental pancreatic grafts. The primary outcomes of endocrine function (assessed by presence of normoglycemia and intravenous glucose tolerance test (IVGTT)) were determined in a DBD autotransplantation model using the tail of the pancreas and anastomosing the splenic vessels to the iliac vessels, with subsequent completion pancreatectomy.

Two experimental perfusates were assessed during HMP: SGF-I (silica gel filtered plasma, osmolality 430 mosm/kg) and SGF-II (similar to SGF-I with the inclusion of mannitol for increased oncotic pressure, osmolality 470–500 mosm/kg). In the cold storage group, subsets of grafts were stored either in Collins solution (osmolality 300mosm/kg) or SGF–I. Pancreases in both experimental groups were preserved over a range of times—24, 48, and 72 h prior to transplantation. A control group underwent transplantation immediately (nonpreserved grafts). Pulsatile, nonoxygenated HMP was delivered with a set mean pressure of 30 mmHg.

The investigators observed that (i) higher HMP pressures (45 and 60 mmHg) resulted in severe edema and eventual cessation of pancreatic flow and that (ii) at lower pressures (15 mmHg), flow could not be established, therefore refining their choice of perfusion pressure to 30 mmHg. Even at 30 mmHg, by 24 h interstitial edema was observed which progressed up to 48 h. Graft failure (development of hyperglycemia) rates ranged between 30% and 40% in the HMP pancreases at 24–48 h compared to 0% in the cold SGF-I group in the same time period. On the positive side, histological assessment of pancreases that functioned for longer than 14 days had similar appearances to cold stored and nonpreserved pancreases as well as similar IVGTT glucose responses. The conclusion of the investigators was to recommend cold storage with high osmolar plasma for less than 48 h over HMP preservation.

Karcz et al. reported their experience of pulsatile HMP in 15 porcine DCD allografts [83]. Average warm ischemia time was 25.13 ± 1.3 min and mean cold ischemia time was 149.6 ± 8.9 min. Pancreases received nonoxygenated HMP via an aortic segment with UW solution for an additional duration of 315 min. Perfusion parameters of pressure, flow, resistance, and temperature were recorded, and wedge biopsies taken before and after perfusion for histopathologic analysis. The investigators recorded a temperature range of 4–10°C, pulsatile perfusion of 58–60 pulses per minute, average optimal resistance index of 0.083 ± 0.042 Hg/mL/min, average flow volume of 147.6 ± 9.969 mL/min, and average systolic perfusion pressure of 21.07 ± 4.261 mmHg. At the end of perfusion there was an increase in pancreas' weight ranging between 3.2% and 18.3%. Histological assessment of preperfusion specimens showed evidence of greater than 20% lobular damage, moderate islet cell damage, inflammation, and edema. There was significant improvement in the postperfusion biopsies noting less sublobular and acinar cell damage, mild islet cell damage, and mild edema and inflammation.

Only a small number of studies have investigated HMP using human pancreases. Using discarded human pancreases Leemkuil et al. [84] compared SCS (n = 10) with oxygenated HMP (n = 10). Both groups underwent preservation for six hours and had equal numbers of DBD (n = 5) and DCD (n = 5) organs. In the HMP group, dual perfusion was delivered via the splenic and superior mesenteric arteries with the portal vein left open to allow passive drainage. The perfusion pressure was set at 25 mmHg for both vessels, temperature range of 4–7°C, and oxygen flow rate of 100 mL/min. Acridine orange was added to the perfusate in six experiments to assess perfusion efficacy, and tissue biopsies were taken to investigate ATP concentration, weight change, fluorescence microscopy, and histologic evaluation. Two HMP pancreases were used to investigate the feasibility of islet isolation post cold perfusion.

Acridine orange was visible in all the biopsies taken from the whole donor pancreases, therefore demonstrating uniform perfusion. The wet to dry ratio did not change after preservation in either HMP or SCS groups. During HMP, ATP concentration increased 6.8-fold in the DCD pancreases and 2.6-fold in the DBD pancreases. There was no evidence of cellular injury or edema in either group on histology. Thiobarbituric acid reactive substance concentration was analyzed in the perfusate to investigate the formation of ROS: this was low in both groups. Finally, islet isolation was feasible with islet cell viability of 90% and 92% after three days in culture. This publication suggests that oxygenated HMP is feasible, safe, and increases pancreatic tissue ATP content, although it does not improve islet cell viability.

Branchereau et al. [85] compared pulsatile HMP to SCS in 11 discarded human pancreases with preservation duration of 24 h. The groups included SCS (n=2), split pancreases (n=2) to enable matched studies, and HMP (n=7). Pancreas and duodenal biopsies provided the primary outcome measure. Pulsatile, nonoxygenated HMP used Perf-Gen™ solution (similar composition to UW) with a 25 mmHg fixed pressure. On macroscopic evaluation there was minimal evidence of edema, and the indices of resistance, initially fluctuating, were stable after the first half of perfusion. Histologically, two of the seven HMP pancreases showed evidence of fat necrosis, which correlated with high BMI donor status. No other lesions were observed after 12 h of HMP compared to findings of ischemic necrosis in the SCS biopsies. At 24 h of HMP there was evidence of focal ischemic necrosis in the acinar tissue in all the biopsies, but appearances of well-preserved islets on staining with insulin, glucagon, and somatostatin antibodies.

Hamaoui et al. [86] described their development of a HMP protocol using porcine pancreases and subsequently applying it to nontransplanted human pancreases. Perfusion was with nonoxygenated UW, supplemented with mannitol as an oncotic agent, and a target maximum pressure of 20 mmHg. Two groups of porcine pancreases were compared: one group underwent 26 h of SCS and the other group that underwent 26 h of SCS followed by five hours of HMP. The efficacy of each intervention was assessed by normothermic reperfusion for two hours using a combination of autologous blood and saline. A glucose challenge was delivered 30 min after starting normothermic reperfusion, to stimulate insulin secretion. For the three human pancreases studied, HMP (same protocol as the porcine grafts) was conducted for two hours prior to normothermic assessment using a Krebs–Henseleit buffer-based solution, supplemented with sodium bicarbonate and oxygenated with 95% oxygen and 5% carbon dioxide mixture. For the first 15 min of normothermic reperfusion, a target pressure of 45 mmHg was maintained, after which pressure and flow were allowed to stabilize independently based on individual organ characteristics.

HMP in both human and porcine pancreases was associated with weight gain of between 3.9% and 14.7% (human) and 15.3% and 27.6% (porcine). During normothermic reperfusion, two-thirds of the pancreases that underwent HMP demonstrated a rise in insulin levels in response to the glucose challenge compared to no response observed in the SCS-alone treated pancreases. In both human and porcine pancreases, the histological appearances after HMP remained similar to pre-HMP appearances. However, histological appearances after normothermic reperfusion showed moderate to severe reperfusion damage in all the investigated pancreases.

The body of evidence described before shows that HMP of pancreases has been tested in a few small and heterogeneous studies. It is not possible on the basis of these studies to deduce the optimum method of perfusion (e.g., pulsatile/continuous, perfusate composition, or the benefit of oxygenation). However, the limited evidence that exists suggest that HMP is probably safe for the pancreas, although some reports of edema suggest that considerable refinement is required with respect to the variables mentioned before. The key question is whether further nonclinical studies are needed before evaluation in a clinical trial.

Normothermic machine perfusion (NMP)

Normothermia is the means to reproduce the physiological milieu. The principal technical objective of all NMP systems is to replicate as closely as possible normal organ metabolism and function of the ex situ organ by providing oxygenation and other essential substrates.

There are a number of theoretical benefits of NMP over HMP: foremost is the avoidance of all the effects and implications of cooling. Although oxygenation delivery during HMP is addressed (at least in part) by the recent introduction of hypothermic oxygenated perfusion (HOPE), this does not address the fact that hypothermia itself is directly injurious to tissues [87]. NMP reduces the effects of IRI by minimizing the exposure to ischemia while the graft remains metabolically active. Maintaining cellular metabolism facilitates objective functional assessments and also permits administration of therapeutic interventions and reconditioning prior to transplantation. NMP has been shown not only to be clinically feasible for livers [52,88], kidneys [53], lungs [54], and hearts [89] but also to be beneficial over SCS in permitting functional assessment [90] and organ repair [91].

Nicholson and Hosgood [53] in the first reported clinical study of renal transplantation after NMP used their protocol to assess its effects in 18 ECD kidneys, comparing these to a control group of 47 ECD kidney that underwent SCS. The kidneys, having undergone NMP for 60 min using ABO-compatible PRBC-based perfusate, experienced significantly less delayed graft function (5.6% versus 36.2%), although there were no differences in graft or patient survival at 12 months.

Nasralla et al. [52] in a multicenter, randomized control trial compared NMP-preserved livers to SCS grafts and observed 50% lower rates of graft injury measured by postoperative transaminase levels, 50% lower organ discard rates (after retrieval), while in the context of 54% longer average preservation times. In the "VITTAL" trial, Mergental el al [90] used NMP as a means to assess high-risk livers declined for transplantation (by all UK transplant centers) by providing objective functional information. In this open, nonrandomized, prospective trial, 31 livers were assessed using criteria of pH, lactate clearance, glucose, bile, and vascular flows during the first four hours of NMP. Predefined criteria were achieved by 71% of the organs (n=22), and these were then transplanted after a median preservation time of 18 h, with 100% 90-day patient and graft survival.

Keshavjee et al. at Toronto General Hospital have demonstrated the benefit of ex situ normothermic lung perfusion to allow successful transplantation of donor lungs that would be otherwise declined for transplantation, showing equivalent long-term outcomes to standard criteria lungs [92]. Adoption of this technology has correlated with a doubling of annual lung transplantation rates (> 200) in Toronto. Using a porcine lung transplant model, the same group has utilized the NMP platform to enable intratracheal IL-10 gene transduction in ex situ donor lungs prior to transplantation, comparing this with in vivo delivery (in the donor). Transplant lung function was superior in the ex situ (*NMP*) group with significantly elevated levels of proinflammatory cytokines posttransplant observed in the in vivo group [91].

To date there is only a small number of studies that have investigated NMP of pancreases and these are mostly limited to experimental preclinical models. The earliest description of pancreas NMP, albeit not in the context of transplantation, was by Babkin and Starling in 1926; this was a physiological model to study secretin effects on pancreatic blood flow [93].

The first report of NMP of the pancreas in the context of transplantation was from Meyer [94] and the University of Minnesota group in 1973. DBD canine pancreaticoduodenal grafts were connected via the celiac artery to the device that delivered a perfusate composed of 1:1 of blood and Ringer's lactate at 37°C, with no further additives. The portal vein was left uncannulated to allow free venous drainage. Perfusion was initially fixed at a pressure of 80 mmHg. Oxygenation was provided by means of an air pump and membrane oxygenator. Controlled amounts of carbon dioxide were introduced to control the pH of the perfusate. During a three-hour perfusion, assessment was performed by assessing insulin release in response to glucose administration as well as pancreas secretion after secretin stimulation. The investigators reported duodenal motility, insulin response to the glucose challenge, and pancreas secretion after secretin stimulation. Macroscopic appearances were reported as normal after three hours of perfusion, although there were some areas of edema on histological assessment.

In 1980 Eckhauser [95] and the Michigan group described the experimental application of NMP of canine DBD pancreases, using a pulsatile perfusion pump and perfusate composed of 300mls of heparinized autologous blood and the addition of balanced electrolytes, bovine serum albumin, and mannitol to achieve an osmolality of 315 mosm/L in a 400 mL circuit. A gas mixture of 95% oxygen and 5% carbon dioxide was delivered by means of a disc oxygenator and an in-line dialysis unit was included to prevent the accumulation of lipase, amylase, and metabolites, and to maintain the electrolyte balance. The flow rate to the graft was maintained at 20 ± 5 mL/min. Initial trial perfusion pressures of greater than 60 to 70 mmHg led to gross pancreatitis and histological appearances of disseminated microvascular thrombosis, therefore lower subsequent pressures ranging from 30 to 50 mmHg were used. Pancreatic biopsies after five hours of NMP were compared to biopsies from in situ perfused pancreases and showed similar appearances to the latter group that were used as the controls. Edema of the pancreas was a consistent finding after NMP, but minimal with perfusions less than four hours, and duodenal contractions were significantly less forceful in perfusions exceeding four hours. Hemolysis was reported as a persistent finding at a rate of 4.8 ± 1.1 mg/dL/hr. The pancreases showed evidence of constant oxygen consumption. Pancreatic secretion responded appropriately to initiation and discontinuation of secretin infusions. Despite dialysis, there was incomplete clearance of metabolites and proteases.

In the more recent era, Barlow et al. reported NMP perfusion of a series of five human pancreases unsuitable for transplantation [13]. The objective was to develop an ex situ perfusion system that enabled objective pancreas assessment in order to improve discard rates. After a median cold ischemia time (CIT) of 13 h 19 min, the organs were perfused via the single limb of the Y graft anastomosed to the splenic and superior mesenteric arteries, with the portal vein left open for free drainage. The duodenum was drained with a wide bore catheter. The organs were perfused for two hours with ABO-compatible packed red blood cells (PRBC) diluted with a colloid, gelofusine for a hematocrit target of 20%. Additives included mannitol, glucose, heparin, and bicarbonate to normalize pH. The custom-built circuit incorporated a centrifugal pump, pressure controlled at 50mmgHg, temperature set at 37°C, and a gas mixture of 95% oxygen/5% carbon dioxide flowing at 0.1 L/min. Outcome measures included blood gas analysis, biochemistry (amylase, lipase, lactate dehydrogenase, and insulin), perfusion parameters (blood flow, pressure, pump speed, and temperature), histology, and stimulated insulin release (glucose and arginine).

One pancreas with a CIT greater than 30 h was excluded from the analysis as both the pancreas and duodenum appeared grossly ischemic during NMP. During perfusion of the remaining grafts, stable arterial flows of 35 ± 2.8 mL/min/100 g (mean) were recorded. High duodenal drainage compromising the circuit volume led to the early termination of the perfusion for one pancreas at 60 min. Amylase and lipase levels increased during the course of NMP in all cases: notably, the level of amylase was highest in an older graft with marked appearances of fat infiltration.

Stimulation of insulin response by glucose was elicited appropriately in only one pancreas, the same organ that had the highest basal insulin levels. Subsequent stimulation with arginine was associated with an insulin level increase in two further pancreases. As noted by the authors, insulin detection is an unreliable marker of viability as its presence may be a sign of beta cell injury. Histological assessment showed focal necrosis and patchy acinar and fat necrosis in all grafts but was most extensively marked in the graft with significant fat infiltration and least apparent in the (younger) graft that showed the best insulin response to glucose stimulation. This work proved that pancreatic NMP is feasible, but also highlighted the potentially injurious nature of the technique, showing the need for considerable refinements.

Kumar et al. focused on determining the optimal perfusion pressures during NMP using a DCD porcine model [96] to compare continuous perfusion pressures of either 20 mmHg (low-pressure) or 50 mmHg (high-pressure). This protocol included 1.5 l heparinized autologous whole blood, cefuroxime and epoprostenol sodium, and a temperature set at 37°C. The proximal aortic segment, pancreatic duct, duodenum, and portal vein were all cannulated to facilitate perfusion and monitoring. Outcome measures included blood gas analysis, biochemistry (electrolytes, glucose, and amylase), and graft biopsies (histology, anticaspase 3, M30 CytoDEATH, and ATP synthase).

After similar durations of cold ischemia (median 127 min and 136 min in the high- and low-pressure groups, respectively), organs were perfused for a median time of three hours (range 2–4 h). Flow rates for the high-pressure group were 141 mL/min versus 40 mL/min in the low-pressure group. The low-pressure group maintained normal gross macroscopic appearances for up to four hours compared to three hours in the high-pressure group.

Amylase levels were significantly lower in the low-pressure group, but pancreatic secretion was otherwise similar. Both groups showed insulin responses to glucose stimulation with no significant differences: pH, lactate, and electrolyte levels were also similar. In both groups hematoxylin and eosin (H&E) staining demonstrated an absence of leukocytes.

Anticaspase 3 and M30 CytoDEATH immunostaining was performed to quantify the extent of cellular death. There was a significantly better cell death profile in the low-pressure group, and immunostaining for presence of ATP synthetase complex V was detected in significantly higher amounts in the low-pressure group suggesting increased ATP replenishment. The investigators concluded that lower perfusion pressures are beneficial and that this may be a consequence of less endothelial sheer trauma and acinar and parenchymal damage during NMP.

Kuan and colleagues [97] compared simultaneous kidney and pancreas NMP to NMP of pancreas alone in a DBD porcine model. The objective of their study was to investigate the added benefit of a kidney as a dialysis organ during NMP. Using autologous whole blood as the perfusate additives included cephazolin, epoprostenol sodium, and heparin, and Hartmann's solution was used to replace duodenal drainage and urine production. The circuit included a roller pump adjusted to achieve a mean arterial pressure of 70–80 mmHg for the pancreases and 90–80 mmHg for the kidneys. Outcome measures included macroscopic appearances (edema, hemorrhage, and congestion), urine production, biochemistry, blood gases, and histology.

Two experiments ($n=2$) were conducted in each group. The first perfusion (pancreas alone) was continued to four hours, although appeared grossly ischemic at 150 min. All other perfusions were run for two hours. The pancreases all appeared moderately/severely edematous, congested, and hemorrhagic after 90 min of perfusion. The kidneys also became edematous and congested after 60–90 min of NMP. There was a progressive drop in hemoglobin which was thought to be secondary to hemolysis (a hypothesis supported by the observation of hematuria in those experiments with kidneys). Notably, the pH progressively increased: biochemistry analysis could not be performed due to significant hemolysis. Histopathology showed worsening acinar necrosis, hemorrhagic congestion, and vessel thrombosis with NMP duration. In contrast, islet cells showed only a small degree of cytoplasmic microvacuolation with no significant evidence of necrosis or inflammation. The authors concluded that the addition of a kidney to pancreas NMP appeared to confer no benefit with respect to acid–base balance. Considering the experience described by this group, it may be that the use of whole blood that included platelets, inflammatory cells, and other inflammatory mediators may have exacerbated the IRI rather than protecting the pancreas from this.

Nassar et al. [98] describe the longest reported duration of pancreas NMP in a series of three declined human DBD pancreases. The average CIT was short at four hours. The perfusate was a combination of PRBC and fresh frozen plasma in a

1:3 ratio. They used a custom-designed liver device that included a pulsatile roller pump to deliver perfusion via the cannulated superior mesenteric, gastroduodenal, and splenic arteries. The portal vein was allowed to drain freely. The pancreatic and duodenal secretions were drained via a catheter introduced through the common bile duct. The duration of perfusion was six hours for two pancreases and 12h for one pancreas. The target flow rate was set at a physiological rate of 55 mL/min/100 g of tissue: the perfusion pressure was recorded at an average of 60 mmHg. Outcome measures included C-peptide level and histologic assessment. The first perfusion was unsuccessful because of high duodenal secretions and hemorrhage which compromised the circuit volume and led to duodenal ischemia. In the other two perfusions, C-peptide levels were detected during NMP but no stimulation tests for insulin release were performed. H&E staining at six hours showed healthy appearing acini and islets, and chromogranin staining also confirmed intact pancreatic islets appearances at six hours NMP. In the longest perfusion of 12h, H&E staining at the end was reported as showing healthy appearances of acini and islets.

In contrast to HMP, NMP offers the valuable opportunity for measuring function and thereby viability assessment. However, the collective experience to date of pancreas NMP suggests that it may be more injurious than beneficial. Nonetheless, the total experience is very limited but should not hinder future research which is needed to move this field into the clinical arena. Much work remains to be done to identify the best perfusion method. Variables include whether the perfusion needs to be pulsatile or continuous, and what pressures should be used. Also, the composition of the perfusate is of vital importance. No commercial devices have been configured to perfuse the pancreas, although it is a reasonable assumption that a variant of a perfusion system that works well for the kidney should be feasible. The addition of therapeutic strategies that block or mitigate the effects of IRI will be particularly desirable in the context of the pancreas. The question of duodenal drainage and the possible ill effects of pancreatic enzyme release into the perfusate are issues which are specific to pancreas perfusion and have not been resolved in the work published to date.

Should there be a strict classification of temperatures in machine preservation?

The experiences described in pancreas HMP and NMP raise the question of an intermediate temperature that combines the benefits of both. Subnormothermic machine perfusion (SNMP, 25–34°C) has been described and tested for livers but not yet in pancreas preservation. In a DBD porcine model, Spetzler and colleagues [99] compared livers that underwent three hours of oxygenated SNMP (33°C) with red blood cells and Steen solution™ to SCS prior to liver transplantation. They observed improved hyaluronic acid levels in the SNMP group (a marker of endothelial cell function) and less histological evidence of bile duct necrosis at the termination of the experiments on day four.

There is emerging evidence that it is the transition between temperatures that is important. "Controlled oxygenated rewarming" (COR) avoids abrupt rewarming that may contribute to reperfusion injury. Minor et al. [100] in a DCD porcine model compared kidneys that underwent COR versus SCS and assessed these during normothermic reperfusion, observing that COR was associated with significantly better creatinine clearance and less inflammation. COR has already been applied clinically with successful transplantation of six DBD livers [101]. COR in itself is not an independent preservation strategy but could be useful for transitioning temperatures to facilitate preservation, physiological assessments, and organ preparation for transplantation. It is a technique that might be valuable as part of an integrated organ recovery protocol that includes a number of variables on an individual organ basis.

Can the benefits of machine perfusion improve islet isolation?

Intuitively, it is logical that the benefits associated with machine perfusion should translate to improved islet isolation outcomes. Islet transplantation carries a much lower risk profile than solid organ transplantation, but one with lower efficiency in terms of insulin independence (while accepting that success in islet transplantation should be measured in terms of abrogation of hypoglycemic unawareness). Not only are there several steps within the isolation process at which the number of viable islets can fall (reducing the yield), but also islets are lost at the time of infusion into the portal venous circulation, as a result of IRI and the phenomenon of "Instant Blood-Mediated Inflammatory Response" (IBMIR). These hurdles mean recipients commonly require more than one islet transplantation and experience rates of insulin independence that are lower than those experienced in whole organ PTx. A few preclinical studies have investigated the feasibility of islet isolation after machine perfusion. Taylor et al. [102] reported that islet isolation after 24-h HMP of porcine pancreases was feasible, and that, although HMP was associated with moderate edema, it had no impact on the function of the isolated islets. Leemkuil et al. [84] showed feasibility of isolation with two DCD allografts after six hours of oxygenated HMP of nontransplanted human pancreases. Islet yield was 3999 IEQ/g tissue and 6452 IEQ/g tissue, with response to glucose stimulation with insulin stimulation index of 2.2 and 1.9. The islets had satisfactory viability of 94% and 98% postisolation which dropped slightly three days postculture to 90% and 92%. Islet transplantation is a clinical service in which there is a large potential demand for donor organs combined with a high organ discard rate, and the prospect of using machine perfusion technology to improve the number and outcome of isolations is very attractive.

The future of perfusion in pancreas transplantation: What needs to happen and what difference will it make?

The pancreas, as the organ with the most severe clinical manifestation of IRI, is the organ in which the introduction of novel IRI strategies is most needed, as well as the one in which early stage clinical trials may be the most informative. Increasing knowledge of the molecular basis of IRI is providing numerous candidate targets for novel strategies. Machine perfusion provides a suitable platform from which to deliver therapeutic interventions. NMP (in particular) provides the potential to deliver much more than simply high-quality preservation (i.e. preventing further deterioration in a stored organ): it should become the means of delivering novel treatments that can also resuscitate the organ (returning it to its preretrieval state of cellular metabolism) and repair it (treat defects that are unrelated to the process of death and organ donation). It is reasonable to assume that a combination of interventions will be required and that some of these will be "generic" (i.e., applicable to all organs) and that others will be "specific" (i.e., indicated by the characteristics of the donor or recipient).

The simplest approach to mitigation of IRI is to minimize exposure to hypoxia, thereby maintaining the oxidative phosphorylation pathway and ATP synthesis, minimizing ROS accumulation and endogenous antioxidant depletion. Clearly, NMP achieves this most effectively, although tissue ATP measurements show that this can also be achieved by oxygenated HMP. Given that limited oxygen-carrying capacity of a perfusate that does not include a specialist oxygen carrier (e.g., red blood cells or perfluorocarbon), this is a situation that is unlikely to occur in NMP circuits but is a real factor in HMP. However, overprovision of oxygen is also very damaging (in some situations even more damaging). Careful design is needed to ensure that the NMP circuits do not allow the very high perfusate pO_2 levels (> 80 kPa) that have been associated with adverse outcomes in liver transplant studies [103].

The number of potential therapeutic targets is growing, and the main challenge is now how to test these. A comprehensive list is given in Table 1. Not every novel treatment will be given ex situ: new antiinflammatory medications are being introduced for the treatment of autoimmune and other conditions and some of these may be of value in IRI. The "repurposing" of treatments designed for other indications is likely to reveal some useful benefits. Other novel strategies may be best delivered ex situ, for reasons of targeting the organ most effectively, avoiding off-target side effects, or simply the cost of treating a whole patient compared to a single organ.

NMP allows organ assessment

Although organ assessment does not strictly come under the heading of "resuscitation," it is nonetheless closely connected: measurement of the status of an organ is necessary in order to monitor the response to resuscitation. The ideal assessment process would meet a number of criteria:

- Sensitive: minimized false negative outputs
- Specific: minimized false positive outputs
- Nonbinary (i.e., continuous variable)
- Rapid (i.e., produces data in clinically relevant timescale)
- Cost effective

To date, the application of perfusion systems to assess organ viability has relied upon a number of individual measures or simple algorithms. Examples include:

- Liver NMP: Lactate clearance, pH, bile biochemistry, bile flow, perfusate transaminase, vascular flows [90]
- Liver HMP: Flavin mononucleotide (discussed later) [118]
- Kidney NMP: perfusate flow, appearance, urine flow [53]
- Kidney HMP: vascular resistance [119]
- Heart NMP: Lactate [89]
- Lung NMP: Gas exchange (pO_2/FiO_2) [54]

Comparable metrics have not yet been established for the pancreas, because there is so little experience of perfusion and effectively none within the clinical arena. The metric used in the preclinical studies described before include lactate dehydrogenase, lactate, glucose, amylase, lipase, insulin, and C-peptide. Broadly, potentially useful biomarkers relate to injury and endocrine function (e.g., insulin response to glucose or arginine stimulation). The last of these has been adopted

TABLE 1 Summary of some therapeutic targets for ex situ pancreas resuscitation and repair from all organs that may be applicable to the pancreas.

Strategy	Therapy and mechanism	Model	Protocol	Outcome	Ref
Antiinflammatory and antiapoptosis	Combination of antiinflammatory agents Alprostadil: Prostaglandin E1; *vasodilator, antiplatelet, fibrinolytic* N-Acetylcysteine: *ROS scavenger* Carbon monoxide: *vasodilation and antiinflammatory* Sevoflurane: anesthetic gas; *endothelial glycocalyx protection*	Porcine	DBD liver allotransplantation model. 3 groups 1. 4h SNMP 33°C with antiinflammatory combination 2. NMP alone 3. SCS Animals were followed up for 3 days posttransplantation.	Lower peak aspartate aminotransferase, Interleukin (IL)-6, (TNF)-α, hyaluronic acid, and increased IL10 levels in the group receiving addition of antiinflammatory strategies	[104]
	APT070 (Mirococept): Membrane targeting C3 convertase inhibitor *complement inhibitor*	Human islets	Human islets were preincubated for an hour with APT070 and exposed to allogeneic whole blood or transplanted underneath the kidney capsule of streptozotocin-induced diabetic humanized mice. Outcome analysis was cytokine levels and islet morphology on immunofluorescence staining.	Compared to controls preincubation with APT070 was associated with less complement deposition, inflammatory cell infiltration, and proinflammatory cytokines. Histological appearance of maintained and intact islets	[105]
	Carbon Monoxide (CO): (CORM)-401: Carbon Monoxide Releasing Molecules	Porcine	DCD (1h warm ischemia time WIT) kidneys received 4h HMP and subsequently treated with CORM-401 or inactive CORM-401 by normothermic perfusion followed by normothermic reperfusion with allogeneic blood for 10h. Outcome measures were urine output, vascular flow rate, selected injury markers, and histology.	CORM-401 reduced proteinuria, kidney injury markers (KIM-1, NGAL), with reduced histological appearances of acute tubular necrosis, apoptosis, and prevented intrarenal hemorrhage and vascular clots.	[106]
	FR167653 (FR): *Antiinterleukin (IL)-1β and antitumor necrosis factor (TNF)-α*	Canine	In vivo isolated tail of pancreas model. Ninety minutes WIT was followed by reperfusion. The FR group had the intervention by continuous intravenous infusion 30min to onset of WIT and until 120min postreperfusion, the control group received normal saline. Outcome measures were dog survival rates up to 72h, biochemistry and histology.	Fewer FR animals died, and overall had lower amylase, lipase, (IL)-1β mRNA expression and less pancreatic tissue damage on histologic assessment.	[107]
Antioxidants	Superoxide dismutase (SOD) and N-(2-mercaptopropionyl) glycine (MPG)	Rat	In vivo study using sodium taurocholate (ST) induced pancreatitis model. Both SOD and MPG were given intravenously prior to ST induction and compared to groups receiving a leukotriene antagonist or saline. Outcome measures were vessel diameter and leucocyte venular endothelial adherence.	SOD and MPG prevented arterial constriction and leucocyte adherence	[108]

Anticoagulation & thrombolysis	Antithrombin III (AT III): *Anticoagulant, antiinflammatory properties*	Rats	Pancreas allotransplantation model, after CIT 12 h included 3 experimental groups (i) receiving AT III intravenously, 30 min before reperfusion; (ii) no AT III; and (iii) healthy control group, i.e., without transplantation. Outcome measures were pancreatic enzymes 4 h post reperfusion, intravital microscopy and immunohistology.	No differences in enzyme levels although significantly higher than healthy controls. In AT III group there was less histologic evidence of damage, higher capillary and venular erythrocyte velocities, and decreased leukocyte–endothelium interaction	[109]
	Thrombalexin (TLN): Endothelial localizing peptide; *cytotopic (cell 'tethered) anticoagulant*	Porcine and human	Porcine kidneys underwent 1.5 h HMP with TLN in the perfusate versus none or inert TLN in the control groups. Subsequently all groups had assessment during 6 h of NMP. Two human kidneys underwent same protocol.	Machine perfusion facilitated TLN tethering in the microvasculature. Compared to controls TLN kidneys had significantly better perfusion parameters, microvascular capillary perfusion, lower levels of lactate and fibrin generation.	[110]
	ONO-3708: Thromboxane A2 receptor antagonist; *Antiplatelet*	Canine	In vivo study using a vascular isolated pancreas with 1 h of ischemia ONO-3708 was given intravenous throughout (pre/during postischemia and reperfusion) versus none as the control.	ONO-3708 reduced lipid peroxide levels in the venous drainage, increased prostaglandin I2, with less amylase levels and pancreatic juice and better histologic appearances.	[111]
Cellular therapies	Mesenchymal Stem Cells (MSC)	Human	5 pairs of DCD human kidneys were normothermically perfused for 24 h on an exsanguinous metabolic support (EMS) tissue engineering platform. One pair received MSC (1×10^8) versus the other pair receiving nothing as the control. Outcome measures were DNA synthesis, cytokine/chemokine synthesis, cytoskeletal regeneration and mitosis.	MSC treatment resulted in less inflammatory cytokine synthesis, increased ATP, and number of cells undergoing mitosis (26%). There was normalization of metabolism and the cytoskeleton. All suggestive of ex vivo regeneration.	[112]
	Adenoviral vector gene delivery	Porcine	Compared normothermic ex vivo lung perfusion (EVLP) and in vivo (donor) intratracheal delivery of an E1-, E3 deleted adenoviral vector encoding either green fluorescent protein (GFP) or interleukin-10 (IL-10) with a control group receiving normal saline only. After 12 h following the intervention the lung transplantation was performed. Outcome measures included posttransplant lung function, transgene expression, and histological assessment.	There was significant transgene expression in both in vivo and EVLP groups. Lung function was excellent in all EVLP groups compared to the in vivo group. Histological assessment of inflammation showed significantly better appearances in the EVLP group.	[91]
	Short interfering RNA (siRNA)	Rat	siRNA against Fas receptor is coated by lipid nanoparticles to enable intracellular endocytosis which is delivered into livers via the portal vein by NMP and HMP. The Fas receptor when binding to its ligand signals hepatocytes to apoptose.	In this efficacy study distribution was observed on fluorescent confocal microscopy encouraging further investigation of FAS silencing (as well as other apoptotic genes and receptors) during ex vivo preservation prior to transplantation	[113]

Continued

TABLE 1 Summary of some therapeutic targets for ex situ pancreas resuscitation and repair from all organs that may be applicable to the pancreas—cont'd

Strategy	Therapy and mechanism	Model	Protocol	Outcome	Ref
Mitochondrial protection	AP39: Mitochondria targeted Synthetic hydrogen sulfide (H_2S) donor	Porcine	Ex vivo DCD kidneys (60 min of WIT) were flushed and stored with 200 nM AP39 and UW for 24 h at subnormothermia 21°C versus kidneys flushed and preserved with UW alone at 21°C and a control UW alone group at 4°C which is standard of care. Outcome measures were assessment of tissue sections by fluorescence microscopy to assess necrosis and nuclei integrity after ethidium homodimer and DAPI staining, respectively	Significantly less tissue necrosis in the AP39 kidneys than the control group and the 21°C group which had the most evidence of necrosis	[114]
Oxygenation	M101: *novel natural oxygen carrier*	Human and rat	-Nonutilized human pancreases had M101 injected into the pancreases and added to the SCS preservation solution for the last 3 h of CIT. -M101 injected into pancreatic duct and added to the preservation solution of rat pancreases for the 6 h of CIT. -Control groups did not have M101	Human pancreases: increased complex I mitochondrial activity, activation of AKT activity (survival marker), and isolated islets had upregulated insulin secretion. Rat pancreases—decreased ROS (oxidative stress), HMGB1 (necrosis), and p38 MAPK (cellular stress pathway). Improved function of isolated islets.	[115]
Vasodilation and vascular targets	Comparison of various vasodilators BQ123: Endothelin-1 antagonist Epoprostenol: prostacyclin analogue Verapamil: calcium channel antagonist	Porcine	DBD liver allotransplantation model with CIT of 2 h. 3 h NMP in four groups comparing ex vivo use of vasodilators 1. BQ123 2. Epoprostenol 3. Verapamil 4. No vasodilator 5. SCS as control (Total CIT 5 h) Outcome measures were liver injury and function parameters during NMP and until day 3 posttransplantation	In the BQ123 group there was significantly higher hepatic artery flow. Aspartate aminotransferase levels and hepatocyte injury were lower in BQ123 and verapamil groups compared to epoprostenol and SCS.	[116]
	Nitric Oxide (NO) synthase/donors Sodium nitroprusside (SNP): vasodilation L-arginine (L-arg): vasodilation	Porcine	In situ study with vascular isolation of the pancreatic tail and cannulation of the splenic vessels postsplenectomy. Three hours of warm ischemia was followed by 6 h reperfusion. SNP was given via splenic artery during reperfusion and L-arg given systemically via as a bolus prior to reperfusion and as an intravenous continuous infusion over 6 h of reperfusion. The two control groups had either (i) no vasodilator delivered or L-NAME (NO synthase inhibitor) given.	Lipase was significantly reduced in the venous effluent of the SNP and L-arg groups compared to controls and L-NAME groups. Vascular resistance was reduced in both NO groups, and tissue oxygen pressures were significantly higher in SNP group. Histology assessment was most severe in the control and L-NAME groups.	[117]

Abbreviations: AT III, Antithrombin III; AP39, Mitochondria targeted Synthetic hydrogen sulfide (H_2S) donor; APT070, Mirococept; BQ123, Endothelin-1 antagonist; CO, Carbon Monoxide; CIT, Cold Ischemia Time; (CORM)-401, Carbon Monoxide Releasing Molecule; EMS, Exsanguinous Metabolic Support; EVLP, ex vivo Lung Perfusion; FR167653, *Antiinterleukin (IL)-1β and antitumor necrosis factor (TNF)-α*; IL-6, Interleukin-6; IL-10, Interleukin-10; L-arg, L-arginine; L-NAME, NO synthase inhibitor; MSC, Mesenchymal Stem Cells; M101, novel natural oxygen carrier; MPG, N-(2-mercaptopropionyl) glycine; ONO-3708, Thromboxane A2 receptor antagonist; siRNA, Short interfering RNA; SNMP, Sub-Normothermic Machine Perfusion; SNP, Sodium Nitroprusside; SOD, Superoxide Dismutase; ROS, Reactive Oxygen Species; TLN, Thrombalexin; WIT, Warm Ischemia Time.

in the clinical practice of islet transplantation, in which insulin response to stimulation is established as a useful (although not infallible) means to determine the quality of isolated islets [120].

To predict newer viability and assessment targets is to consider the pancreas as a whole and in its functional components, endocrine, exocrine, vascular, and intestinal. Using perfusate samples to identify site-specific injury markers would be valuable similar to utility of the kidney damage marker, kidney ischemia molecule-1 (KIM-1) which is released from the proximal tubular cells and elevated hyaluronic acid indicating sinusoidal endothelial cell damage in the liver. Flavin mononucleotide (FMN) is an emerging injury biomarker with majority of its evidence being investigated in HMP organ assessment [121]. FMN is a cofactor of the mitochondrial complex I and an acceptor of electrons from NADH for the maintenance of energy metabolism. FMN dissociates from complex I under conditions of reverse electron transfer when the mitochondria oxidizes succinate accumulated during ischemia [122]. The presence of FMN in HMP perfusate has been identified as a marker of mitochondrial injury and quantified to predict posttransplantation liver function (lactate clearance and coagulation factors) and outcomes (primary nonfunction and severe allograft dysfunction) [118].

Cell-free DNA (cfDNA) is a cell injury marker of increasing interest. At the time of cell death, short DNA fragments are released and can be detected and quantified, with increasing levels in proportion to the rate of cell death. Progress is being made to localize the origin of cfDNA by cell or tissue type by methylation patterns [123]. cfDNA has been investigated as a biomarker of injury in ex situ normothermic machine perfusion of lungs and correlates with the degree of allograft dysfunction [124]. Although there is no evidence regarding cfDNA in the setting of pancreas machine perfusion, it has been studied as a marker of beta cell loss in islet transplantation and observed to be predictive at 24h posttransplant for 1-month islet engraftment, defined as reduced insulin requirement/independence and higher stimulated C-peptide levels [125]. Isolation and analysis of total cfDNA during NMP should be very feasible, especially given that the organ is isolated from the potentially confounding effects of other organs and physiological cfDNA clearance mechanisms.

Imaging can provide invaluable information. Orthogonal polarization spectral imaging and side stream dark field imaging are portable devices that use the absorption of hemoglobin to visualize the microcirculation using a polarized light technique negating the need for contrast agents. It provides information including total and perfused vessel density and the proportion of perfused different sized vessels categorized by diameter. As noted earlier, the pancreas has arteriovenous shunts at the lobular level: the implication of this is that gross organ perfusion may not reflect microcirculatory perfusion, and therefore a strategy to assess microcirculation patency is particularly valuable. Orthogonal polarization spectral imaging and side stream dark field imaging have been used successfully to analyze the pancreatic microcirculation including the human in vivo pancreas in a few studies [126].

Magnetic resonance imaging (MRI) and spectroscopy (MRS) are in widespread clinical use. For the ex situ pancreas, it would be particularly useful to repurpose existing protocols to diagnose and quantify the severity of pancreatitis [127], pancreatic steatosis and fibrosis [128]. Weis et al. [129] used ^{1}H MRS to estimate lipid concentration in the nonadipose pancreatic cells in 11 nontransplanted SCS-preserved human pancreases, showing this to be a potentially useful means of assessment of allograft quality prior to whole pancreas transplantation or islet isolation.

MRI has been used to assess dynamically the metabolic changes in livers and kidneys during hypothermic and normothermic machine perfusion. Young et al. [130] investigated the use of the 3T MRI scanner to perform metabolic and structural assessments of a NMP-perfused porcine liver. After a period of warm ischemia achieved by stopping the oxygen supply, production rates of metabolites ATP, inorganic phosphate, lactate, pyruvate, alanine, and bicarbonate were evaluated; the perfusion of the liver tissue could also be monitored. Changani et al. [131] developed a rapid protocol to measure the regeneration of phosphate nucleotides using ^{31}P MRS by an oxygenated HMP porcine liver. ATP regeneration rates were shown to decrease with prolonged storage and their protocol produced images that reflected the degree of hepatic vascular perfusion.

Similar MR assessment has been reported of the kidney by Mariager et al. [132]. Hyperpolarized MRI and MRS were used to compare delivered pyruvate turnover in vivo versus ex situ in NMP porcine kidneys as a strategy to assess dynamic metabolism in machine-perfused grafts. Metabolic function was reduced compared to in vivo kidneys; the investigators concluded that the combination of MR imaging and NMP are valuable tools for ex situ assessments. The ability to image machine-perfused organs with MR enables noninvasive monitoring of the effect of interventions. It is certainly desirable that MR and other imaging technologies are integrated with the development of novel pancreas perfusion designs.

Perfusion dynamic parameters (e.g., pressure/resistance/flow) may also prove useful for pancreas assessment. The response of an isolated perfused organ to perfusion is complex, and in the absence of autonomic innervation and systemic vasoactive hormones may be more sensitive to endogenous endothelial factors such as nitric oxide (NO). Injury to the endothelium which can occur during ischemia leads to a reduced production of the endothelial-derived substances NO and prostacyclin which are vasodilators, and the upregulation of vasoactive endothelin-1 which, unopposed, causes an increase in vascular resistance and reduced flow.

Concluding remarks

The pancreas is in many ways the most challenging of the abdominal organs to transplant, and this is certainly the case with respect to organ retrieval and preservation. Not only is the pancreas vulnerable to preservation injury, but also the response to that injury is typically more severe in its consequence than the kidney or liver. Ischemia–reperfusion injury of the pancreas can be detected after most transplants and can be monitored by radiological and biochemical means. Although active resuscitation of the pancreas is not yet a reality, and indeed machine perfusion methods have barely been tested, it is in the pancreas that we might expect to see the greatest benefit of novel interventions designed to downregulate the inflammatory response. Indeed, because it is so sensitive, we might even see pancreas transplant studies become the preferred environment in which to test new therapeutic approaches. With a plethora of novel strategies to test in the coming years, the resulting pancreas transplant renaissance would be greatly welcomed.

References

[1] Krieger NR, Odorico JS, Heisey DM, D'Alessandro AM, Knechtle SJ, Pirsch JD, et al. Underutilization of pancreas donors. Transplantation 2003;75(8):1271–6.
[2] Fridell JA, Rogers J, Stratta RJ. The pancreas allograft donor: current status, controversies, and challenges for the future. Clin Transplant 2010;24(4):433–49.
[3] Leemkuil M, Leuvenink HGD, Pol RA. Pancreas transplantation from donors after circulatory death: an irrational reluctance? Curr Diab Rep 2019;19(11):129.
[4] D'Alessandro AM, Stratta RJ, Sollinger HW, Kalayoglu M, Pirsch JD, Belzer FO. Use of UW solution in pancreas transplantation. Diabetes 1989;38(Suppl. 1):7–9.
[5] Wahlberg JA, Love R, Landegaard L, Southard JH, Belzer FO. 72-Hour preservation of the canine pancreas. Transplantation 1987;43(1):5–8.
[6] Warshaw AL, O'Hara PJ. Susceptibility of the pancreas to ischemic injury in shock. Ann Surg 1978;188(2):197–201.
[7] Fernandez-del Castillo C, Harringer W, Warshaw AL, Vlahakes GJ, Koski G, Zaslavsky AM, et al. Risk factors for pancreatic cellular injury after cardiopulmonary bypass. N Engl J Med 1991;325(6):382–7.
[8] National Health Service Blood and Transplantation. Organ donation and transplantation activity report 2019/20, 2020. [Internet]. Available from: https://nhsbtdbe.blob.core.windows.net/umbraco-assets-corp/19481/activity-report-2019-2020.pdf.
[9] OPTN. Annual report of the U.S. organ procurement and transplantation network and the scientific registry of transplant recipients: 1994–2020 [Internet]. UNOS; 2021. Available from http://optn.transplant.hrsa.gov.
[10] Sung RS, Zhang M, Schaubel DE, Shu X, Magee JC. A reassessment of the survival advantage of simultaneous kidney-pancreas versus kidney-alone transplantation. Transplantation 2015;99(9):1900–6.
[11] Totsuka E, Fung JJ, Lee MC, Ishii T, Umehara M, Makino Y, et al. Influence of cold ischemia time and graft transport distance on postoperative outcome in human liver transplantation. Surg Today 2002;32(9):792–9.
[12] Hawthorne WJ, Davies S, Mun H-c, Chew YV, Williams L, Anderson P, et al. Successful islet outcomes using Australia-wide donors: a national centre experience. Metabolites 2021;11(6), 360.
[13] Barlow AD, Hosgood SA, Nicholson ML. Current state of pancreas preservation and implications for DCD pancreas transplantation. Transplantation 2013;95(12):1419–24.
[14] van Loo ES, Krikke C, Hofker HS, Berger SP, Leuvenink HG, Pol RA. Outcome of pancreas transplantation from donation after circulatory death compared to donation after brain death. Pancreatology 2017;17(1):13–8.
[15] Proneth A, Schnitzbauer AA, Schenker P, Wunsch A, Rauchfuss F, Arbogast H, et al. Extended pancreas donor program-the EXPAND study: a prospective multicenter trial testing the use of pancreas donors older than 50 years. Transplantation 2018;102(8):1330–7.
[16] Schneeberger S, Biebl M, Steurer W, Hesse UJ, Troisi R, Langrehr JM, et al. A prospective randomized multicenter trial comparing histidine-tryptophane-ketoglutarate versus University of Wisconsin perfusion solution in clinical pancreas transplantation. Transpl Int 2009;22(2):217–24.
[17] Boggi U, Vistoli F, Del Chiaro M, Signori S, Croce C, Pietrabissa A, et al. Pancreas preservation with University of Wisconsin and Celsior solutions: a single-center, prospective, randomized pilot study. Transplantation 2004;77(8):1186–90.
[18] Nicoluzzi J, Macri M, Fukushima J, Pereira A. Celsior versus Wisconsin solution in pancreas transplantation. Transplant Proc 2008;40(10):3305–7.
[19] Axelrod DA, Sung RS, Meyer KH, Wolfe RA, Kaufman DB. Systematic evaluation of pancreas allograft quality, outcomes and geographic variation in utilization. Am J Transplant 2010;10(4):837–45.
[20] Blok JJ, Kopp WH, Verhagen MJ, Schaapherder AF, de Fijter JW, Putter H, et al. The value of PDRI and P-PASS as predictors of outcome after pancreas transplantation in a large European pancreas transplantation center. Pancreas 2016;45(3):331–6.
[21] Mittal S, Lee FJ, Bradbury L, Collett D, Reddy S, Sinha S, et al. Validation of the pancreas donor risk index for use in a UK population. Transpl Int 2015;28(9):1028–33.
[22] Farney AC, Rogers J, Stratta RJ. Pancreas graft thrombosis: causes, prevention, diagnosis, and intervention. Curr Opin Organ Transplant 2012;17(1):87–92.
[23] Muthusamy ASR, Giangrande PLF, Friend PJ. Pancreas allograft thrombosis. Transplantation 2010;90(7):705–7.
[24] Troppmann C. Complications after pancreas transplantation. Curr Opin Organ Transplant 2010;15(1):112–8.
[25] Sakorafas GH, Tsiotos GG, Sarr MG. Ischemia/reperfusion-induced pancreatitis. Dig Surg 2000;17(1):3–14.

[26] Martin JL, Costa ASH, Gruszczyk AV, Beach TE, Allen FM, Prag HA, et al. Succinate accumulation drives ischaemia-reperfusion injury during organ transplantation. Nat Metab 2019;1(10):966–74.
[27] Zhang Y, Zhang M, Zhu W, Yu J, Wang Q, Zhang J, et al. Succinate accumulation induces mitochondrial reactive oxygen species generation and promotes status epilepticus in the kainic acid rat model. Redox Biol 2020;28, 101365.
[28] Panconesi R, Flores Carvalho M, Mueller M, Dutkowski P, Muiesan P, Schlegel A. Mitochondrial reprogramming—what is the benefit of hypothermic oxygenated perfusion in liver transplantation? Transplantology 2021;2(2):149–61.
[29] Chouchani ET, Pell VR, Gaude E, Aksentijević D, Sundier SY, Robb EL, et al. Ischaemic accumulation of succinate controls reperfusion injury through mitochondrial ROS. Nature 2014;515(7527):431–5.
[30] Petrenko A, Carnevale M, Somov A, Osorio J, Rodriguez J, Guibert E, et al. Organ preservation into the 2020s: the era of dynamic intervention. Transfus Med Hemother 2019;46(3):151–72.
[31] Benz S, Schnabel R, Morgenroth K, Weber H, Pfeffer F, Hopt UT. Ischemia/reperfusion injury of the pancreas: a new animal model. J Surg Res 1998;75(2):109–15.
[32] Benz S, Bergt S, Obermaier R, Wiessner R, Pfeffer F, Schareck W, et al. Impairment of microcirculation in the early reperfusion period predicts the degree of graft pancreatitis in clinical pancreas transplantation. Transplantation 2001;71(6):759–63.
[33] Schaser KD, Puhl G, Vollmar B, Menger MD, Stover JF, Köhler K, et al. In vivo imaging of human pancreatic microcirculation and pancreatic tissue injury in clinical pancreas transplantation. Am J Transplant 2005;5(2):341–50.
[34] Sunamura M, Yamauchi J, Shibuya K, Chen HM, Ding L, Takeda K, et al. Pancreatic microcirculation in acute pancreatitis. J Hepatobiliary Pancreat Surg 1998;5(1):62–8.
[35] Sweiry JH, Mann GE. Pancreatic microvascular permeability in caerulein-induced acute pancreatitis. Am J Physiol 1991;261(4 Pt. 1):G685–92.
[36] Cuthbertson CM, Christophi C. Disturbances of the microcirculation in acute pancreatitis. Br J Surg 2006;93(5):518–30.
[37] Vollmar B, Menger MD. Microcirculatory dysfunction in acute pancreatitis. A new concept of pathogenesis involving vasomotion-associated arteriolar constriction and dilation. Pancreatology 2003;3(3):181–90.
[38] Navina S, Acharya C, DeLany JP, Orlichenko LS, Baty CJ, Shiva SS, et al. Lipotoxicity causes multisystem organ failure and exacerbates acute pancreatitis in obesity. Sci Transl Med 2011;3(107):107ra10.
[39] Liong SY, Dixon RE, Chalmers N, Tavakoli A, Augustine T, O'Shea S. Complications following pancreatic transplantations: imaging features. Abdom Imaging 2011;36(2):206–14.
[40] Nadalin S, Girotti P, Königsrainer A. Risk factors for and management of graft pancreatitis. Curr Opin Organ Transplant 2013;18(1):89–96.
[41] Troppmann C, Gruessner AC, Dunn DL, Sutherland DE, Gruessner RW. Surgical complications requiring early relaparotomy after pancreas transplantation: a multivariate risk factor and economic impact analysis of the cyclosporine era. Ann Surg 1998;227(2):255–68.
[42] Klöppel G, Detlefsen S, Feyerabend B. Fibrosis of the pancreas: the initial tissue damage and the resulting pattern. Virchows Arch 2004;445(1):1–8.
[43] Cornateanu SM, O'Neill S, Dholakia S, Counter CJ, Sherif AE, Casey JJ, et al. Pancreas utilisation rates in the UK—an 11-year analysis. Transpl Int 2021;34(7):1306–18.
[44] Mathur A, Marine M, Lu D, Swartz-Basile DA, Saxena R, Zyromski NJ, et al. Nonalcoholic fatty pancreas disease. HPB (Oxford) 2007;9(4):312–8.
[45] Acharya C, Navina S, Singh VP. Role of pancreatic fat in the outcomes of pancreatitis. Pancreatology 2014;14(5):403–8.
[46] Papachristou GI, Papachristou DJ, Avula H, Slivka A, Whitcomb DC. Obesity increases the severity of acute pancreatitis: performance of APACHE-O score and correlation with the inflammatory response. Pancreatology 2006;6(4):279–85.
[47] Shin KY, Lee WS, Chung DW, Heo J, Jung MK, Tak WY, et al. Influence of obesity on the severity and clinical outcome of acute pancreatitis. Gut Liver 2011;5(3):335–9.
[48] Alempijevic T, Dragasevic S, Zec S, Popovic D, Milosavljevic T. Non-alcoholic fatty pancreas disease. Postgrad Med J 2017;93(1098):226–30.
[49] Oxford English Dictionary. Resuscitation. 3rd. Oxford University Press; 2020.
[50] Jochmans I, Brat A, Davies L, Hofker HS, van de Leemkolk FEM, Leuvenink HGD, et al. Oxygenated versus standard cold perfusion preservation in kidney transplantation (COMPARE): a randomised, double-blind, paired, phase 3 trial. Lancet 2020;396(10263):1653–62.
[51] van Rijn R, Schurink IJ, de Vries Y, van den Berg AP, Cortes Cerisuelo M, Darwish Murad S, et al. Hypothermic machine perfusion in liver transplantation—a randomized trial. N Engl J Med 2021;384(15):1391–401.
[52] Nasralla D, Coussios CC, Mergental H, Akhtar MZ, Butler AJ, Ceresa CDL, et al. A randomized trial of normothermic preservation in liver transplantation. Nature 2018;557(7703):50–6.
[53] Nicholson ML, Hosgood SA. Renal transplantation after ex vivo normothermic perfusion: the first clinical study. Am J Transplant 2013;13(5):1246–52.
[54] Cypel M, Yeung JC, Liu M, Anraku M, Chen F, Karolak W, et al. Normothermic ex vivo lung perfusion in clinical lung transplantation. N Engl J Med 2011;364(15):1431–40.
[55] Ardehali A, Esmailian F, Deng M, Soltesz E, Hsich E, Naka Y, et al. Ex-vivo perfusion of donor hearts for human heart transplantation (PROCEED II): a prospective, open-label, multicentre, randomised non-inferiority trial. Lancet 2015;385(9987):2577–84.
[56] Oniscu GC, Randle LV, Muiesan P, Butler AJ, Currie IS, Perera MTPR, et al. In situ normothermic regional perfusion for controlled Donation after circulatory death—the United Kingdom experience. Am J Transplant 2014;14(12):2846–54.
[57] Hessheimer AJFC. Normothermic regional perfusion in solid organ transplantation. In: Advances in extracorporeal membrane oxygenation, vol. 3. IntechOpen; 2019.
[58] Rojas-Peña A, Sall LE, Gravel MT, Cooley EG, Pelletier SJ, Bartlett RH, et al. Donation after circulatory determination of death: the University of Michigan experience with extracorporeal support. Transplantation 2014;98(3):328–34.
[59] Matsumoto S, Kuroda Y. Perfluorocarbon for organ preservation before transplantation. Transplantation 2002;74(12):1804–9.

[60] Kuroda Y, Kawamura T, Suzuki Y, Fujiwara H, Yamamoto K, Saitoh Y. A new, simple method for cold storage of the pancreas using perfluorochemical. Transplantation 1988;46(3):457–60.
[61] Fujino Y, Kuroda Y, Suzuki Y, Fujiwara H, Kawamura T, Morita A, et al. Preservation of canine pancreas for 96 hours by a modified two-layer (UW solution/perfluorochemical) cold storage method. Transplantation 1991;51(5):1133–5.
[62] Matsumoto S, Kandaswamy R, Sutherland DE, Hassoun AA, Hiraoka K, Sageshima J, et al. Clinical application of the two-layer (University of Wisconsin solution/perfluorochemical plus O2) method of pancreas preservation before transplantation. Transplantation 2000;70(5):771–4.
[63] Papas KK, Hering BJ, Gunther L, Rappel MJ, Colton CK, Avgoustiniatos ES. Pancreas oxygenation is limited during preservation with the two-layer method. Transplant Proc 2005;37(8):3501–4.
[64] Yotsumoto G, Jeschkeit-Schubbert S, Funcke C, Kuhn-Régnier F, Fischer JH. Total recovery of heart grafts of non-heart-beating donors after 3 hours of hypothermic coronary oxygen persufflation preservation in an orthotopic pig transplantation model. Transplantation 2003;75(6):750–6.
[65] Saad S, Minor T, Kötting M, Fu ZX, Hagn U, Paul A, et al. Extension of ischemic tolerance of porcine livers by cold preservation including postconditioning with gaseous oxygen. Transplantation 2001;71(4):498–502.
[66] Rolles K, Foreman J, Pegg DE. A pilot clinical study of retrograde oxygen persufflation in renal preservation. Transplantation 1989;48(2):339–42.
[67] Minor T, Klauke H, Isselhard W. Improved preservation of the small bowel by luminal gas oxygenation: energetic status during ischemia and functional integrity upon reperfusion. Transplant Proc 1997;29(7):2994–6.
[68] Suszynski TM, Rizzari MD, Scott 3rd WE, Tempelman LA, Taylor MJ, Papas KK. Persufflation (or gaseous oxygen perfusion) as a method of organ preservation. Cryobiology 2012;64(3):125–43.
[69] Scott 3rd WE, O'Brien TD, Ferrer-Fabrega J, Avgoustiniatos ES, Weegman BP, Anazawa T, et al. Persufflation improves pancreas preservation when compared with the two-layer method. Transplant Proc 2010;42(6):2016–9.
[70] Scott 3rd WE, Weegman BP, Ferrer-Fabrega J, Stein SA, Anazawa T, Kirchner VA, et al. Pancreas oxygen persufflation increases ATP levels as shown by nuclear magnetic resonance. Transplant Proc 2010;42(6):2011–5.
[71] Moers C, Smits JM, Maathuis M-HJ, Treckmann J, van Gelder F, Napieralski BP, et al. Machine perfusion or cold storage in deceased-donor kidney transplantation. N Engl J Med 2009;360(1):7–19.
[72] Zimmer H-G. The isolated perfused heart and its pioneers. Phys Ther 1998;13(4):203–10.
[73] Carrel A, Lindbergh CA. The culture of whole organs. Science 1935;81(2112):621–3.
[74] Hamaoui K, Papalois V. Machine perfusion and the pancreas: will it increase the donor Pool? Curr Diab Rep 2019;19(8):56.
[75] Kuan KG, Wee MN, Chung WY, Kumar R, Mees ST, Dennison A, et al. Extracorporeal machine perfusion of the pancreas: technical aspects and its clinical implications—a systematic review of experimental models. Transplant Rev (Orlando) 2016;30(1):31–47.
[76] Prudhomme T, Kervella D, Le Bas-Bernardet S, Cantarovich D, Karam G, Blancho G, et al. Ex situ perfusion of pancreas for whole-organ transplantation: is it safe and feasible? A systematic review. J Diabetes Sci Technol 2020;14(1):120–34.
[77] Treckmann J, Moers C, Smits JM, Gallinat A, Maathuis MH, van Kasterop-Kutz M, et al. Machine perfusion versus cold storage for preservation of kidneys from expanded criteria donors after brain death. Transpl Int 2011;24(6):548–54.
[78] Jochmans I, Moers C, Smits JM, Leuvenink HG, Treckmann J, Paul A, et al. Machine perfusion versus cold storage for the preservation of kidneys donated after cardiac death: a multicenter, randomized, controlled trial. Ann Surg 2010;252(5):756–64.
[79] Watson CJ, Wells AC, Roberts RJ, Akoh JA, Friend PJ, Akyol M, et al. Cold machine perfusion versus static cold storage of kidneys donated after cardiac death: a UK multicenter randomized controlled trial. Am J Transplant 2010;10(9):1991–9.
[80] Tersigni R, Toledo-Pereyra LH, Pinkham J, Najarian JS. Pancreaticoduodenal preservation by hypothermic pulsatile perfusion for twenty-four hours. Ann Surg 1975;182(6):743–8.
[81] Brynger H. Twenty-four-hour preservation of the duct-ligated canine pancreatic allograft. Eur Surg Res 1975;7(6):341–54.
[82] Florack G, Sutherland DE, Heil J, Squifflet JP, Najarian JS. Preservation of canine segmental pancreatic autografts: cold storage versus pulsatile machine perfusion. J Surg Res 1983;34(5):493–504.
[83] Karcz M, Cook HT, Sibbons P, Gray C, Dorling A, Papalois V. An ex-vivo model for hypothermic pulsatile perfusion of porcine pancreata: hemodynamic and morphologic characteristics. Exp Clin Transplant 2010;8(1):55–60.
[84] Leemkuil M, Lier G, Engelse MA, Ploeg RJ, de Koning EJP, t Hart NA, et al. Hypothermic oxygenated machine perfusion of the human donor pancreas. Transplant Direct 2018;4(10), e388.
[85] Branchereau J, Renaudin K, Kervella D, Bernadet S, Karam G, Blancho G, et al. Hypothermic pulsatile perfusion of human pancreas: preliminary technical feasibility study based on histology. Cryobiology 2018;85:56–62.
[86] Hamaoui K, Gowers S, Sandhu B, Vallant N, Cook T, Boutelle M, et al. Development of pancreatic machine perfusion: translational steps from porcine to human models. J Surg Res 2018;223:263–74.
[87] Nagpal BM, Sharma R. Cold injuries: the chill within. Med J Armed Forces India 2004;60(2):165–71.
[88] Ravikumar R, Jassem W, Mergental H, Heaton N, Mirza D, Perera MT, et al. Liver transplantation after ex vivo normothermic machine preservation: a phase 1 (first-in-man) clinical trial. Am J Transplant 2016;16(6):1779–87.
[89] Iyer A, Gao L, Doyle A, Rao P, Cropper JR, Soto C, et al. Normothermic ex vivo perfusion provides superior organ preservation and enables viability assessment of hearts from DCD donors. Am J Transplant 2015;15(2):371–80.
[90] Mergental H, Laing RW, Kirkham AJ, Perera MTPR, Boteon YL, Attard J, et al. Transplantation of discarded livers following viability testing with normothermic machine perfusion. Nat Commun 2020;11(1):2939.
[91] Yeung JC, Wagnetz D, Cypel M, Rubacha M, Koike T, Chun Y-M, et al. Ex vivo adenoviral vector gene delivery results in decreased vector-associated inflammation pre- and post-lung transplantation in the pig. Mol Ther 2012;20(6):1204–11.

[92] Divithotawela C, Cypel M, Martinu T, Singer LG, Binnie M, Chow CW, et al. Long-term outcomes of lung transplant with ex vivo lung perfusion. JAMA Surg 2019;154(12):1143–50.
[93] Babkin BP, Starling EH. A method for the study of the perfused pancreas. J Physiol 1926;61(2):245–7.
[94] Meyer W, Castelfranchi PL, Schulz LS, Ruiz OJ, Hendrickx J, Aquino CJ, et al. Physiologic studies during perfusion of the isolated canine pancreas. The endocrine and exocrine behavior. Eur Surg Res 1973;5(2):105–15.
[95] Eckhauser F, Knol JA, Porter-Fink V, Lockery D, Edgcomb L, Strodel WE, et al. Ex vivo normothermic hemoperfusion of the canine pancreas: applications and limitations of a modified experimental preparation. J Surg Res 1981;31(1):22–37.
[96] Kumar R, Chung WY, Runau F, Isherwood JD, Kuan KG, West K, et al. Ex vivo normothermic porcine pancreas: a physiological model for preservation and transplant study. Int J Surg 2018;54(Pt. A):206–15.
[97] Kuan KG, Wee MN, Chung WY, Kumar R, Mees ST, Dennison A, et al. A study of normothermic hemoperfusion of the porcine pancreas and kidney. Artif Organs 2017;41(5):490–5.
[98] Nassar A, Liu Q, Walsh M, Quintini C. Normothermic ex vivo perfusion of discarded human pancreas. Artif Organs 2018;42(3):334–5.
[99] Spetzler VN, Goldaracena N, Echiverri J, Kaths JM, Louis KS, Adeyi OA, et al. Subnormothermic ex vivo liver perfusion is a safe alternative to cold static storage for preserving standard criteria grafts. Liver Transpl 2016;22(1):111–9.
[100] Minor T, Sutschet K, Witzke O, Paul A, Gallinat A. Prediction of renal function upon reperfusion by ex situ controlled oxygenated rewarming. Eur J Clin Invest 2016;46(12):1024–30.
[101] Hoyer DP, Mathe Z, Gallinat A, Canbay AC, Treckmann JW, Rauen U, et al. Controlled oxygenated rewarming of cold stored livers prior to transplantation: first clinical application of a new concept. Transplantation 2016;100(1):147–52.
[102] Taylor MJ, Baicu S, Greene E, Vazquez A, Brassil J. Islet isolation from juvenile porcine pancreas after 24-h hypothermic machine perfusion preservation. Cell Transplant 2010;19(5):613–28.
[103] Watson CJE, Kosmoliaptsis V, Randle LV, Gimson AE, Brais R, Klinck JR, et al. Normothermic perfusion in the assessment and preservation of declined livers before transplantation: hyperoxia and vasoplegia-important lessons from the first 12 cases. Transplantation 2017;101(5):1084–98.
[104] Goldaracena N, Echeverri J, Spetzler VN, Kaths JM, Barbas AS, Louis KS, et al. Anti-inflammatory signaling during ex vivo liver perfusion improves the preservation of pig liver grafts before transplantation. Liver Transpl 2016;22(11):1573–83.
[105] Xiao F, Ma L, Zhao M, Smith RA, Huang G, Jones PM, et al. APT070 (mirococept), a membrane-localizing C3 convertase inhibitor, attenuates early human islet allograft damage in vitro and in vivo in a humanized mouse model. Br J Pharmacol 2016;173(3):575–87.
[106] Bhattacharjee RN, Richard-Mohamed M, Sun Q, Haig A, Aboalsamh G, Barrett P, et al. CORM-401 reduces ischemia reperfusion injury in an ex vivo renal porcine model of the Donation after circulatory death. Transplantation 2018;102(7):1066–74.
[107] Yamamoto H, Sugitani A, Kitada H, Arima T, Nishiyama K, Motoyama K, et al. Effect of FR167653 on pancreatic ischemia-reperfusion injury in dogs. Surgery 2001;129(3):309–17.
[108] Kusterer K, Poschmann T, Friedemann A, Enghofer M, Zendler S, Usadel KH. Arterial constriction, ischemia-reperfusion, and leukocyte adherence in acute pancreatitis. Am J Physiol 1993;265(1 Pt. 1):G165–71.
[109] Hackert T, Werner J, Uhl W, Gebhard M-M, Büchler MW, Schmidt J. Reduction of ischemia/reperfusion injury by antithrombin III after experimental pancreas transplantation. Am J Surg 2005;189(1):92–7.
[110] Hamaoui K, Gowers S, Boutelle M, Cook TH, Hanna G, Darzi A, et al. Organ pretreatment with cytotopic endothelial localizing peptides to ameliorate microvascular thrombosis and perfusion deficits in ex vivo renal hemoreperfusion models. Transplantation 2016;100(12):e128–39.
[111] Kuroda T, Shiohara E, Haba Y, Hanazaki K. The effect of a thromboxane A2 receptor antagonist (ONO 3708) on ischemia-reperfusion injury of the dog pancreas. Transplantation 1994;57(2):187–94.
[112] Brasile L, Henry N, Orlando G, Stubenitsky B. Potentiating renal regeneration using mesenchymal stem cells. Transplantation 2019;103(2):307–13.
[113] Gillooly AR, Perry J, Martins PN. First report of siRNA uptake (for RNA interference) during ex vivo hypothermic and normothermic liver machine perfusion. Transplantation 2019;103(3):e56–7.
[114] Juriasingani S, Akbari M, Chan JY, Whiteman M, Sener A. H2S supplementation: a novel method for successful organ preservation at subnormothermic temperatures. Nitric Oxide 2018;81:57–66.
[115] Lemaire F, Sigrist S, Delpy E, Cherfan J, Peronet C, Zal F, et al. Beneficial effects of the novel marine oxygen carrier M101 during cold preservation of rat and human pancreas. J Cell Mol Med 2019;23(12):8025–34.
[116] Echeverri J, Goldaracena N, Kaths JM, Linares I, Roizales R, Kollmann D, et al. Comparison of BQ123, epoprostenol, and verapamil as vasodilators during normothermic ex vivo liver machine perfusion. Transplantation 2018;102(4):601–8.
[117] Benz S, Obermaier R, Wiessner R, Breitenbuch PV, Burska D, Weber H, et al. Effect of nitric oxide in ischemia/reperfusion of the pancreas. J Surg Res 2002;106(1):46–53.
[118] Muller X, Schlegel A, Kron P, Eshmuminov D, Würdinger M, Meierhofer D, et al. Novel real-time prediction of liver graft function during hypothermic oxygenated machine perfusion before liver transplantation. Ann Surg 2019;270(5):783–90.
[119] Patel SK, Pankewycz OG, Nader ND, Zachariah M, Kohli R, Laftavi MR. Prognostic utility of hypothermic machine perfusion in deceased donor renal transplantation. Transplant Proc 2012;44(7):2207–12.
[120] Papas KK, Suszynski TM, Colton CK. Islet assessment for transplantation. Curr Opin Organ Transplant 2009;14(6):674–82.
[121] Schlegel A, Muller X, Mueller M, Stepanova A, Kron P, de Rougemont O, et al. Hypothermic oxygenated perfusion protects from mitochondrial injury before liver transplantation. EBioMedicine 2020;60, 103014.
[122] Ten V, Galkin A. Mechanism of mitochondrial complex I damage in brain ischemia/reperfusion injury. A hypothesis. Mol Cell Neurosci 2019;100, 103408.

[123] Akirav EM, Lebastchi J, Galvan EM, Henegariu O, Akirav M, Ablamunits V, et al. Detection of β cell death in diabetes using differentially methylated circulating DNA. Proc Natl Acad Sci U S A 2011;108(47):19018–23.

[124] Kanou T, Nakahira K, Choi AM, Yeung JC, Cypel M, Liu M, et al. Cell-free DNA in human ex vivo lung perfusate as a potential biomarker to predict the risk of primary graft dysfunction in lung transplantation. J Thorac Cardiovasc Surg 2021;162(2):490–499.e2.

[125] Gala-Lopez BL, Neiman D, Kin T, O'Gorman D, Pepper AR, Malcolm AJ, et al. Beta cell death by cell-free DNA and outcome after clinical islet transplantation. Transplantation 2018;102(6):978–85.

[126] von Dobschuetz E, Biberthaler P, Mussack T, Langer S, Messmer K, Hoffmann T. Noninvasive in vivo assessment of the pancreatic microcirculation: orthogonal polarization spectral imaging. Pancreas 2003;26(2):139–43.

[127] Arvanitakis M, Delhaye M, De Maertelaere V, Bali M, Winant C, Coppens E, et al. Computed tomography and magnetic resonance imaging in the assessment of acute pancreatitis. Gastroenterology 2004;126(3):715–23.

[128] Yoon JH, Lee JM, Lee KB, Kim S-W, Kang MJ, Jang J-Y, et al. Pancreatic steatosis and fibrosis: quantitative assessment with preoperative multiparametric MR imaging. Radiology 2016;279(1):140–50.

[129] Weis J, Ahlström H, Korsgren O. Proton MR spectroscopy of human pancreas allografts. Magn Reson Mater Phys Biol Med 2019;32(4):511–7.

[130] Young LA, Ceresa CD, Miller JJ, Valkovic L, Voyce D, Tunnicliffe EM, Ellis J. Assessing metabolism and function of normothermically perfused ex vivo livers by multinuclear MR imaging and spectroscopy. ISMRM; 2018.

[131] Kumar Changani K, Fuller BJ, Bryant DJ, Bell JD, Ala-Korpela M, Taylor-Robinson SD, et al. Non-invasive assessment of ATP regeneration potential of the preserved donor liver: a ³¹P MRS study in pig liver. J Hepatol 1997;26(2):336–42.

[132] Mariager CØ, Hansen ESS, Bech SK, Munk A, Kjærgaard U, Lyhne MD, et al. Graft assessment of the ex vivo perfused porcine kidney using hyperpolarized [1-13C]pyruvate. Magn Reson Med 2020;84(5):2645–55.

Chapter 4

Normothermic regional perfusion for whole pancreas and islet transplantation

Andrew I. Sutherland[a,b], Aaron Hui[c,d], and Gabriel Oniscu[b,e]

[a]*Edinburgh Transplant, Royal Infirmary of Edinburgh, Edinburgh, Scotland,* [b]*University of Edinburgh, Edinburgh, Scotland,* [c]*Transplant and General Surgery, Monash Health, Melbourne, VIC, Australia,* [d]*Transplant and General Surgery, St. Vincents Health, Melbourne, VIC Australia,* [e]*Royal Infirmary of Edinburgh, Edinburgh, Scotland*

Introduction

Since the first pancreas transplant in 1966, there have now been over 45,000 pancreas transplants worldwide [1]. Numbers have increased to a plateau in 2004 and the 1-year graft survival has increased from 37% in 1983 to greater than 90% in 2021 [2]. Similarly, the long-term pancreas graft function has improved, especially when performed as part of a simultaneous kidney and pancreas transplant. Islet Transplantation, with the first successful long-term islet transplant program established in Edmonton, Canada has over the past 2 decades become established in transplant centers throughout the world and is considered the "standard of care" in patients with Type 1 diabetes and severely impaired hypoglycemia awareness. Despite these successes, the number of people with diabetes continues to rise and there remains a significant shortage of good quality donor pancreases for transplantation. One under-utilized source of pancreases is from donors after circulatory arrest (DCD). Although DCD pancreases have been shown to have good outcomes in both solid organ and islet transplants only a relatively small number of countries have established programs. This is largely due to concern about additional damage from warm ischemia time during the "agonal phase." Novel organ perfusion technologies have been developed over the past two decades which have helped mitigate some of these risks in liver and kidney transplants. The use of organ perfusion technology in pancreas transplants is, however, still largely limited to the experimental setting. Normothermic regional perfusion, a process wherein DCD donors, warm oxygenated blood is recirculated in the donor, is now part of established practice in a number of European countries. This has significantly improved outcomes in DCD livers and kidney transplantation [3,4] and more recently thoracoabdominal NRP has been introduced for the procurement of DCD hearts [5]. This chapter expands on this technique, detailing how it is performed and how organs are assessed, as well as the early results on outcomes in transplantation including pancreas and islet cell transplantation.

Donors after circulatory death

DCD (Deceased after Circulatory death) refers to the retrieval of organs for the purpose of transplantation from patients whose death is diagnosed and confirmed using cardio-respiratory criteria. This contrasts with donors that are referred for organ donation based on brain stem death testing. Patients are referred through the DCD pathway usually after suffering a catastrophic brain injury that does not meet brain stem death criteria. DCD donors are classified under two principal types; uncontrolled or controlled donors. This classification was further defined in 1995 as the *Maastricht Classification* to help sub categorize the DCD donors based on organ viability/ischaemic insult and graft survival. This has unified the naming of DCD across European and Western transplant centers (see Table 1) allowing better graft follow up and uniformity of data in published literature [6].

In the United Kingdom, Maastricht III and IV donors are only permitted to proceed to organ donation, due to the current ethical and legal restrictions restricting access to these patients. In some European countries (e.g., Spain and France) there are a larger focus on Maastricht I and II donors, which can reflect the different numbers of DCD donor pools from each country and also organ outcomes. In countries that allow the use of Maastricht I and II donors, premortem interventions are also permitted. These include procedures such as femoral cannulation and heparinisation prior to WLST (withdrawal of life-saving treatment) which may also influence outcomes when compared to donors from other countries. Such pre-mortem interventions have allowed centers in Spain for example, to achieve greater utilization of

Pancreas and Beta Cell Replacement. https://doi.org/10.1016/B978-0-12-824011-3.00004-7

TABLE 1 Modified Maastricht classification of DCD and the locations where mainly practiced.

Category	Description	Type of DCD	Locations practiced
I	Dead on arrival	Uncontrolled	ED in a transplant center
II	Unsuccessful resuscitation	Uncontrolled	ED in a transplant center
III	Anticipated cardiac arrest	Controlled	ICU and ED
IV	Cardiac arrest in a brain-dead donor	Controlled	ICU and ED
V	Unexpected arrest in ICU patient	Uncontrolled	ICU in a transplant center

ICU, intensive care unit; *ED*, emergency department.
Source: Otero A, Gomez-Gutierrez M, Suarez F, et al. Liver transplantation from Maastricht category 2 non-heart beating donors. Transplantation 2003; 76: 1068–1073.

novel technologies such as NRP and refine techniques in "uncontrolled DCD" donors, dramatically increasing the donor pool and revitalizing interest and research into DCD donors [6,7].

Organs from controlled/uncontrolled DCD donors are exposed to a greater duration of warm ischemia than those from comparable DBD donors. This is at its most profound between the onset of asystole and establishing organ cold perfusion with a preservation fluid. A better measure of ischaemic injury is the so-called functional warm ischemia time, which is considered to begin when the patient's systolic arterial pressure decreases below 50 mmHg, the arterial oxygen saturation decreases below 70% or both. This period from WLST to point of cardiac arrest is known as the *agonal phase* which is the most unpredictable time in the retrieval process. Times vary from a few minutes up to hours and play a major influence in the utilization rates of organs that may have come from an otherwise suitable donor. Ischaemic injury increases the risks of primary graft failure, delayed graft function, and other ischaemic complications (e.g., biliary structures in liver grafts), and is a considerable concern to retrieval and implantation teams [7,8]. This is of particular importance in the pancreas where any periods of instability expose the potential donor graft to the deleterious effects of hypoxia and increase the rate so end-organ damage and chances of vascular thrombosis and graft pancreatitis in the recipient. As a consequence, transplanting teams are often cautious in accepting organs from older potential DCD donors or those with co-morbidities such as diabetes mellitus, hypertension, and peripheral vascular disease that may compound such ischaemic damage expected from a DCD donor (Table 2).

The length of the agonal phase and the total cold ischaemic time (CIT) have a major influence on whether organs are accepted and used for transplantation. The pancreas is very susceptible to the deleterious effects of ischemia. Even in cold storage, the process of anaerobic metabolism occurs with the accumulation of toxic metabolites which amplify post-transplantation reperfusion injury in donor organs. The result of prolonged ischaemic time is graft pancreatitis/graft thrombosis and higher rates of graft pancreatectomy.

With the advent of static cold storage (SCS) by Collins et al in the 1960s which demonstrated cheap and acceptable outcomes for kidneys over prolonged periods of storage time [9]; perfusion preservation technology had largely taken a back seat to this more cost-effective method. Evidence has demonstrated that even with a short warm ischaemic time, SCS methods do not prevent the ongoing effects of ischemia, and this becomes more important when large distances are involved between donor and recipient centers. Multiple studies into liver/kidney/lung grafts have demonstrated that the

TABLE 2 DCD pancreas donor selection criteria in the United Kingdom.

Age 5–60 years
Body mass index <30
Absence of contraindications for organ donation (infection, malignancies, etc.)
Treatment withdrawal to circulatory arrest <60 min
Adequate in situ perfusion

longer the CIT the greater chance of primary nonfunction and delayed graft function [10]. The increasing need for donor organs has seen increased use of DCD organs and even more marginal donors, driving renewed investment and interest in perfusion preservation methods for this donor population.

DCD pancreas and islet cell transplant

DCD donors continue to be a very under-used resource for pancreas and islet cell transplants worldwide. In the US less than 3% of pancreas transplants are from DCD donors [11]. In Europe, although 18 out of 35 countries have established DCD programs, only five centers currently transplant pancreases from DCD donors [12a]. Between 2008 and 2016, out of 473 pancreas transplants performed from DCD donors, 85% were performed in the UK, with Belgium and the Netherlands performing 8% and 6% respectively, and Spain and Switzerland performing less than 1% between them [12a]. However, even in the UK, this accounted for only 10% of pancreases transplanted during that period. Although DCD pancreas utilization has continued to increase in the UK (it is now about 30%), this has still not reached parity with the proportion of DCD donors (40%), and there is a significant variation in utilization between different units within the UK [12b]. It should also be noted that in the three countries with the most DCD transplants in Europe, over 99% of DCD donors are from controlled DCD, not uncontrolled donors [12a]. The reluctance to utilize DCD pancreases is largely driven by concern about the additional warm ischemia time causing an increased risk of complications including graft pancreatitis and graft thrombosis. However, this reluctance seems to be at odds with the published outcomes from DCD donors. Although several papers have reported an increased rate of graft thrombosis in DCD pancreases [13–15], they have also demonstrated broadly equivalent 5-year graft survival compared to DBD donors. There is, however, evidence that centers are more selective in their criteria for DCD donors, and certainly, in the largest series [13] the DCD donors had significantly lower mean age and BMI. Nevertheless, these studies demonstrate that excellent graft outcomes can be achieved with DCD donors, albeit with restricted criteria. A meta-analysis of four studies found that there was no significant difference in graft survival between DBD and DCD donors up to 10 years while confirming that graft thrombosis was 1.67 times higher in DCD organs (95% CI, 0.74–1.31, $P=0.006$) [16]. Interestingly in this meta-analysis, there was not a large difference in mean donor age between DCD and DBD (26 years versus 27). A recent registry study by Callaghan et al. [17] of the British experience of DCD kidney-pancreas transplant, also found no differences in 5-year graft survival between DCD and DBD donors (79.5% versus 80.4%; $P=0.86$). The DCD donors were younger and had lower BMIs. However, following univariate and multivariate analysis there was no statistically significant association between increasing donor age and pancreas CIT on the survival of pancreases transplanted from DCD donors. This differs from Chen et al [18] analysis of US data that donor age is the most important predictor of long-term graft function in DCD donors. However, Chen et al also found no negative association between prolonged CIT and graft survival in SPK transplants from DCD donors, which goes against conventional wisdom regarding minimizing risk from DCD donors. The authors of both these studies agree that greater statistical power is needed to delineate the predictive effect of CIT on DCD donor pancreas transplant outcomes.

Islet transplantation differs from solid organ pancreas transplant in that there is a second opportunity to assess organ quality prior to transplantation by way of checking islet yield, purity, and viability. However, this further quality evaluation has not led to wider adoption of using DCD pancreases for islet cell transplants. In the UK, despite having a long-established DCD pancreas program, less than 10% of islet transplants are from DCD donors [18] This may be because islet yield, viability, and purity does not necessarily give a functional assessment of the islets, especially in terms of long-term function and does not necessarily reflect the results of isolation outcomes for a significantly lower transplant. The early results in uncontrolled DCD donors in Japan were encouraging in terms of islet yield and early outcomes prior to transplant [19,20]. However, this was based on the early intervention of the DCD donor compared to other jurisdictional criteria that require cessation of heartbeat prior to intervention. Additionally, when DBD donors were approved in Japan, a retrospective analysis comparing DBD and DCD islet transplants showed that although yield and viability were similar, long-term outcomes were inferior for DCD donor pancreases [21]. De Paep et al. described the Belgian experience and found that similar beta cell yields could be achieved with DCD donors if the agonal phase was less than 10 min [22]. De Konnig and team, however, found significantly lower yields from DCD donors, and this was not related to warm ischemia time [23]. Despite the lower yields from DCD donor pancreases, beta-cell function, measured by glucose-stimulated insulin secretion, was similar between the DCD and DBD groups. Furthermore, following transplantation, there was no difference in c-peptide (area under the curve) following mixed meal tolerance tests between DCD and DBD graft recipients [24]. This differs from our own experience in Edinburgh where we analyzed the yields from DCD and DBD donors over the last 10 years [25]. There were no significant differences in mean yield (248,800 vs 213,713, $P=0.18$), viability (81.9% vs 81.4%, $P=0.88$), and purity (69.3% vs 71.6%, $P=0.58$), between DBD and DCD donors respectively. Understanding the long-term outcomes is more challenging as only a minority of recipients received both of their islets transplants from DCD donors. However, when we compared the outcomes of recipients who received at least one islet transplant from a DCD donor to those that received both their routine and priority islet transplants from DBD donors and there were no significant differences in one-year graft survival (84% vs 87%), stimulated c-peptide (671 vs 580 pmol/L, $P=0.54$), number of severe hypoglycaemic events/year (0.04 vs 0.28, $P=0.25$), HbA1C (56 vs 52 mmol/mol, $P=0.25$) and median beta score (3 vs 3, $P=0.21$). The only difference we found was that awareness of hypoglycemia was superior in the DBD group compared to the DCD group (Gold Score 2 vs 4, $P<0.01$).

In some ways in the field of pancreas and islet transplantation, donor organ shortage is not the major issue, it is organ utilization. The limiting factors are the uncertainty of organ assessment prior to transplantation with regard to steatosis and the likelihood of post-transplant pancreatitis. This is further compounded by current limitations to accurately estimate b-cell viability and function of the organ, all of which are heavily influenced by donor details. Normothermic regional perfusion gives a further "window" with which to evaluate organ quality and function prior to deciding to utilize the pancreas.

Normothermic regional perfusion: Methodology

Before the withdrawal of supportive care for the DCD donor, it is important that the retrieval team has established the normothermic extra-corporeal circuit. Although there is some variation in the circuits used by different centers, the principles for each NRP circuit are the same, relying on an extra-corporeal centrifugal pump, membrane oxygenator, an external heat exchanger to maintain blood temperature close to 37.3 degrees Celsius, and an external reservoir to collect blood and store perfusate. In the UK a leucocyte filter is used but is left out of the circuit in the first 2–3 min to minimize blood clotting that may occur. Circuits are primed with heparin except where a heparin-coated circuit is used. The prime solution comprises sodium lactate and succinylated gelatin and contains antimicrobials, sodium bicarbonate, and heparin. For thoracoabdominal NRP mannitol is also added.

In some European countries that have pioneered DCD-NRP antemortem interventions are permitted. In Spain/Netherlands/Italy Maastricht III donors are permitted to be cannulated via the femoral vessels or have vascular sheath introduced to facilitate femoral vessel cannulation. In addition to this IV, heparin is administered in ITU when the withdrawal of treatment occurs. In Spain, for Maastricht 2 donors femoral cannulation can be performed by paramedic staff, and IV heparinisation is given before the donor has arrived after unsuccessful CPR [24,26,27]. All antemortem interventions are not currently permitted in the UK (Fig. 1).

In the UK, following the withdrawal of treatment and confirmation of cardiac death, donors are rapidly transferred to the operating theater and placed supine. A rapid midline laparotomy is performed, and the distal aorta (or common iliac artery) is cannulated and in most instances, an intra-aortic balloon is placed concurrently; this allows for occlusion of the supradiaphragmatic aorta and commencement of the perfusion pump before a thoracotomy is performed

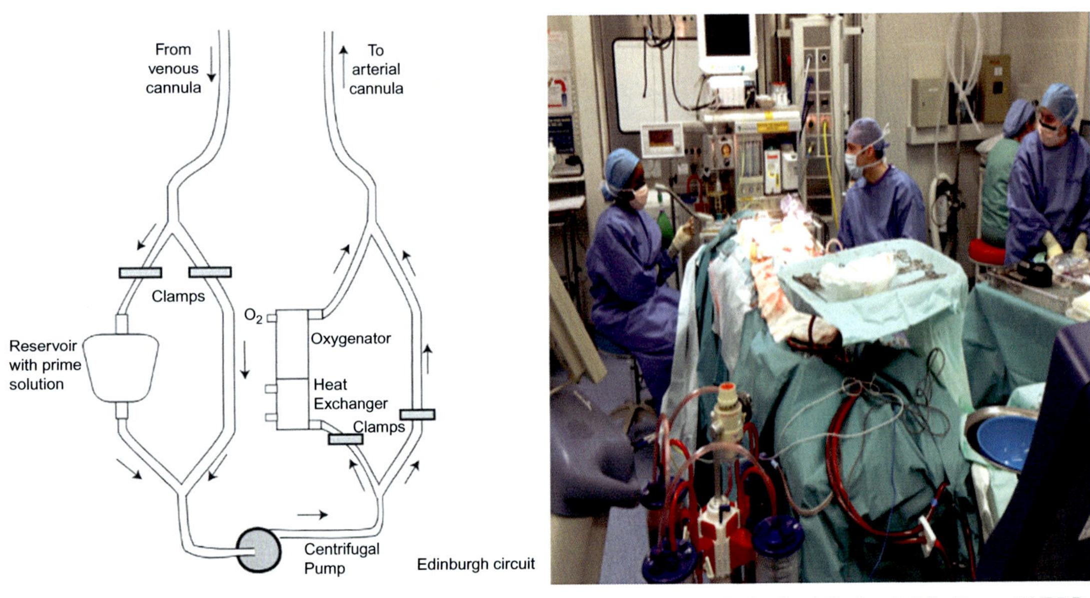

FIG. 1 Schematic of NRP set-up in practice. *(Modified from Oniscu G., Randle L.V., Muisean P., Butler A.J., Currie I.S., Perera M.T.P.R., et al. In situ normothermic regional perfusion for controlled donation after circulatory death. The United Kingdom experience. Am J Transplant 2014;14:2846–2854.)*

reducing warm ischaemic time. Inferior vena cava cannulation is performed by using a patented (drainage 3-way need product information) system to allow for greater drainage of the renal/liver and pancreas outflow tracts; it is reinforced circumferentially to prevent the "collapse" of the IVC ensuring uninhibited drainage. Once the cannulae are connected to the external circuit a closed circuit is achieved and NRP is commenced. Once the NRP circuit has been established the retrieving surgeon often performs a thoracotomy and places a descending thoracic aortic clamp supplementing the aortic balloon placed to isolate the abdominal circulation and to prevent any cerebral perfusion.

The pump flow should be maintained above 2 L/min with a maximum of 4 L. Continuous oxygen/air mixture is supplied to the circuit to maintain blood gas parameters within normal limits; no volatile agents are given. The NRP circuit is initially established bypassing the oxygenator (and reservoir, where present) to prevent clotting on their large surface area until the heparin in the circuit had mixed with the circulating blood. After 1-min blood flow is diverted through the oxygenator and reservoir, to continue oxygenated perfusion for 2 h.

DCD-NRP involving cardiothoracic organs

Thoracoabdominal NRP can be performed either by cannulating the ascending aorta and IVC via the right atrial appendage, or more commonly cannulating the abdominal aorta and IVC, with a clamp placed across the origins of the brachiocephalic trunk, left common carotid, and left subclavian arteries; Current preference for DCD heart recovery involves abdominal aortic cannulation since that affords better access for clamping the arch vessels to prevent cerebral perfusion.

Where thoracoabdominal-NRP is performed and cardiac output restored, NRP is stopped at 30 to 60 min, and the heart is allowed to support the limited thoracoabdominal circulation while its function was evaluated. If the cardiac function is considered inadequate and failing to sustain an adequate mean arterial pressure, extracorporeal perfusion is recommenced.

Thoracoabdominal NRP adds an additional layer of complexity and donor risk as once the cardiac organs are retrieved (within 30 min of commencing NRP), the thoracic cavity can remain in the circuit due to the presence of shunts and collateral vascular supply, it is often advisable to have additional surgical team members to help in the inevitable additional bleeding that is encountered with the donor when heart and lungs are retrieved.

Organ assessment during normothermic regional perfusion

During abdominal NRP parameters are set to maintain a pump flow of 2–4 L/min. A continuous maintenance pressure of 60–65 mmHg in the arterial cannula, and a temperature of 37 °C; bicarbonate is usually administered just after NRP has started to maintain pH 7.35–7.45, and a hematocrit > 25% is targeted. Blood samples from the circuit are obtained just after starting NRP and at least every 30 min for biochemistry analysis, serum lactate levels, and hematocrit values. The ALT should be < 3 times the upper limit of normal at the initiation of NRP and < 4 times the upper limit of normal at the end of NRP when selecting the liver grafts. In the UK organs are allocated in keeping with the current allocation criteria and the use of NRP does not affect the offering system (Figs. 2 and 3).

The UK experience published was by Onicsu et al. in 2014 [26] involved all pioneering centers that were implementing NRP. Although centers differed in minor aspects of setup, biochemical analysis while on NRP was relatively uniform with alanine transferase (ALT) and Lactate being used as surrogate markers for liver function. Donor livers that did not see significant improvement in either of these two parameters were not used from the implant and donated for clinical research (see Fig. 4). In addition to the biochemical analysis, macroscopic appearances of the liver, bowel perfusion with absence or presence of peristalsis, the appearance of the gall bladder mucosa and bile duct were all used in graft assessment. In this study, a total of 21 DCD donors proceeded and NRP was utilized, in total 11 livers were utilized, 38 kidneys and three pancreata were from a donor pool of 21 (see Table 3). The current duration of optimal NRP has not yet been established. Some Spanish Centres advocate for 4 h as the optimum time to reverse ischemic injury, however, the UK implements 2 h as being sufficient to allow organ assessment and aid recovery, a time supported by more recent studies. A two-hour NRP period is more favorable in regard to logistical support, theater allocation without compromising organ quality.

Unlike liver grafts, there is no standardized method for assessing pancreas function on NRP and recipient centers are still reliant on donor and recipient information when deciding on accepting donor pancreata.

Mechanistic insights into how normothermic regional perfusion works

When considering how normothermic regional perfusion works it is difficult to separate the whole from the sum of the constituent parts. The concept of warm, oxygenated perfusion is not new, indeed organ preservation using machine perfusion preceded static cold perfusion. With the advent of normothermic machine perfusion came the concept of

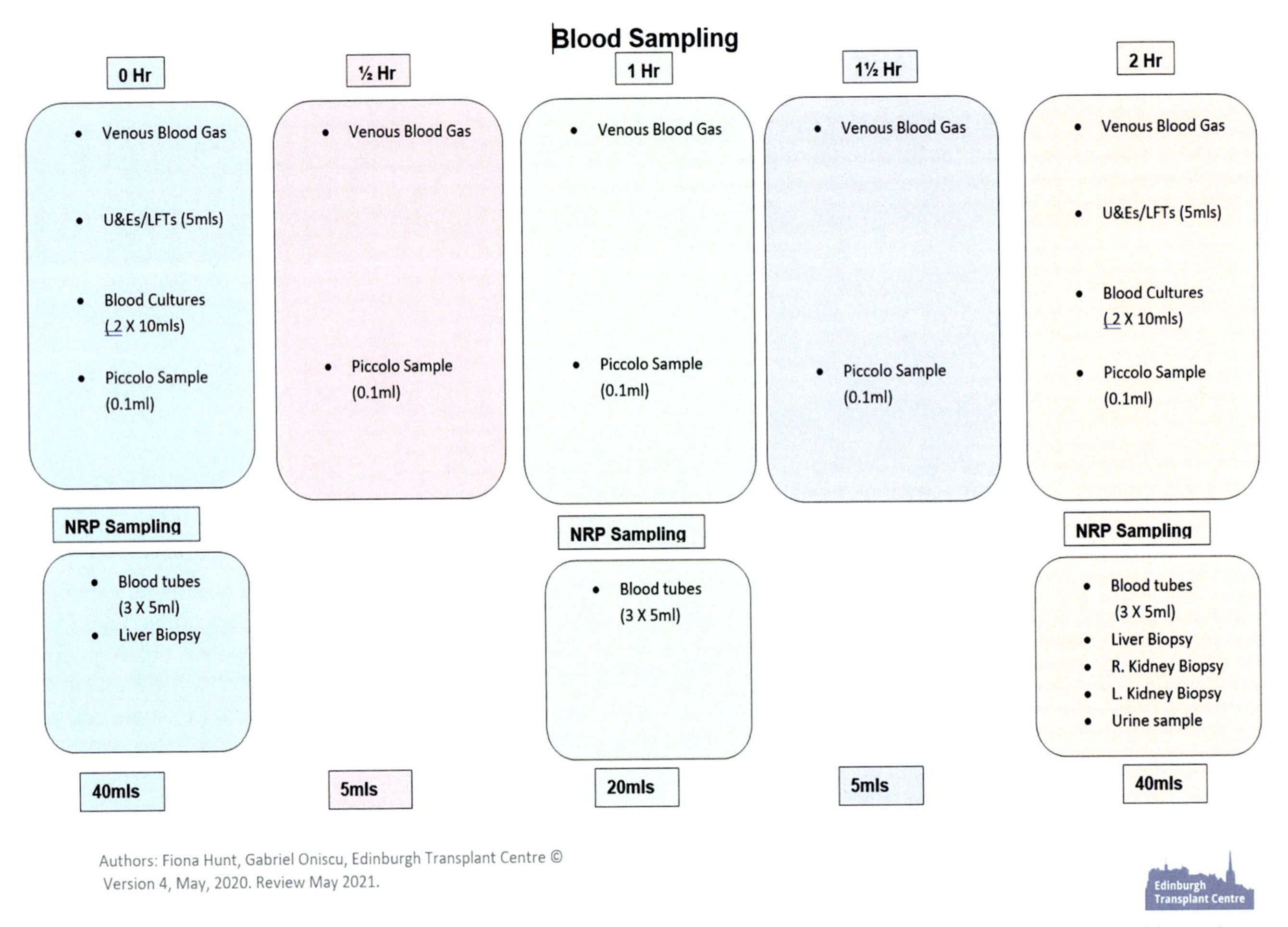

FIG. 2 Blood sampling is taken at scheduled intervals during the Normothermic Regional Perfusion (NRP process to give subjective evidence of organ function).

"ischaemic reconditioning," the process whereby organs have cellular energy substrates restored, reducing injury following reperfusion. There is also evidence to suggest that ischaemic preconditioning may occur, the process whereby a short period of ischemia protects organs from a subsequent prolonged period of ischemia. However, the effect of both these processes cannot be separated from the role played by organ assessment. As discussed above, monitoring of physiological parameters and biochemical markers plays a critical role in deciding whether organs are used. Thus, the improvements that have been seen in graft outcomes [24,26] are likely the combination of all these factors (Fig. 5). The contribution of ischaemic preconditioning and reconditioning are discussed below.

Ischaemic preconditioning

Ischaemic preconditioning is the process whereby a short period of ischemia protects organs from a subsequent prolonged period of ischemia. There is an early phase of preconditioning which is apparent immediately after reperfusion and lasts for approximately 2h, there then follows a longer more prolonged period of protection that is apparent 24–72h later. There is preliminary evidence that both these processes may be triggered by the agonal phase during normothermic regional perfusion. A key mediator of the early phase of protection is adenosine [27]. Garcia-Valdecaseas' group has demonstrated in a porcine model of NRP that adenosine can re-produce the protective effect of NRP in porcine livers and that inhibition of adenosine-A2 receptors abrogates the protective effect of NRP [27,28] (Fig. 6).

The delayed phase of preconditioning is regulated by genes induced during the initial period of ischemia. One of the principal mediators of this response is Hypoxia Inducible Factor (HIF). HIF is continuously degraded by prolyl hydroxylation and proteasomal degradation during normoxia. However, in hypoxia or ischemia, the HIF pathway is immediately activated causing the up-regulation of more than 100 hypoxia-regulated genes. Among these are several genes, including

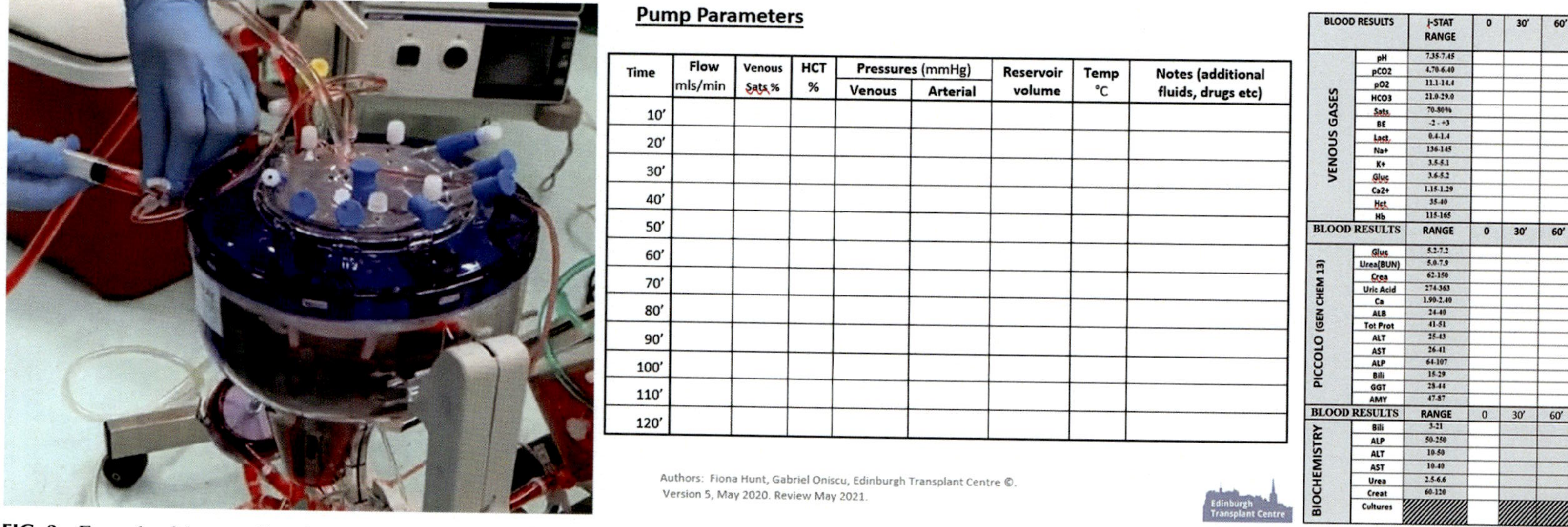

Pump Parameters

Time	Flow mls/min	Venous Sats %	HCT %	Pressures (mmHg)		Reservoir volume	Temp °C	Notes (additional fluids, drugs etc)
				Venous	Arterial			
10′								
20′								
30′								
40′								
50′								
60′								
70′								
80′								
90′								
100′								
110′								
120′								

Authors: Fiona Hunt, Gabriel Oniscu, Edinburgh Transplant Centre ©.
Version 5, May 2020. Review May 2021.

BLOOD RESULTS		i-STAT RANGE	0	30′	60′	90′	120′
VENOUS GASES	pH	7.35-7.45					
	pCO2	4.70-6.40					
	pO2	11.1-14.4					
	HCO3	21.0-29.0					
	Sats	70-80%					
	BE	-2 - +3					
	Lact	0.4-1.4					
	Na+	136-145					
	K+	3.5-5.1					
	Gluc	3.6-5.2					
	Ca2+	1.15-1.29					
	Hct	35-40					
	Hb	115-165					
BLOOD RESULTS		**RANGE**	**0**	**30′**	**60′**	**90′**	**120′**
PICCOLO (GEN CHEM 13)	Gluc	5.2-7.2					
	Urea(BUN)	5.0-7.9					
	Crea	62-150					
	Uric Acid	274-363					
	Ca	1.90-2.40					
	ALB	24-40					
	Tot Prot	41-51					
	ALT	25-43					
	AST	26-41					
	ALP	64-107					
	Bili	15-29					
	GGT	28-44					
	AMY	47-87					
BLOOD RESULTS		**RANGE**	**0**	**30′**	**60′**	**90′**	**120′**
BIOCHEMISTRY	Bili	3-21					
	ALP	50-250					
	ALT	10-50					
	AST	10-40					
	Urea	2.5-6.6					
	Creat	60-120					
	Cultures						

Edinburgh Transplant Centre

FIG. 3 Example of the recordings for pump parameters and blood biochemistry during Normothermic Regional Perfusion (NRP).

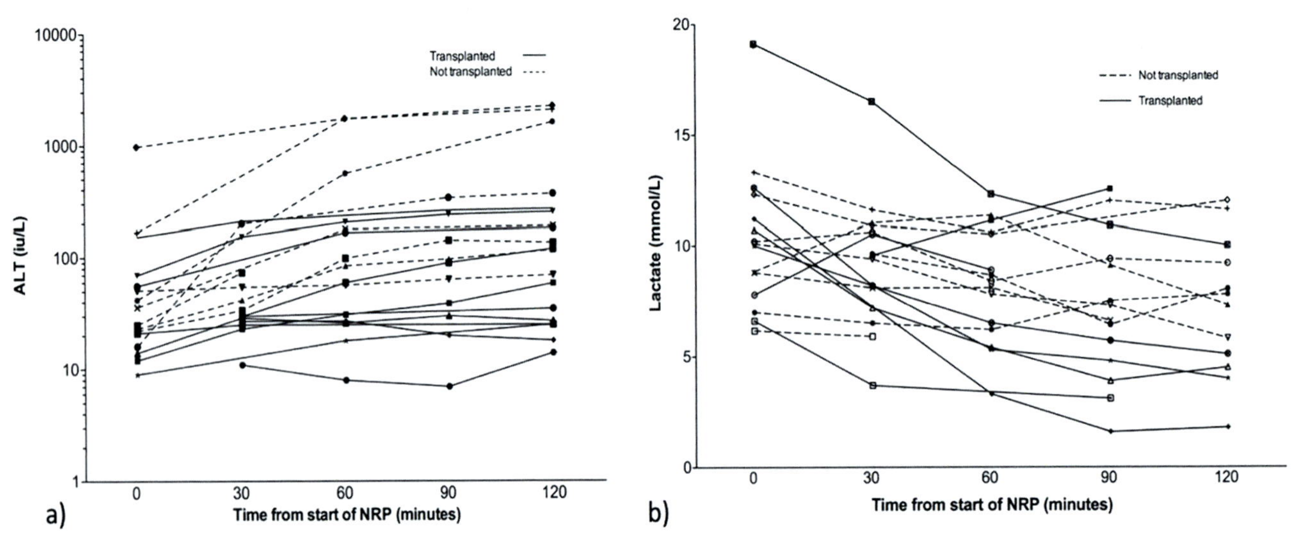

FIG. 4 (A) Changes in donor alanine transferase (ALT) (IU/L). (B) Changes in the lactate during normothermic regional perfusion (NRP) according to liver utilization (logarithmic scale). *(Source: Oniscu G, Randle L. V., Muisean P, Butler A.J., I.S Currie, M.T.P.R Perera, J.L. Forsythe and C.J.E Watson.: In situ normothermic regional perfusion for controlled donation after circulatory death. The United Kingdom experience Am J Transplant: 2014; 14 2846–2854.)*

TABLE 3 Individual center normothermic regional perfusion retrieval and organ transplant activity.

	Number of			
Transplant center	Donors	Livers	Kidneys	Pancreata
Birmingham	3	2	5[a]	–
Cambridge	9	4	16[b]	2
Edinburgh	9	5	17[c]	1[d]
All	21	11	38	3

[a] *One donor had a previous nephrectomy.*
[b] *Three double kidney transplants, two discarded, two combined pancreas and kidney transplants.*
[c] *One double kidney transplant and one discarded.*
[d] *One pancreas was used for research, one pancreas for islet isolation with insufficient yield, and one pancreas for islets.*
Modified from Gabriel C. Oniscu; Jennifer Mehew; Andrew J. Butler; Andrew Sutherland; Rohit Gaurav; Rachel Hogg; Ian Currie; Mark Jones; Christopher J E Watson, Improved organ utilisation and better transplant outcomes with in situ Normothermic Regional Perfusion in controlled donation after circulatory death, Am J Transplant 2017;17:2165-2172.

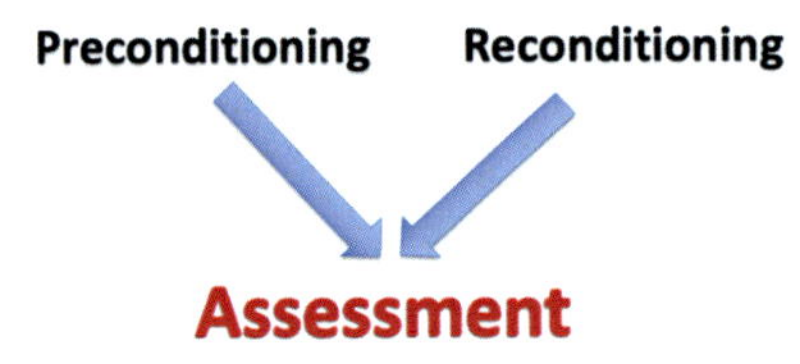

FIG. 5 NRP mechanism: Although the underlying mechanism of NRP is unclear it is likely to be a combination of preconditioning, reconditioning, and organ assessment.

HO-1, VEGF, iNOS, and EPO that have been found to protect organs from ischemia reperfusion injury [24–27]. Evidence of HIF up-regulation during NRP has been shown by Kerforne et al. in a porcine model of NRP [6]. HIF1a expression was shown to be present in NRP kidneys but not those that did not receive NRP. HIF target genes including VEGF, EPO, HO-1, and Glut-1 were found to be up-regulated in the kidney after 2 h, 4 h, and 6 h of NRP. In this model of allotransplantation, the post-transplant function was improved following 4 and 6 h NRP but not 2 h. Although this suggests HIF may play a role in NRP induced preconditioning, at this stage it is only an association. In our own work, he has investigated whether HIF

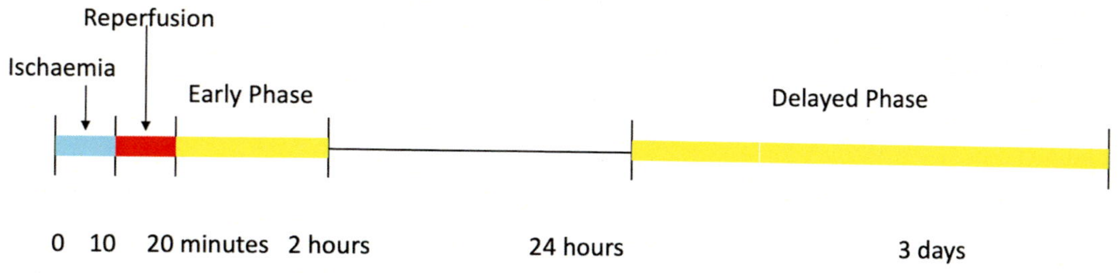

FIG. 6 Ischaemic preconditioning is the process whereby a short period of ischemia protects organs from a subsequent prolonged period of ischemia. There is an early phase of preconditioning which is apparent immediately after reperfusion and lasts for approximately 2 h, there then follows a longer more prolonged period of protection that is apparent 24–72 h later.

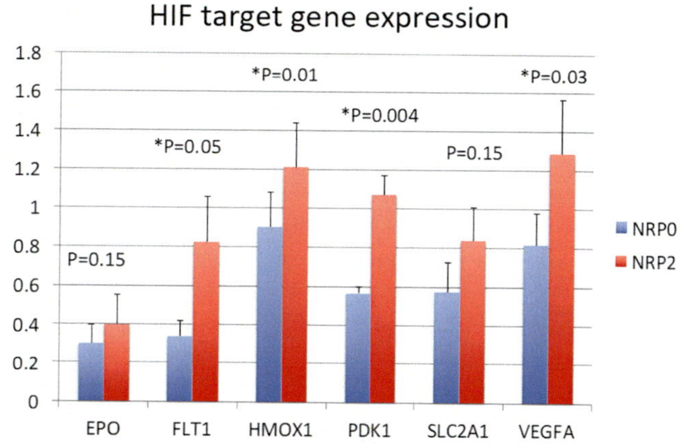

FIG. 7 HIF targert gene expression in liver at the beginning of NRP (NRP 0) and after 2 h of NRP (NRP2). Transplant International, 2017, Vol. 30(Suppl. 2), p. 537.

target genes are up-regulated in the clinical NRP setting. We examined HIF target gene expression in the liver at 0 and 2 h of NRP. Even in this short period, there was evidence of significant up-regulation (Fig. 7).

Physiological effects of NRP and evidence of reconditioning

Ischaemic reconditioning is the principal concept behind the benefits of NMP and NRP and other perfusion technologies, over traditional methods of SCS. During the hemodynamic instability occurring in the agonal phase of DCD, the donor experiences several physiological processes that are deleterious to the donor organs. The overall effect is tissue ischemia and accumulation of toxic metabolites that are the results of anaerobic metabolism. In cells deprived of oxygenated blood, cellular respiration slows down with irreversible damage occurring rapidly within minutes in sensitive tissues such as the pancreas [29]. Timely reestablishment of blood flow is essential to salvage the ischemic tissue, however, reperfusion itself can paradoxically cause further damage to the ischemic tissue, which characterizes "ischemia-reperfusion injury." It is the ATP degradation during warm ischemia that leads to the progressive accumulation of xanthine and hypoxanthine which are important sources of superoxide radical at organ reperfusion [29]. NRP has been shown in experimental models and clinical studies to restore ATP levels (Fig. 8) and reduce nucleotide degradation products, as well as increase the concentrations of endogenous antioxidants [30,31].

Ischemic reperfusion injury is one of the primary concerns of transplant surgeons, none more so than pancreatic transplant surgeons because it is the main cause of graft pancreatitis post-implantation and the most common cause of pancreas graft failure. Stopping or reducing the effects of ischaemic reperfusion injury has become a focus and aim of newer therapies. NRP offers the potential of reversing or reducing the initial effects of ischemia as well as the secondary effects of ischemia-reperfusion injury.

In summary normothermic preservation entails several concepts that reduce overall oxygen deprivation in the transplant organs and may reverse the initial effects of ischaemic reperfusion injury:

- The organ is continuously perfused with oxygenated blood/oxygen carrier before being implanted into the recipient and therefore not exposing the recipient to toxic anaerobic metabolites.
- Medications and nutrients can be provided at body temperature during storage (+/transport).

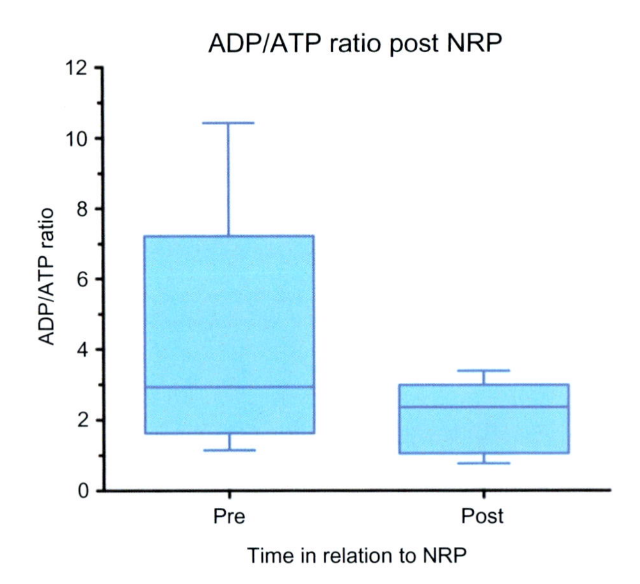

FIG. 8 ADP/ATP ration before and after 2h normothermic regional perfusion. *(Modified from Butler AJ, Randle LV, Watson CJ. Normothermic regional perfusion for donation after circulatory death without prior heparinization. Transplantation. 2014;97(13):1272–8. Doi: https://doi.org/10.1097/TP.0000000000000082. PMID: 24646774.)*

This has the following advantages:

- Enabling normal cellular metabolism with the recovery of cellular energy status and removal of toxic metabolites.
- Allowing repair of reversible injury with normal cellular processes.
- Facilitating functional testing of the organ before transplantation through the measurement of perfusion and biochemical parameters during preservation.
- Potentially reducing the hemodynamic on the recipients during reperfusion.

Both normothermic machines and in situ perfusion apply the same concepts of continual organ preservation by supplying "normal" cellular substrates to the potential transplant grafts. The outcome from both machine perfusion and in situ is promising with the advantage of NRP being able to perfuse multiple organs at one time compared with individual perfusion with the ex-situ machines.

Donor selection for DCD-NRP: Transplant organ utilization

Currently, in the UK, donor selection criteria for DCD-NRP are the same as for DCD-only donors. Functional warm ischemia time (WIT) systolic BP<50mmHg to start of perfusion of less than 30min for the liver and pancreas, and 1h for kidneys. Organs are not currently considered for implant if WIT passes these time limits, however, some centers will allow for some flexibility for kidneys and livers with these timings if the donor criteria are favorable and if NRP is involved, but currently, no guidelines for this are established [22,32].

The need to increase the donor pool and the increased use of extended criteria organ donors has driven innovation in perfusion technology particularly in the area of uncontrolled donors [33,34]. This has evolved due to the high discard rate of organs offered via the DCD pathway and the poorer outcomes of organs when compared to organs obtained from the DBD pathway. Currently, DCD donors account for roughly 40% of all organ donations in the UK. There is reluctance by centers in the UK in particular the pancreas from DCD donors due to the reduced organ recovery rates as compared to the DBD (deceased after brain death) donors, with a national average decline rate of 45% (23%–100%) across all centers (see Table 4) [33]. The increased use of DCD donors is demonstratable across most developed nations. In The United States, DCD donors have come to represent the fastest-growing constituent of the donor pool, increasing from approximately 1% of all deceased donors in 1996 to 11% of deceased donors in 2008. In some United Network for Organ Sharing (UNOS) regions with limited standard criteria donors, DCD comprises up to 16%–21% of the total donor pool [34].

TABLE 4 DCD donor pancreas offer decline rates by the transplant center, 1 April 2017 and 31 March 2020.

Centre	Code	2017/18 N	2017/18 (%)	2018/19 N	2018/19 (%)	2019/20 N	2019/20 (%)	Overall N	Overall (%)
Cambridge	A	7	(0)	10	(30)	9	(33)	26	(23)
Cardiff	B	5	(60)	3	(33)	5	(40)	13	(46)
Edinburgh	C	6	(50)	5	(100)	4	(25)	15	(60)
Guy's	D	7	(29)	12	(33)	11	(36)	30	(33)
Manchester	E	25	(16)	28	(43)	15	(53)	68	(35)
Newcastle	F	3	(100)	5	(100)	4	(100)	12	(100)
Oxford	G	16	(38)	19	(42)	18	(50)	53	(43)
WLRTC	H	7	(71)	2	(50)	2	(0)	11	(55)
UK		**76**	**(34)**	**84**	**(46)**	**68**	**(46)**	**228**	**(42)**

Centre has reached the upper 99.8% confidence limit
Centre has reached the upper 95% confidence limit
Centre has reached the lower 95% confidence limit
Centre has reached the lower 99.8% confidence limit

Source: NHSBT Annual report on Pancreas and islet cell transplantation: Published September 2020.

Evidence for use of NRP

Despite cDCD donors making up around 40% of all organ donors in the UK and concerted public health efforts to promote organ donation through DCD pathways, there has not been the expected increase in organ acceptance in this group due to concerns of organ assessment and implant performance. There is a particularly high attrition rate from the number of organs offered for transplantation to those transplanted, for liver and pancreas with only 29% and 20% of the respective offered DCD organs being transplanted. The US reports similar discard rates for DCD pancreata with 25.4% discarded after retrieval with the most common reasons reported as damage and anatomical abnormalities as compared to only 60% from DBD pancreas donors. NRP has demonstrable short- and long-term benefits in both kidney and liver grafts with early adopting countries now implementing DCD-NRP as the standard retrieval method for DCD donors (France and Spain), with increasing use in the UK and Italy.

Oniscu et al. reported the effects on organ utilization with the implementation of the NRP in the UK. Out of the 4712 donors considered in the analysis period, NRP was used for 159 donors (3%). Fig. 9 demonstrates individual organ utilization

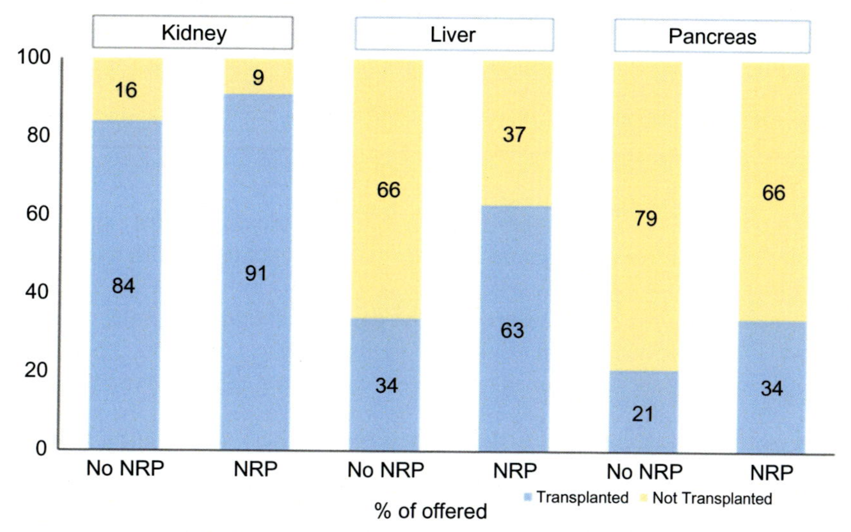

FIG. 9 Individual % of organ utilisation for DCD donors compared to DCD-NRP. *(Modified from Oniscu et al. (2021) in print.)*

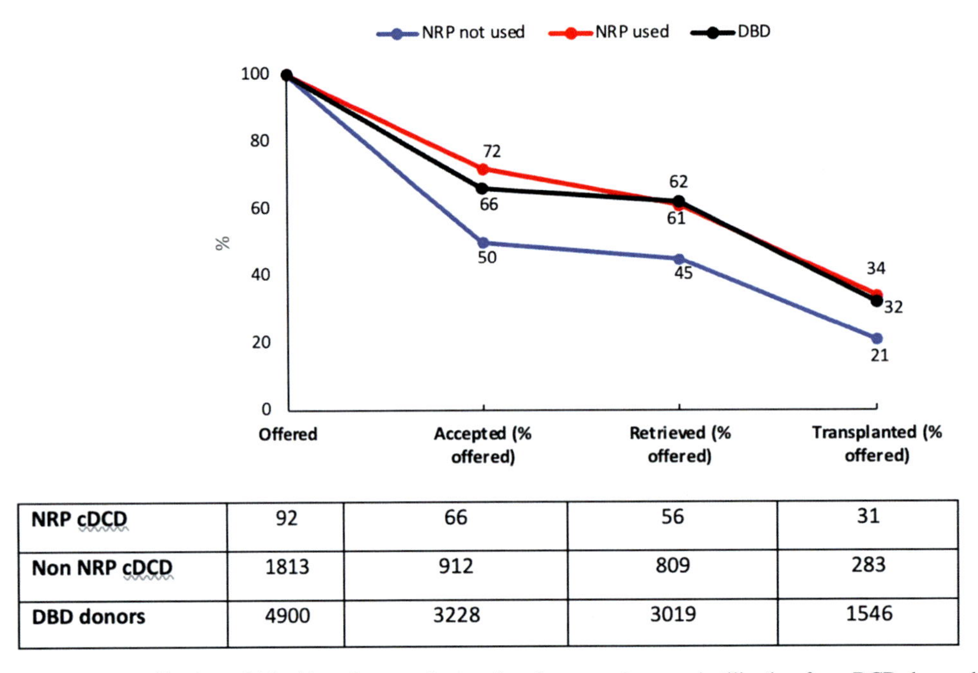

NRP cDCD	92	66	56	31
Non NRP cDCD	1813	912	809	283
DBD donors	4900	3228	3019	1546

FIG. 10 Demonstrates pancreas utilization which although not as impressive, there was increased utilization from DCD donors by an odds ratio of 1.7 when compared to non-NRP DCD donors. See Table 5. Although NRP numbers are small in this study it is some evidence that the benefits of NRP are slowly becoming accepted across transplanting units in the UK. This is in part due to the known beneficial effects already been demonstrated in the liver and renal grafts. These organs allow real-time in situ assessment of function and biochemistry which is critical in deciding on organ acceptance and implantation. *(Modified from Oniscu et al. (2021).)*

TABLE 5 The effect of NRP on the probability of pancreas transplantation resulting from a DCD pancreas offer adjusted for the Pancreas Donor Risk Index (PDRI) (p-value for Likelihood ratio test).

	N	Odds ratio (95% CI)	*P* value
PDRI Quartile			<0.0001
Q3	477	3.280 (1.943–5.535)	
Q4 (lowest quality)	476	1.000	
NRP use			0.0442
Yes	92	1.676 (1.022–2.746)	
No	1813	1.000	

Source: Oniscu et al. (2021).

according to the use of NRP. There was a significant increase in the utilization rates of all abdominal organs. The most noteworthy increase in utilization rates was observed for the liver, with an unadjusted utilization rate for NRP donors nearly double that for non-NRP donors (63% versus 34%) (Fig. 10).

These methods of assessment are currently not afforded to the pancreas and transplanting centers are still heavily reliant on donor details and retrieval back table anatomical assessment in making decisions around pancreas implantation. Despite this, the Edinburgh group looked at pancreas utilization rates over the past 5 years for DCD-NRP donors. During the analysis period, 366 DCD retrievals were performed, 41 DCD donors proceeded where the pancreas had been offered and accepted. There were 11 NRP-DCD retrievals and 30 standard DCD retrievals. In the 11 DCD-NRP retrievals performed, seven pancreases were utilized as whole organs as simultaneous pancreas-kidney transplants, two for islet transplant, and one for simultaneous kidney and islet transplant, one pancreas was reviewed by the recipient center and deemed not suitable after back table dissection. In the DCD alone group, 27 were not accepted for either pancreas or islet transplant. Three were deemed suitable for the implant (1 SPK and 2 islets). In this small series, DCD-NRP was shown to increase the utilization of pancreas grafts compared to DCD alone (91% vs. 10%, $P<0.0001$) [35].

Liver transplant outcomes from DCD-NRP

The reluctance to accept DCD organs is not limited to the pancreas. It is well recognized that livers grafts accepted from traditional DCD donors have higher rates of primary nonfunction (PNF) and more commonly increased rates of ischaemic biliary complications [36,37]. This is in particular reference to non-anastomotic stricture or "ischaemic cholangiopathy." Some single center studies have demonstrated a 13% incidence in ischaemic complications in DCD livers compared to the reported rate of 1% in DBD [26,36,37]. The largest review into ischaemic lesions from DCD donors published in 2016 by Hessheimer et al. [36] reported rates of 20%–30% of ischaemic complications in DCD liver after long-term follow-up.

NRP-DCD was first developed for Maastricht II donors in Spain and reported by Fondevila et al. in 1997. [38] There group published the first studies on the use of the recirculation of oxygenated blood that was used at 37°C to improve the viability of porcine livers damaged by warm ischemia. This original line of investigation revealed that normothermic extracorporeal membrane oxygenation (NECMO) could partially restore the cellular energy load lost during a period of cardiac arrest. When NECMO was compared with total body cooling after cardiac arrest, it was shown that NECMO was significantly superior in improving the histological ischaemic lesions and posttransplant graft function. They reported that the beneficial effects of NECMO were demonstrated to lie in its ability to turn the initial period of cardiac arrest into one of ischemic preconditioning. Based on this experimental work, the Barcelona Hospital Clinic implemented a clinical protocol in 2002 to transplant livers arising from uncontrolled DCD by systematically maintaining donors with NECMO prior to embarking on the organ retrieval process [38,39].

In 2007 Fondelvila et al. described the first series of DCD-NRP versus DBD in clinical practice. In this study, there is a 2:1 ratio of DBD to DCD donors in the comparison group, out of the 44 DCD donor livers in the study 11 DCD-NRP were ultimately transplanted. Follow-up data out to 24 months demonstrated comparable grafts function to DBD donors. This study although small numbers demonstrated the clinical application and viability in the clinical practice and provide the benchmark for further investigation of DCD donation in other developed countries.

In the UK NRP has been implemented since 2003 and on the back of the Spanish trials and the increasing number of favorable publications, there has been general acceptance among the transplant community that NRP is a viable method for increasing liver grafts to the donor pool and a method for providing superior outcomes compared to traditional DCD donors for liver grafts [24,37,38].

The use of antemortem intervention is prohibited in the UK under current law, which makes some of the Spanish experience difficult to apply to UK donors. Oniscu et al. described the UK experience of NRP across three pioneering transplanting centers (Edinburgh/Birmingham/Cambridge), with the use of Maastricht III donors (as opposed to Maastricht II from Spain). Results in this study reported overall reported similar findings of increased organ recovery and more importantly the viability of the NRP as a process in Maastricht III donors. In the UK "controlled' withdrawal occurs most commonly in the ITU and without the addition of antemortem interventions, there is a greater agonal phase which has led to concerns about clot formation within donor venous systems that may prevent NRP from being established whilst additionally adding to the ischaemic injury.

Watson et al. [38] then followed up on this study to directly compare the outcome of the contemporaneous comparator group of all the DCD livers transplanted in the study centers over the same seven-year study period to DCD-NRP livers from two contributing transplanting centers (Edinburgh/Cambridge). The two centers contributed equally to this study with the results demonstrating both increased graft and patient survival when death was censored (See graphs A, B, C in Fig. 11), it is important to note however that the mean age of the donors in the DCD-NRP group was significantly lower than the DBD group, a finding consistently seen in clinical practice in the UK. The most notable finding of the study was the markedly reduced rate of ischaemic cholangiopathy seen between the two groups. This is consistent with the data on the Spanish series and was so successful that out of the 43 patients that received a liver in the DCD-NRP group no ischaemic cholangiopathy was reported. (See Table 6). In Spain, the success in reducing ischaemic cholangiography and improved graft outcomes paired with greater clinical and technical experience has made DCD-NRP the accepted standard when donors are put forward through the DCD pathway. As this method is adopted by more centers it is believed that the UK and other developed countries will follow suit.

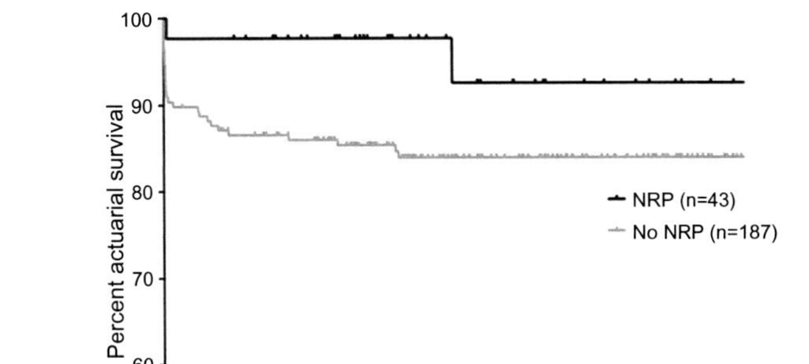

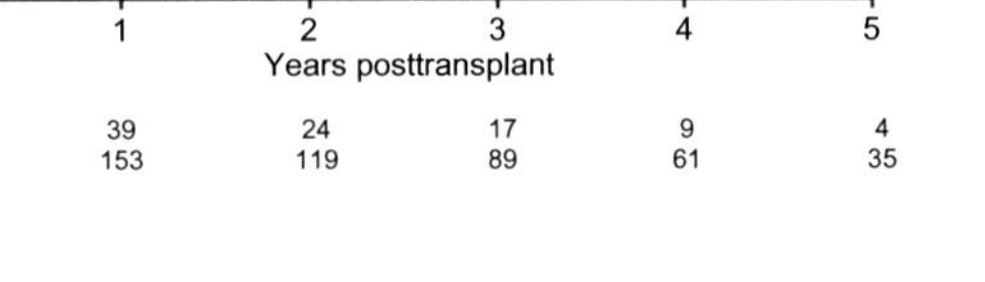

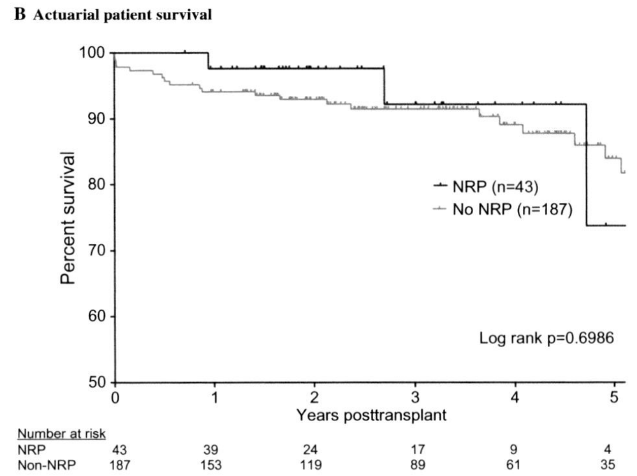

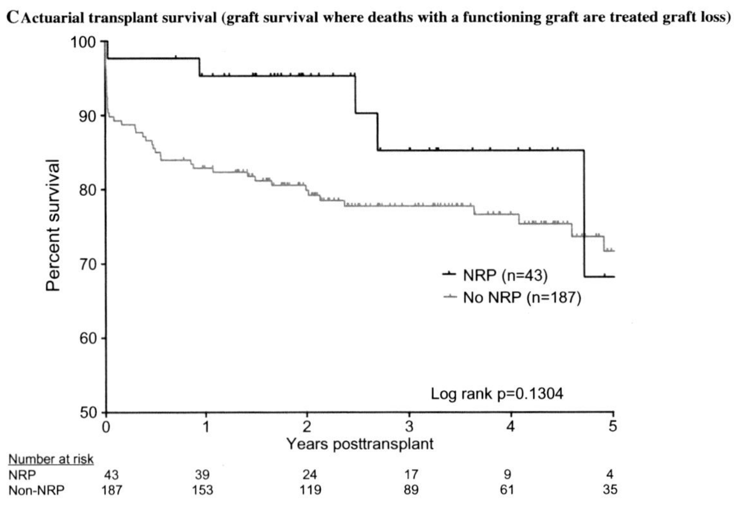

FIG. 11 Kaplain Meier plots showing (A) death-censored graft survival, (B) patient survival, (C) graft survival where deaths with functioning grafts were treated as graft loss.

TABLE 6 Modified summary of categorical donor and transplant-related variables for 231 DCD donor livers.

		Ischemic cholangiopathy		
Categorical variable	**Level**	**Yes (n = 47)**	**No (n = 183)**	**Fishers exact P-value**
Locality of donor	Local	6(12.8%)	56(30.6%)	0.03
	Regional	17(36.2%)	61(33.3%)	
	National	24(51.1%)	66(36.1%)	
Blood group	Identical	46(97.9%)	181(98.9%)	0.5
compatibility	Compatible	1(2.1%)	2(1.1%)	
Perfusion used	Comparator	47(100%)	140(76.5%)	<0.0001
	NRP	0(0%)	43(23.5%)	

Modified from Watson C.J.E., Hunt F., Messer S., Currie I., Large S., Sutherland A., et al. In situ normothermic perfusion of livers in controlled circulatory death donation may prevent ischemic cholangiopathy and improve graft survival. The. Am J Transplant 2019;19:1745–1758.

Kidney transplant outcomes from DCD-NRP

Although the renal parenchyma appears to be much more resilient to the ischaemic injury that occurs during DCD organ donation, there is still a higher reported rate of DGF and PNF from DCD donors compared to DBD donors [39]. Kidney preservation from both hypothermic and normothermic modalities has been in practice for quite some time with varying degrees of success. The earliest reports came from cold machine perfusion in the USA for transport of large distances between donor and recipient centers during the development of cold preservation fluids Beltzer et al. 1968. This study demonstrated the viability of transport and preservation systems even with the modest engineering systems of the 1960s. Progress in machine preservation had been largely confined to single centers with some reluctance to take up the expensive technology across transplant communities [4,40]. However, the need to increase the donor pool renewed interest in this field, progressing from machine perfusion to in situ machine perfusion (both Hypothermic/normothermic) with an increasing number of studies on kidney function in both animals and human models.

Valero et al. [41] reported on DCD-NRP kidneys in the first series directly demonstrating this preservation method against in situ cold perfusion and traditional DCD organ procurement. Each method was compared for graft function out to 5 years, demonstrated quicker graft recovery, and improved 5-year graft survival in the DCD-NRP (Normothermic recirculation) group. (See Table 7).

Minabres et al. [42] results on DCD versus DCD-NRP graft outcomes also provided further evidence suggesting NRP has advantageous effects on kidney donor function. The group demonstrated over a two-year period from the inception of their program very similar outcomes between DCD-NRP kidneys and DBD kidneys. Although statistical significance was not achieved for any of the analyzed variables, reduced in-hospital stay and immediate graft function was reported in nearly all the kidneys from the NRP group (see Table 8).

Pancreas outcomes and future directions for NRP and the pancreas

Due to the benefits seen with NRP in liver allograft transplantation and encouraging evidence in renal transplantation, there is a belief within the transplant community that we will potentially see similar benefits in the pancreas as the technology becomes more widely available and accepted. The data, however, for pancreas and islet transplantation remains very limited. Nevertheless, the data outlined above from the UK experience Oniscu et al. [26] suggests that the pancreas utilization from DCD donors has already been increased by using NRP. In terms of pancreas transplant outcomes only a few case studies and small series have been published. In Spain, five cDCD-NRP pancreases were transplanted between 2015 and 2017 with 100% early graft survival [40]. The Cambridge group recently published their 10-year experience of cDCD pancreas transplants, including 13 DCD-NRP pancreas transplants. Similar outcomes across graft survival for both kidney and pancreas transplants were demonstrated and although the overall findings of the paper were to demonstrate that DCD pancreas is a viable option to increase the donor pool, an important comment was several pancreases accepted using NRP were from marginal donors, that would have otherwise rejected based on donors' details. Interestingly three of the 13

TABLE 7 Donor's cause of death, cold and warm ischemia time, recipients characteristics and HLA-A, B, DR matching according to perfusion technique used for organ procurement. Values expressed as mean and standard deviation and as number of cases (M male, F female, PLT polytrauma).

	In situ	Total body cooling	Normothermic recirculation
Donor's cause of death: PLT/Non PLT	9/13[a]	1/3	0/4
Cold ischemia time (h)	18.5 ± 6.2	15.3 ± 4.6	17.8 ± 6.7
Warm ischemia time (min)	58 ± 41[a]	81 ± 15	82 ± 11
Recipients age (years)	36.7 ± 16.4	36 ± 5.9	46.7 ± 8.1
Recipients sex (M/F)	26/14	3/5	7/1
HLA-A, B, DR matching	1.9 ± 1.3	1.6 ± 1.1	1.3 ± 1.6

[a] *P=0.02 compared to other groups.*
Modified from R. Valero, Catiana Cabrer, Frederic Oppenheime, Esteve Trias, Jacinto Sanchez-Ibaiiez, Francisco M. De Cabo, Aurora Navarro, David Paredes, Antonio Alcaraz Rafael GutiCrrez, Marti Manyalich. Normothermic recirculation reduces primary graft dysfunction of kidneys obtained from non-heart-beating donors. Transpl Int; 2000 (13); 303–310.

TABLE 8 Kidney recipient details and outcomes.

	DCD ($n = 37$)	DBD ($n = 36$)	P
Age (years)	57 (47–63)	55.5 (49–61)	0.838
Gender (male)	26 (70.6%)	23 (60.8%)	0.562
HLA incompatibility > 3	19 (73.1%)	31 (88.6%)	0.179
FWIT in donors (min)	12 (10–19)	–	
Cold ischemic time (min)	960 (475–1290)	1140 (900–1290)	0.350
History of previous transplantations	5 (13.5%)	8 (22.8%)	0.331
DGF	10 (27%)	12 (33.3%)	0.557
No. of dialysis sessions	2 (2–3)	2.5 (2–3)	0.674
Patients with < 3 dialysis sessions[a]	6 (60%)	6 (50%)	0.691

[a] *Those who had DGF.*
Results are presented as absolute number (percentage) or median (IQR). DCD, donation after circulatory death; DBD, donation after brain death; FWIT, functional warm ischemic time; DGF. delayed graft function.
Modified from Minambres E., Suberviola B., Dominguez-Gil B., Rodrigo E., Ruiz-San Millan J.C., Juan J.C.R.-S., et al. Improving the outcomes of organs obtained from controlled donation after circulatory death donors using abdominal normothermic regional perfusion. Am J Transplant 2017;17:2165–2172.

DCD-NRP pancreas grafts used had functional warm ischemia greater than 60 min and would not have been utilized based on current UK DCD pancreas acceptance criteria [17]. It is inevitable as the NRP experience grows so will the interest in research into NRP variables. Many of the current papers on NRP have focused on graft outcomes understandably. But as with machine perfusion, there is a lot of information that needs to be further analyzed to achieve the greatest results from NRP and therefore greater number of organs to the donor pool. As the NRP numbers increase so will the data on DCD-NRP pancreas and will hopefully deal with the current lack of large-scale data in this field.

The benefits of NRP over other forms of perfusion preservation are that the organs can be functionally assessed in situ. This current benefit has only been reported in liver allografts as continual BSL/ASL/Bilirubin and lactate are used for surrogate markers of liver function. Kidney function can be directly assessed by measuring urine output, but this has yet to be used in donor organ acceptance criteria but proves useful in anecdotal reports from NRP centers. This is the area that would most benefit the implanting pancreas surgeon in making a decision about acceptance or rejection of DCD pancreases. Any advances in aiding pancreatic functional assessment that provides subjective values would be invaluable. Currently, there is a reliance on subjective information, clinical retrieval photos, donor data history, and surrogates of pancreas function such as amylase around the time of death. These can easily represent other physiological processes that are not related to the pancreas and may deter the implanting surgeon away from an otherwise suitable pancreas [17,41].

There is a larger body of research into current NMP models which focuses on variables such as pump speed, continuous versus pulsatile flow, and the differences in hypothermic/normothermic climates on the porcine and discarded human pancreas. All of which have shown promise in the early data that an ex-situ perfusion is a viable option for the pancreas, without any method being favored or beneficial in outcomes. Many of these variables have not yet been studied in NRP, but it is felt that the "normal" homeostatic environment provided by NRP is the best current mode of organ preservation via perfusion. This is inferred from the liver trials from Spain, which have varied the time on NRP and demonstrated that 1 h is enough to achieve the benefits of organ reconditioning, as long as continuous perfusion pressure and oxygen levels are maintained [40–42]. We currently perform a two-hour minimum target for NRP in the UK but what time is the most beneficial/optimal time for the pancreas performance is not yet known.

The use of in situ assessment lends itself to the current trends of using biomarkers to assess function and viability. Although candidate markers such as microRNA and methylated DNA have been tested in the experimental setting, no one has yet to be able to demonstrate their utility in clinical trials. NRP assessment allows a short window to use such biomarkers to measure pancreatic/B cell function and viability when assessing an organ's suitability for pancreas transplantation. As these biomarkers become available, they have the potential to greatly increase the utility of NRP.

NRP requires additional resources as well as poses several logistical challenges. As these challenges are met and NRP is introduced to more centers, the body of evidence for NRP will grow. Whereas there seem to be clear benefits for liver transplantation, the use of NRP for pancreas and islet transplant is still in its naissance. The encouraging early results as well as the need for a more in-depth assessment of DCD pancreases indicate a likely greater role for NRP in DCD pancreas procurement in the future. As we transplant more pancreases after NRP and our understanding increases, there is the potential to utilize significantly more DCD pancreases as well as improve outcomes.

References

[1] Gruessner AC, Gruessner RW. Pancreas transplantation of US and Non-US cases from 2005 to 2014 as reported to the United Network for Organ Sharing (UNOS) and the International Pancreas transplant Registry (IPTR). Rev Diabet Stud 2016;13:35–58. https://doi.org/10.1900/RDS.2016.13.35.

[2] Gruessner A, Sutherland DE. Pancreas transplantation in the United States (US) and non-US as reported to the United Network for Organ Sharing (UNOS) and the International Pancreas Transplant Registry (IPTR). Clin Transpl 1996;47–67.

[3] Maathuis MH, Manekeller S, van der Plaats A, et al. Improved kidney graft function after preservation using a novel hypother- mic machine perfusion device. Ann Surg 2007;246:982–8.

[4] Leemkuil M, Lier G, Engelse MA, et al. Hypothermic oxygenated machine perfusion of the human donor pancreas. Transplant Direct 2018;4, e388.

[5] Messer S, Page A, Axell R, Berman M, Hernández-Sánchez J, Colah S, Parizkova B, Valchanov K, Dunning J, Pavlushkov E, Balasubramanian SK, Parameshwar J, Omar YA, Goddard M, Pettit S, Lewis C, Kydd A, Jenkins D, Watson CJ, Sudarshan C, Catarino P, Findlay M, Ali A, Tsui S, Large SR. Outcome after heart transplantation from donation after circulatory-determined death donors. J Heart Lung Transplant 2017;36(12):1311–8. https://doi.org/10.1016/j.healun.2017.10.021. Epub 2017 Oct 24. Erratum in: J Heart Lung Transplant. 2018 Apr;37(4):535 29173394.

[6] Otero A, Gomez-Gutierrez M, Suarez F, et al. Liver transplantation from Maastricht category 2 non-heartbeating donors. Transplantation 2003;76:1068–73.

[7] Scalea JR, Redfield RR, Arpali E, Leverson GE, Bennett RJ, Anderson ME, Kaufman DB, Fernandez LA, D'Alessandro AM, Foley DP, Mezrich JD. Does DCD donor time-to-death affect recipient outcomes? Implications of time-to-death at a high-volume center in the United States. Am J Transplant 2017;17:191–200.

[8] Heylen L, Jochmans I, Pirenne J, Spranger B, Samuel U, Tieken I, Naesens M. The duration of asystolic ischemia determines the risk of graft failure after circulatory-dead donor kidney transplantation: a Eurotransplant cohort study. Am J Transplant 2018;18:881–9.

[9] Fondevila C, Hessheimer AJ, Ruiz A, Calatayud D, Ferrer J, Charco R, Fuster J, Navasa M, Rimola A, Taura P, Gine P, Manyalich M, Garcıa-Valdecasas JC. Liver transplant using donors after unexpected cardiac death: novel preservation protocol and acceptance criteria. Am J Transplant 2007;7:1849–55.

[10] Ravaioli M, De Pace V, Comai G, Capelli I, Baraldi O, D'Errico A, Bertuzzo VR, Del Gaudio M, Zanfi C, D'Arcangelo GL, Cuna V, Siniscalchi A, Sangiorgi G, La Manna G. Preliminary experience of sequential use of normothermic and hypothermic oxygenated perfusion for donation after circulatory death kidney with warm ischemia time over the conventional criteria—a retrospective and observational study. Transpl Int 2018;31:1233–44.

[11] Kandaswamy R, Stock PG, Gustafson SK, Skeans MA, Urban R, Fox A, et al. OPTN/SRTR 2017 annual data report: pancreas. Am J Transplant 2019;19:124–83.

[12] (a) Lomero, et al. Donation after circulatory death today: an updated overview of the European landscape. Transpl Int 2020;33:76–88. (b) NHSBT. Annual report on pancreas and islet transplantation; 2019. [22nd May 2020]; Available from: https://nhsbtdbe.blob.core.windows.net/umbraco-assets-corp/16776/nhsbt-pancreas-and-islet-transplantation-annual-report-2018-2019.pdf.

[13] Muthusamy AS, Mumford L, Hudson A, Fuggle SV, Friend PJ. Pancreas transplantation from donors after circulatory death from the United Kingdom. Am J Transplant 2012;12(8):2150–6. https://doi.org/10.1111/j.1600-6143.2012.04075.x. 22845910.

[14] Salvalaggio PR, Davies DB, Fernandez LA, Kaufman DB. Outcomes of pancreas transplantation in the United States using cardiac-death donors. Am J Transplant 2006;65(Pt 1):1059–65.

[15] Bellingham JM, Santhanakrishnan C, Neidlinger N, et al. Donation after cardiac death: a 29-year experience. Surgery 2011;150:692–702.

[16] Shahrestani S, Webster AC, Lam VW, Yuen L, Ryan B, Pleass HC, Hawthorne WJ. Outcomes from pancreatic transplantation in donation after cardiac death: a systematic review and meta-analysis. Transplantation 2017;101(1):122–30. https://doi.org/10.1097/TP.0000000000001084. 26950713.

[17] Callaghan CJ, Ibrahim M, Counter C, Casey J, Friend PJ, Watson CJE, Karydis N. Outcomes after simultaneous pancreas-kidney transplantation from donation after circulatory death donors: a UK registry analysis. Am J Transplant 2021;1–11.

[18] Chen J, Mikhail DM, Sharma H, et al. Donor age is the most import- ant predictor of long term graft function in donation after cardiac death simultaneous pancreas-kidney transplantation: a retrospective study. Am J Surg 2019;218(5):978–87.

[19] Saito T, Gotoh M, Satomi S, Uemoto S, Kenmochi T, Itoh T, et al. Islet transplantation using donors after cardiac death: report of the Japan islet transplantation registry. Transplantation 2010;90(7):740–7. https://doi.org/10.1097/TP.0b013e3181ecb044. 20811319.

[20] Anazawa T, Saito T, Goto M, Kenmochi T, Uemoto S, Itoh T, et al. Long-term outcomes of clinical trans-plantation of pancreatic islets with uncontrolled donors after cardiac death: a multicenter experience in Japan. Transplant Proc 2014;46(6):1980–4. https://doi.org/10.1016/j.transproceed.2014.06.006. 25131088.

[21] Ito T, Kenmochi T, Kurihara K, Kawai A, Aida N, Akashi Y, et al. The effects of using pancreases obtained from brain-dead donors for clinical islet transplantation in Japan. J Clin Med 2019;8(9):1430. https://doi.org/10.3390/jcm8091430. 31510059.

[22] De Paep DL, Van Hulle F, Ling Z, Vanhoeij M, Pirenne J, Keymeulen B, Pipeleers D, Jacobs-Tulleneers-Thevissen D. Lower beta cell yield from donor pancreases after controlled circulatory death prevented by shortening acirculatory warm ischemia time and by using IGL-1 cold preservation solution. PLoS One 2021;16(5). https://doi.org/10.1371/journal.pone.0251055, e0251055. 33939760. PMC8092795.

[23] Doppenberg JB, Nijhoff MF, Engelse MA, de Koning EJP. Clinical use of donation after circulatory death pancreas for islet transplantation. Am J Transplant 2021. https://doi.org/10.1111/ajt.16533. Epub ahead of print 33565712.

[24] Dorweiler B, Pruefer D, Andrasi TB, Maksan SM, Schmiedt W, Neufang A, Vahl CF. Ischemia-reperfusion injury pathophysiology and clinical implications. Eur J Trauma Emerg Surg 2007;33:600–12.

[25] Sutherland A, Forbes S, Timpson A, Irvine L, Duncan K, Anderson D, Barclay J, Casey J. Islet cell isolation and transplantation from DCD donors in Scotland. EPITA; 2021.

[26] Oniscu G, Randle LV, Muisean P, Butler AJ, Currie IS, Perera MTPR, Forsythe JL, Watson CJE. In situ normothermic regional perfusion for controlled donation after circulatory death. The United Kingdom experience. Am J Transplant 2014;14:2846–54.

[27] Rojas-Pena A, Sall LE, Gravel MT, Cooley EG, Pelletier SJ, Bartlett RH. Donation after circulatory determination of death: the University of Michigan experience with extra-corporeal support. Transplantation 2014;98:328–334.22.

[28] Kerforne T, Allain G, Giraud S, Bon D, Ameteau V, Couturier P, Hebrard W, Danion J, Goujon J-M, Thuillier R, Hauet T, Barrou B, Jayle C. Defining the optimal duration for normothermic regional perfusion in the kidney donor: a porcine preclinical study. Am J Transplant 2019;19:737–51.

[29] Nasralla D, Coussios CC, Mergental H, et al. A randomized trial of normothermic preservation in liver transplantation. Nature 2018;557(7703):50–6.

[30] Maheshwari A, Maley W, Li Z, Thuluvath PJ. Biliary complications and outcomes of liver transplantation from donors after cardiac death. Liver Transpl 2007;13:1645–53.

[31] Chan EY, Olson LC, Kisthard JA, et al. Ischemic cholangiopathy following liver transplantation from donation after cardiac death donors. Liver Transpl 2008;14:604–10.

[32] Andres A, Kin T, O'Gorman D, Livingstone S, Bigam D, Kneteman N, Senior P, Shapiro AMJ. Clinical islet isolation and transplantation outcomes with deceased cardiac death donors are similar to neurological determination of death donors. Transpl Int 2016;29:34–40.

[33] NHS Blood and Transplant. Organ donation and transplantation activity report 2018/19. Available at: http://www.organdonation.nhs.uk/statistics/transplant_activity_report/current_activity_reports/ukt/activity_report_2019/20, 2020. Accessed: November 20, 2021.

[34] Oniscu GC, Mehew J, Butler AJ, Sutherland A, Gaurav R, Hogg R, Currie I, Jones M, Watson CJE. Improved organ utilisation and better transplant outcomes with *in situ* Normothermic Regional Perfusion in controlled donation after circulatory death. Am J Transplant 2017;17:2165–72.

[35] Jochmans I, Akhtar MZ, Nasralla D, Kocabayoglu P, Boffa C, Kaisar M, Brat A, O'Callaghan J, Pengel LHM, Knight S, Ploeg RJ. Past, present, and future of dynamic kidney and liver preservation and resuscitation. Am J Transplant 2016;16:2545–55.

[36] Hessheimer AJ, Cardenas A, Garcia-Valdecasas JC, Fondevila C. Can we prevent ischemic-type biliary lesions in donation after circulatory determination of death liver transplantation? Liver Transpl 2016;22(7):1025–33.

[37] Fondevila C. Is extracorporeal support becoming the new standard for the preservation of DCD grafts? Am J Transplant 2010;10:1341–2. United Network for Organ Sharing. Available at: http://www.UNOS.org.2010. [Accessed 15 February 2010].

[38] Watson CJE, Hunt F, Messer S, Currie I, Large S, Sutherland A, Crick K, Wigmore SJ, Fear C, Cornateanu S, Randle LV, Terrace JD, Upponi S, Taylor R, Allen E, Butler AJ, Oniscu GC. In situ normothermic perfusion of livers in controlled circulatory death donation may prevent ischemic cholangiopathy and improve graft survival. The. Am J Transplant 2019;19:1745–58.

[39] Branchereau J, Renaudin K, Kervella D, et al. Hypothermic pulsatile perfusion of human pancreas: preliminary technical feasibility study based on histology. Cryobiology 2018;85:56.

[40] Belzer FO, Ashby BS, Gulyassy PF, Powell M. Successful seventeen-hour preservation and transplantation of human-cadaver kidney. N Engl J Med 1968;278:608–10.

[41] Valero R, Cabrer C, Oppenheime F, Trias E, Sanchez-Ibaiiez J, De Cabo FM, Navarro A, Paredes D, GutiCrrez AAR, Manyalich M. Normothermic recirculation reduces primary graft dysfunction of kidneys obtained from non-heart-beating donors. Transpl Int 2000;13:303–10.

[42] Minambres E, Suberviola B, Dominguez-Gil B, Rodrigo E, Ruiz-San Millan JC, Juan JCR-S, Ballesteros MA. Improving the outcomes of organs obtained from controlled donation after circulatory death donors using abdominal normothermic regional perfusion. Am J Transplant 2017;17:2165–72.

Chapter 5

Allogeneic islet isolation: Methods to improve islet cell transplantation with new technologies in organ transplant retrieval and isolation techniques

Appakalai N. Balamurugan[a,b,c], Krishna Kumar Samaga[a], Siddharth Narayanan[a], Ahad Ahmed Kodipad[a], Sri Prakash L. Mokshagundam[c,d], and Jaimie D. Nathan[b,c]

[a]*Center for Clinical and Translational Research, Abigail Wexner Research Institute, Nationwide Children's Hospital, Columbus, OH, United States,* [b]*Department of Surgery, Nationwide Children's Hospital, Columbus, OH, United States,* [c]*Department of Pediatrics, College of Medicine, The Ohio State University, Columbus, OH, United States,* [d]*Department of Endocrinology, University of Louisville, Louisville, KY, United States*

Introduction

For the subgroup of patients with type 1 diabetes mellitus and severe hypoglycemic unawareness, isolation of islets from a deceased donor pancreas with intrahepatic transplantation of allogeneic islets can improve their severely problematic hypoglycemia, stabilize glycemic lability, and maintain on-target glycemic control, leading to improved quality of life, often with insulin independence [1]. Allogeneic islet transplantation (IT) also benefits from the significant number of donor pancreases that are unsuitable for whole organ transplantation but may be potentially utilized for islet cell isolation and transplantation. IT is also considered for subgroups of patients with debilitating pancreatitis where they require a native pancreatectomy and autologous IT (auto-IT, reviewed in detail elsewhere [2–4]). The overlaps between allo- and auto-IT are well identified, such that most facilities, equipment, and techniques can be utilized for both processes. Several clinical IT programs across the globe have the means to provide allo- and auto-IT services [2–5]. Detailed and widespread research on various aspects of IT has resulted in significant improvements of techniques and long-term clinical outcomes for patients [1,4–6]. This chapter focuses on the latest developments within allo-IT regarding donor selection, pancreas retrieval, organ preservation, and islet isolation and purification techniques.

Allogeneic islet transplantation

Allo-IT has become a viable therapy demonstrating clinical outcomes on par with its predecessor, whole pancreas transplantation, particularly with respect to insulin independence and graft and patient survival [1,5]. Common indications for allo-IT include type 1 diabetes mellitus for >5 years, negative stimulated C-peptide levels, hypoglycemic unawareness, severe hypoglycemic episodes despite optimal therapy, and glycemic lability [7]. Multiple studies have reported positive outcomes after IT in patients with labile diabetes [5,7,8]. This therapy is highly successful at resolving hypoglycemia, but more than one pancreas donor may be required for insulin independence [2]. Therefore it is essential to select appropriate candidates for whom the procedure will be of greatest benefit, given the shortage of potentially transplantable organs.

IT entails a series of essential interdependent steps, including donor pancreas procurement, islet isolation, purification, culture, and infusion. A thorough and systematic approach toward this process is required to maximize islet yield and quality. Organ donation, a multistep process, is usually coordinated by several organ procurement organizations. A detailed assessment of donor characteristics, including age, body mass index (BMI), and absence of diabetes in the donor (hemoglobin A1c <6.5%), is evaluated, as many of these variables are known to affect islet yield. Furthermore, several improvements in the formulation and use of pancreatic tissue-dissociation enzyme blends and purification protocols have considerably improved the success of islet isolation and helped increase the donor pool.

Pancreas and Beta Cell Replacement. https://doi.org/10.1016/B978-0-12-824011-3.00008-4

Clean room facility and equipment

Human islet isolation for clinical IT should be performed in a current good manufacturing process (cGMP) facility. The islet isolation procedure involves numerous steps that follow regulatory standards to prevent contamination. Factors such as islet yield, viability, and overall functionality are stringently assessed to satisfy the release criteria for transplantation. An isolation facility provides a strictly controlled environment where the islets can be safely and effectively purified from the donor pancreas (Fig. 1). An ideal clean room facility must permit effective workflows between various stations that are established for the process. The US Food and Drug Administration also requires these isolation units to adhere to regulatory guidelines to be able to provide a clean, safe, and effective working environment for the manufacture of islet cells or any cellular product. Several additional measures such as numerous standardized barriers, controlled air flow/direction, graded pressure, temperature, humidity, effective validated cleaning systems, and monitoring programs are implemented in the design of an appropriate clean room to minimize the risk of contamination [9].

Donor selection

Donor selection and pancreas procurement for islet isolation involve precise care, detailed scrutiny, and competent handling—analogous to whole organ transplantation. One of the principal components that determine the success of IT is the optimal physiology of the donor pancreas. Donor characteristics play a major role in final islet yield; therefore, choosing an appropriate organ is a vital aspect of allo-IT.

Several donor characteristics are likely to predict the quantity and quality of the islet preparations, with some of these factors correlating with clinical outcomes [10–12]. Single-center and multicenter retrospective studies have outlined several donor factors that could influence islet isolation outcomes, including donor age, gender, height, weight, cause of death, BMI, length of hospitalization, use of vasopressors, and blood glucose levels [13–20]. Apart from these, other influential variables are the experience of the procuring surgical team, the techniques used to retrieve and perfuse the pancreas, and the cold ischemic time (CIT) [9]. Additionally, the choice of tissue-dissociation enzyme blends, culture media, and culture conditions also has a strong impact on the overall numbers and quality of the resultant islets [9].

Donor weight, height, and BMI appear to be the most relevant variables correlating with islet yield and quality [12,19]. Obese (high BMI), young (age < 45 years) donor pancreases may be most optimal for allo-IT, as they tend to provide high islet yields and high-quality islets, respectively [1,21–23]. The former is also particularly useful for IT, as high-quality organs may be utilized that would be otherwise discarded when considered for whole organ transplantation.

A positive correlation has been observed with donor age and islet yield, with an optimal age of 30–50 years [4,17,19,22]. Islet yield is proportional to the size of the pancreas [19,24]. A juvenile donor pancreas often results in inadequate islet yield, as evident by the smaller size [25,26]. Despite the higher islet yield obtained from pancreases from older donors, the insulin secretion abilities are significantly lower than in organs from younger donors [5]. However, younger donor islet isolations result in large quantities of mantled islets, which are difficult to purify, resulting in a lower recovery rate [22,25]. Kin et al. analyzed 359 donor pancreases [24] and showed that (a) male donors have larger pancreases than female donors [27,28];

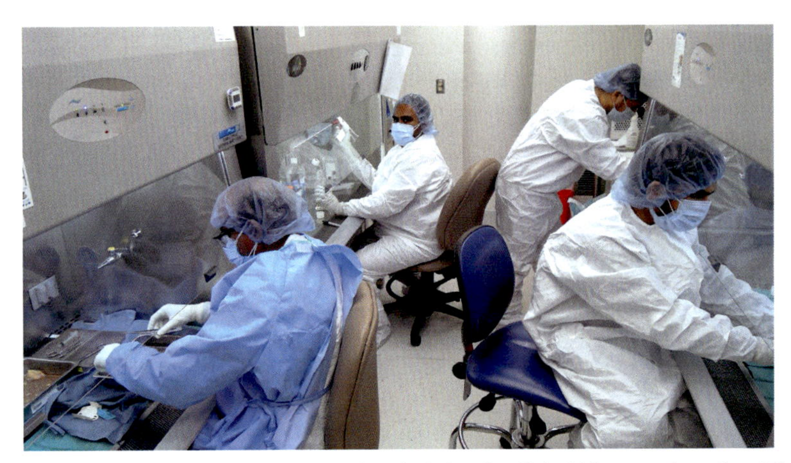

FIG. 1 A characteristic example of a clean room facility for pancreas islet isolation for clinical islet transplantation. All the steps for the islet isolation follow regulatory standards including the use of facemasks, full containment overalls, hoods, cover-shoes, and gloves to reduce the introduction of contaminants into the facility. The entire procedure is carried out within closed systems and performed inside of Class II biological safely cabinets as shown.

(b) male donors have a higher probability of yielding adequate islets [29,30]; (c) pancreas weight increases with age, reaching a plateau in the fourth decade; and (d) BMI correlates with pancreas weight, but body surface area is a better predictor of pancreas weight than BMI. BMI as a measure of obesity fails to distinguish fat mass from lean mass. To account for the disparity, Dybala and colleagues suggest using BMI and body surface area with percentage body fat to precisely assess potential pancreas donors that may result in better potential islet yields [31].

Donor BMI is an important factor for determining isolation success, with a significant positive association with islet yield [32]. Donors with BMI > 25 kg/m^2 have a higher probability of yielding adequate islets [30,33]. Pancreases from such donors are characterized by a larger amount of fatty pancreas texture, islet hypertrophy, and greater insulin production to regulate higher metabolic demands, which contribute to a higher islet mass (Fig. 2A). Consistent with previous studies, Sakuma et al. validated that high BMI donors and large pancreas size are important for successful human islet isolation, yielding higher islet isolation success rates [34].

Individuals with type 2 diabetes are not considered for pancreas donation. However, it is not always evident if a donor has type 2 diabetes since the disease is insidious and can go undiagnosed for many years. Whereas an older donor with higher BMI is preferred for IT [35,36], such donors are at risk of type 2 diabetes. Because of its high specificity in detecting chronic hyperglycemia, hemoglobin A1c is used to screen donors when type 2 diabetes is suspected. The islet isolation facility at the University of Alberta rejects donors with a hemoglobin A1c >6.5% [37].

As evidenced by several studies, higher glycemic values during donor management are detrimental to islet recovery [17,19,20]. However, considering the pathophysiology of brain death and pharmacology of drugs administered during management of brain death, the donor's blood glucose levels are not a true indicator of the donor's glucose metabolism. Lakey et al. observed that elevated blood glucose during hospitalization and higher levels of vasopressors (> 15 μg/kg/min dopamine or > 5 μg/kg/min norepinephrine) correlated with poorer isolation outcomes [17]. Kaddis et al. observed that islet yield was negatively affected by hormonal medications [33]; they also noted that normal amylase levels, absence of organ edema, and no administration of fluid/electrolyte medications improved outcomes. Since islets are particularly vulnerable to ischemia, CIT should be minimized to < 12 h [4].

Several screening parameters have been suggested to account for the high degree of variability in the donor spectrum. Based on stringent donor pancreas acceptance criteria, the available pancreas is assessed and scored such that a donor with a higher score would likely result in a better transplantation outcome. Lakey et al. developed one of the first scoring systems to evaluate critical factors that influence islet isolation outcomes from brain-dead organ donors [17]. Several other scoring

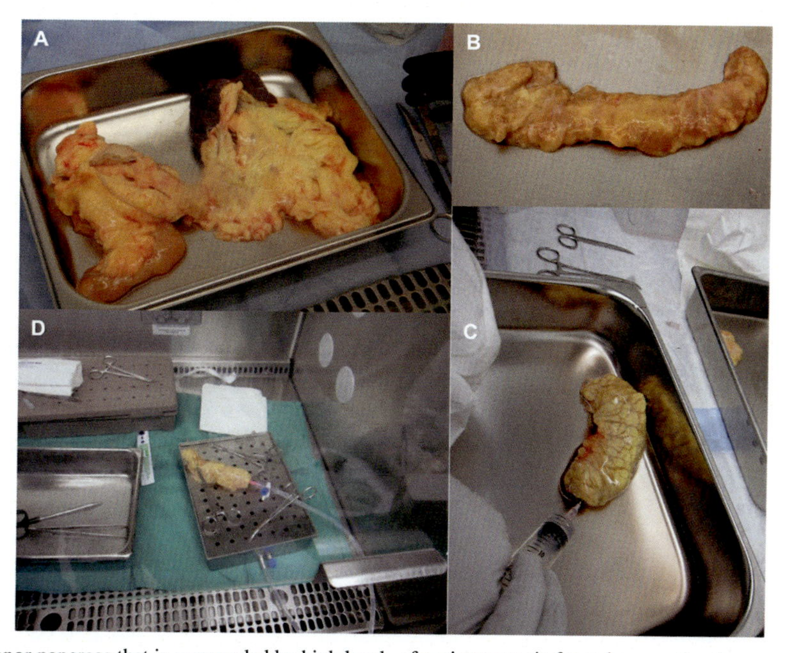

FIG. 2 (A) An image of a donor pancreas that is surrounded by high levels of peri-pancreatic fat and connective tissues making it difficult to clean and dissect the pancreas. It also makes it more difficult to effectively decontaminate the pancreas for islet isolation. (B) An image showing a trimmed organ that is ready for cannulation. (C) The pancreas is usually dissected into the head and body-tail lobes, cannulated, and then manually infused with a pancreatic tissue-dissociation enzyme (collagenase–neutral protease) blend. The efficacy of enzymatic distention is pivotal in maximizing islet yields from donor pancreases. (D) An image showing the pump infusion method of enzyme distention.

systems have since been developed, including the Islet Donor Score [38], the North American Islet Donor Score [39], and a scoring system by O'Gorman et al. [40]. While such tools are useful for subjective assessment and theoretically better selection of potential pancreases for islet isolation, these algorithms or scores serve only as a guide. Although donor score prediction is helpful, it does not always determine the IT outcome.

Several studies [13,14,16–20] have not considered transplantation outcome while predicting the donor score. Despite using a donor score of 79 as the most appropriate cutoff value, the sensitivity and specificity for predicting successful islet isolations were only 43% and 82%, respectively. Older donors with higher BMI could be termed ideal when considering islet outcome alone; however, when considering the transplantation outcome, the quality of islets from these donors might not be adequate. An effective approach to remedy these issues could be to come up with a robust donor score prediction system that considers islet isolation as well as IT outcomes [41].

Currently, the donor source for IT comprises deceased donors without diabetes [1]. It has been suggested that adequate donor selection and improved preservation techniques can lead to enhanced pancreas utilization and outcome [42]. Recent guidelines have been proposed on the procurement of pancreases from donation after circulatory death donors (DCD) for use in transplantation, including a limit on warm ischemic time (<30–60 min) [1]. There is a lacuna in comparative data on clinical outcomes with the standard use of brain-death donors. While associated with a higher risk of complications, use of DCD pancreases has shown similar results compared with donation after brain death donors (DBD), when accumulation of other risk factors is avoided [42]. A recent review summarized outcomes of three new reports comparing DCD with DBD pancreas transplantation and suggested that when it comes to risk factors and predicting outcomes, DCD seems to play a less important role than initially thought [42].

Another recent systematic review identified 18 studies comparing DBD and DCD pancreas donors. With groups comprising more than 23,609 transplant recipients, they showed that antemortem interventions including heparinization are beneficial, as is the initiation of perfusion in the intensive care unit and the early transport of the donor to the operating room before cessation of ventilation [43]. Takita et al. reported that cerebrovascular accident/stroke as a cause of death and intracranial hemorrhage/stroke as a mechanism of death were significantly associated with unsuccessful islet isolation [40]. Due to the insufficient number of organs retrieved from DBD, pancreases are also procured from DCD donors for IT [44]. Contrary to other clinical settings, which have not proved promising yet, experimental isolations using DCD donor pancreases have resulted in an islet yield comparable to their brain-dead counterparts [45]. To be able to achieve this and make the most of available DCD donors, islet groups in Japan have devised techniques for organ preservation using the Kyoto preservation solution and the two-layer method [46], thus optimizing retrieval of organs from these donors. They reported that IT from DCD donors had 76.5%, 47.1%, and 33.6% overall graft survival at 1, 2, and 3 years, respectively, whereas corresponding graft survival after multiple transplantations was 100%, 80.0%, and 57.1%, respectively [22]. They further reported that all the recipients were independent of insulin for 14, 79, and 215 days [47] and remained free of severe hypoglycemia. The caveat to this is that the DCD donors in Japan are retrieved in a different manner than most other countries as they can intervene at an earlier stage of the organ donor process with no significant donor down time (cessation of cardiac output) as compared to other countries where the DCD donor has a stipulated period of cardiac asystole.

Pancreas procurement

The success of human pancreatic islet isolation depends largely on the techniques used during pancreas procurement and the quality of the gland. Islets are highly susceptible to ischemic damage [48]. Warm ischemia of the pancreas of any duration during organ procurement may be detrimental to subsequent islet yield and the functional viability of the islets [48,49]. The procurement, preservation, and transport of the deceased donor pancreas for IT requires the same care and speed as that for a pancreas used for whole organ transplantation, with cold ischemic time (CIT) kept to <12 h [1]. In vitro studies corroborate clinical findings to indicate that prolonged cold storage beyond 10–12 h drastically affects the success of single-donor IT [48].

In order to maintain a shorter CIT, one study suggested that local organ procurement may lead to more islet isolation success than distant procurement [49]. Procurement techniques and the maintenance of a low temperature of the pancreas are critical to subsequent islet recovery and function. Lakey et al. suggested that sustaining a low pancreas temperature during procurement through the addition and replenishment of iced saline slush surrounding the anterior and posterior aspects of the pancreas greatly improves islet yield and the functional viability of the isolated islets and therefore improves success in clinical IT [45].

En bloc procurement is preferred over in vivo dissection. Ideally, the pancreas should be removed with the spleen and a stapled cuff of proximal and distal duodenum. Adhering to the *no touch* technique and maintaining the capsule of the pancreas intact are essential, as it is important to ensure complete distension of the gland during the enzyme infusion step of the islet isolation [9]. A damaged pancreatic capsule leads to enzyme leakage and loss of ductal integrity; a poorly distended pancreas drastically affects islet yield for clinical transplantation [48], as does major injury to the pancreas [50]. En bloc procurement also reduces potential complications related to the level of the retrieving surgeon's skills and abilities [51,52]. Once the donor pancreatectomy is initiated, the pancreas is flushed and cooled with preservation solutions via an intravascular flush. Rapid mobilization of the spleen to the midline after cross-clamping the aorta and embedding the entire pancreas in iced saline slush were shown to double islet yield and significantly improve islet viability.

After the donor pancreas is completely removed, the organ is immediately immersed in cold organ preservation solution, most commonly University of Wisconsin (UW) or histidine–tryptophan–ketoglutarate (HTK) solution. The pancreas is then cold packed (according to United Network of Organ Sharing policy) and transported to the islet isolation facility. Thus three factors are paramount: (a) very careful and atraumatic handling of the pancreas, (b) rapid in situ cooling to minimize warm ischemia and stabilize endogenous enzyme activity prior to islet isolation, and (c) immediate transfer of the pancreas to the isolation facility to minimize CIT [48].

Organ preservation

Multiple techniques have been developed and tested in order to reduce ischemia–reperfusion injury after transplantation. These techniques mostly focus on maintaining cellular adenosine triphosphate (ATP) by delivery of high oxygen concentrations to the tissue [42]. Primary biochemical injury to the pancreas is caused from loss of ATP by lack of oxidative phosphorylation, which is slowed by cooling to reduce cellular metabolism [53].

The two-layer method (TLM), developed in 1988 by Kuroda, has been shown by several groups to help improve oxygenation during cold storage, which then leads to improved islet yield [50,54–57]. Maintenance of ATP production in pancreases stored at the interface between perfluorocarbon and UW solution was observed as a result of oxygenation [58]. Many centers reported improvement in islet isolation outcomes for pancreases preserved in TLM over UW alone [55,59]. A recent meta-analysis concluded that TLM was indeed beneficial for pancreases with prolonged CIT prior to human islet isolation; however, the benefits of the TLM for short-term preservation were inconclusive [60]. These conclusions on TLM have been contested, with some studies demonstrating no beneficial effect of the TLM [61,62].

Oxygen perfusion or persufflation and maintenance of a hyperoxic environment could lead to more favorable results in terms of islet viability [63] and are therefore considered a promising technique. It has been shown that with this relatively simple technique, oxygen may be delivered to organs during cold storage, whereby ATP levels are replenished and maintained with associated reduction of oxidative stress and lipid peroxidation [42]. The persufflation technique has been shown to be more advantageous than the TLM [64], with no additional loss of islet function or viability upon extended preservation time [65].

Evidence suggests that hypothermic machine perfusion of the human pancreas is also a safe preservation method that minimizes tissue injury and edema formation and aids renewal of ATP [42]. However, the main limitation is the lack of a posttransplant evaluation reperfusion model in a normothermic environment.

An in situ regional cooling system for kidney procurement has been adapted for use in pancreas procurement to prevent warm ischemic damage by infusion of a hypothermic preservation solution [52,66]. Static cold storage is preferred for pancreas preservation over machine perfusion (typically used for kidney preservation) due to the complexities and higher failure rates with the machine perfusion [52,67]. Ductal injection/perfusion during pancreas procurement may facilitate enzymatic digestion during the isolation procedure [68].

Preservation solutions have been developed to manage ischemic insults [69]. With an increase in the number of perfusion solutions for whole organ transplant and IT, several studies have evaluated islet isolation outcomes using various solutions such UW, HTK, Celsior solution, Institute Georges Lopez-1, and ET-Kyoto solution, to name a few. Studies have shown that the UW, HTK, CS Celsior solution, and Institute Georges Lopez-1 have similar preservation properties [70]. No study has evaluated the effect of different preservation solutions on DCD pancreas transplants; different treatments might be necessary due to differences in injury during the DBD and DCD procedures [69]. Colloid-free Celsior solution, while comparable to UW for cold storage for whole organ transplantation, is inferior to UW when used for preservation prior to islet isolation [52]. HTK is a common alternative to UW and reports comparable results [52,71,72]. However, some evidence suggests that use of HTK as a preservation solution could be linked to reduced graft survival in solid-organ pancreas transplantation [73], though there is little data on this same effect for IT. Some evidence suggests that ductal perfusion with UW solution during procurement has a positive impact on islet yield [74]. Solution de conservation d'organes et de tissus (SCOT) has shown potential as an alternative to UW due to its immunoprotective effects on islet cells, yielding reduced graft immunogenicity [52,75]; cell swelling and pancreas edema are not significantly different compared to UW following cold storage [52,75]. SCOT has shown improved islet morphology and islet yield [75]. A recent and most informative study from the Collaborative Islet Transplant Registry reported that among 1017 pancreases retrieved between 2007 and 2010, >42% of donor pancreases were perfused with UW solution, 7% with HTK, and only 2.3% with Celsior solution, with the remainder not reporting what perfusate was used [12].

Human islet isolation procedure

Adequate laboratory preparation prior to the islet isolation procedure is critical so the team can begin as soon as the pancreas arrives at the facility. A singular, pivotal factor that predicts insulin independence after transplantation is the islet cell mass (IEQ/kg) acquired from the isolation procedure. The islet isolation procedure is divided into the following essential steps: ductal cannulation, enzyme distention, tissue digestion, tissue recombination, islet purification, islet culture and viability assessment, and final transplant preparation [48].

Most often the islet isolation procedure is performed in a clean room (cGMP) facility. It is vital to keep the pancreas cold from the time of organ removal to the initiation of enzymatic digestion. The pancreatic capsule is kept intact and uninjured for successful enzyme distention and to reduce the likelihood of enzyme leakage during ductal perfusion.

Efficient and uniform delivery of the tissue-dissociation enzyme blends with a specific ratio/dose of collagenase and neutral protease enzymes to the entire pancreas is essential to achieve a good isolation outcome. Enzymatic tissue dissociation separates the exocrine and endocrine components of the pancreas. Perfusion is done by either hand syringe injection or by using a semiautomated system with a peristaltic pump connected to the main pancreatic duct of a whole or segmented (head and body/tail sections) pancreas (Fig. 2B and D). The infused enzymatic blend digests the extracellular matrix of the pancreas to release islets and exocrine cells [76–78]. Effective islet cell isolation depends on adequate distension of the pancreas with the tissue-dissociation enzyme blends to maximize digestion efficiency. Difficulties with the distention phase need to be addressed promptly to ensure that the enzyme blend distributes well throughout the pancreas. Leaks may be addressed using hemostats, and ductal occlusions may be bypassed with catheters of a higher gauge.

Once the pancreas has been distended, a quick, final trimming helps remove excess surface fat and connective tissue that may interfere with the digestion phase. The pancreas is then cut into small pieces with a diameter of 2–5 cm and placed in the Ricordi chamber along with any enzyme solution remaining after the distention phase. The following section describes the development of pancreatic tissue-dissociation enzyme blends and their further refinement, which has led to significant improvements in islet yields over the past decade.

Tissue-dissociation enzyme blends for human islet isolation

Liberase-HI was the enzyme of choice for many years around the world for clinical IT until it was withdrawn from the market in 2007 due to safety concerns (use of bovine brain-derived materials were used for culture during the fermentation process). At that point, the only other available enzyme for clinical isolation was the purified clinical-grade SERVA/Nordmark collagenase and neutral protease enzymes, supplied in separate vials [79]. Lacking viable alternatives, many centers struggled to establish standardized isolation protocols with these enzymes, which resulted in lower islet yields, eventually leading to fewer IT procedures [3,12]. Identifying a suitable replacement tissue-dissociation enzyme blend became a priority for the continued success of IT programs across the globe [3].

Szot et al. successfully used the SERVA/Nordmark enzyme blend to isolate islets for clinical allotransplantations [80]. The University of Minnesota group reported that the same enzyme could also be successfully utilized for autologous islet isolation from pancreases with chronic pancreatitis [81]. The results from the autologous islet isolations using the SERVA/Nordmark blend were comparable to those using traditional Liberase-HI in terms of islet yield and clinical outcomes [81]. But the same outcomes were not achieved with brain-dead donor pancreases for allo-IT.

To improve the percentage of isolated islets that went on to be used for transplantation, VitaCyte CI*zyme* Collagenase HA was used in place of the SERVA/Nordmark collagenase. Biochemical comparison of these collagenase products showed that SERVA/Nordmark enzyme contained a higher fraction of truncated C1 collagenase than in Collagenase HA, which contained primarily intact C1 collagenase. The use of SERVA/Nordmark enzyme blends affected islet yields as it poorly digested the tissue, resulting in mantle islets (islets embedded in exocrine tissue), which in turn resulted in low islet yields [82]. Balamurugan et al. demonstrated unique differences between the C1 collagenase found in purified collagenase products manufactured by three main suppliers, VitaCyte LLC, Roche, and SERVA/Nordmark [76]. The results emphasized the importance of intact C1 collagenase for successful human islet isolation and transplantation [76].

Interestingly, nonsimultaneous administrations of the same enzymes have also been shown to improve isolation outcomes [83]. Brandhorst et al. performed detailed enzyme activity analyses for different enzyme lots and determined that high trypsin-like activity is a key parameter for the efficacy of pancreatic tissue-dissociation enzyme blends [84]. To identify suitable enzyme combinations, Balamurugan et al. evaluated many different enzyme products containing different levels of intact or truncated collagenases, used in combination with thermolysin or neutral protease [82]. The group incorporated biochemical analysis of the collagenases as part of the evaluation criteria and identified a new blend containing intact C1 and C2 collagenases and neutral protease, both from *C. histolyticum*, and called it New Enzyme Mixture (NEM) [77]. Isolations with the NEM were considerably more effective than any other enzyme combinations, recovering consistently high yields of quality islets from human pancreases [76]. Additionally, clinical allo-IT and auto-IT with islets isolated with NEM achieved greater success than previous transplantations [85].

During the course of enzyme standardization, Roche introduced Liberase MTF, a new enzyme blend preparation of purified collagenase and purified thermolysin. High-performance liquid chromatography analysis confirmed that the product had intact C1 collagenase. This enzyme has been utilized successfully for clinical islet isolations [86,87]. Currently, the three different enzyme combinations (from VitaCyte LLC, Roche, and SERVA/Nordmark) are extensively utilized for clinical preparations. Despite the steady progress made, only an average of 50% of the total isolations have yielded sufficient islet numbers for transplantation [12].

Recombinant enzymes for human islet isolation

Manufacturing purified recombinant collagenases may overcome problems commonly encountered with purified natural collagenase. rC1and rC2 collagenases are expressed from single genes in the *Escherichia coli* strain containing low protease activity. This minimizes the proteolysis of collagenase enzymes, leading to the generation of intact rC1 and rC2 collagenases. Clostripain contamination is eliminated, improving the control of protease activity during the islet isolation procedure. Balamurugan et al. successfully utilized purified rC1 and rC2 collagenases and assessed four different collagenase formulations to recover islets from human pancreases ($n=12$; three per group) [85]. Varying amounts of rC1 and rC2 collagenase activity, supplemented with a fixed amount of BP protease activity, were tested using a statistically designed experiment in a two-split pancreas model. A lower-dose recombinant collagenase in combination with BP protease was successfully used to recover >5000 IEQ/g pancreas. Results from this study indicate the effectiveness of this animal-free enzyme mixture to recover islets at doses of collagenase that are approximately 50% of those commonly used for human islet isolation.

Loganathan et al. used low doses of a collagenase–protease mixture to successfully isolate islets, with good functionality, from whole human pancreases [21]. Compared with islet isolation performed using a normal collagenase–protease mixture (60:40 C1–C2 collagenase mixture containing about 530 mg of collagenase to digest 100 g pancreas) with equivalent (12 WU/g) or standard (20 WU/g) amounts, the lower-dose mixture demonstrated better islet recovery, tissue digestion efficiency, insulin secretion, and transplantation outcomes. This was the first report showing an improvement in islet functionality with a low-dose enzyme. Overall, the quality and functionality assessments suggest that this recombinant dose has the potential to be used for human islet isolations to improve clinical outcomes.

A purified enzyme comprising an optimized fixed ratio (60:40) of C1–C2 collagenase was developed. A broader C1–C2 ratio can be used for human islet isolation than has been used in the past. Recombinant collagenase is an effective replacement for the natural enzyme, and Loganathan et al. determined that high islet yield can be obtained even with low doses of rC1–rC2, which is beneficial for the survival of islets [88]. Low doses of rC1–rC2 collagenase can be successfully utilized to obtain a higher islet yield than obtained with the standard dose of collagenase.

Enzyme combinations for human islet isolation from young donor pancreases

Islets isolated from young donor pancreases remain mantled and embedded by surrounding acinar cells, resulting in large pellet size after enzymatic digestion of the pancreas [25]. Though the surrounding layer of acinar cells may protect islet integrity [89], mantled islets are often lost during the purification process [90]. The release of mantled islets from young donor pancreases is likely due to the composition of the extracellular matrix. Accepting this assumption, one approach to reducing the percentage of mantled islets is to modify the composition of collagenase–protease mixtures used to free cells from the extracellular matrix.

Loganathan et al. identified the unique role of protease in digesting the peri-islet extracellular matrix proteins to recover mantle-free islets from young donor pancreases. They developed a novel approach to this problem by using a "trisected pancreas" model to compare the effectiveness of different enzyme mixtures [21]. Results from this study have immense clinical benefits; utilizing young donor pancreases increases the number of acceptable donors for allo-IT, and based on earlier reports [15,91] could improve clinical outcomes [21].

Cost-effective enzyme combinations for human islet isolation

In addition to clinical use, human islets are routinely isolated from brain-dead donor pancreases for various research studies. Currently, most of these laboratories utilize the highly purified enzyme blends to isolate islets. The major drawback in utilizing these enzyme blends for research preparation is that the cost can be prohibitive. At the same time, the crude enzymes available from Sigma-Aldrich and Worthington have been used for both human and other animal pancreatic islet isolations, but these have not gained widespread application due to factors such as low digestion efficacy related to enzyme impurity, imbalanced combination of key active components, significant batch-to-batch and vial-to-vial enzyme variation, and high endotoxin levels.

Loganathan et al. successfully utilized a new enriched collagenase product called the Defined Enriched Collagenase 800 (>85% purity) from VitaCyteLLC to improve the islet isolation process, mainly focusing on pancreases utilized for research purposes [92]. This new product was tested at three different islet processing centers to confirm the efficacy of low-cost enriched collagenase on islet yield and function. The use of low-cost enriched collagenase products with biochemical characteristics comparable to purified collagenase enzymes was shown to be as effective as higher-priced products in human islet isolation.

Purification

Islet purification is the process of separating islets from exocrine tissue to reduce the volume of tissue to be infused into the patient and to reduce the potential for exocrine cell contamination [93]. Typically, islet isolations from brain-dead donor pancreases result in a large tissue volume. Smaller tissue volumes are much easier to infuse into the patient and pose less risk of portal vein thrombosis during infusion [94,95].

After digestion, the remaining pancreatic tissue is collected in recombination flasks. The collected cells are washed several times to eliminate waste such as dead cells or remnant enzyme solution. Islets and exocrine cells have different cellular densities, which allow them to be separated through centrifugation by density gradients (Fig. 3). Islet and exocrine cell densities are dependent on a variety of factors, including differences in donor characteristics, preservation method and effectiveness, CIT prior to isolation, and osmolality of the gradient solution [96–98]. Several islet purification methods have been reported, including traditional methods such as handpicking, serial sieving, and discontinuous density gradient using tubes and bottles [93], but density gradient centrifugation using the COBE 2991 Cell Processor remains the only consistently successful method for large-scale clinical use for islet separation from the digested tissue [99].

Density gradients have been used for islet purification on the basis that islets and exocrine tissue are separated by density, observable by the cells floating or sinking in the gradient [93]. Density gradient selection is one of the primary ways to improve islet purity and yield. A better gradient will provide better separation of the islet and exocrine tissue. There are both continuous and discontinuous density gradients. A discontinuous gradient is formed by layering varying densities of solution on top of one another to create steps of density. A continuous gradient is similar, but the density

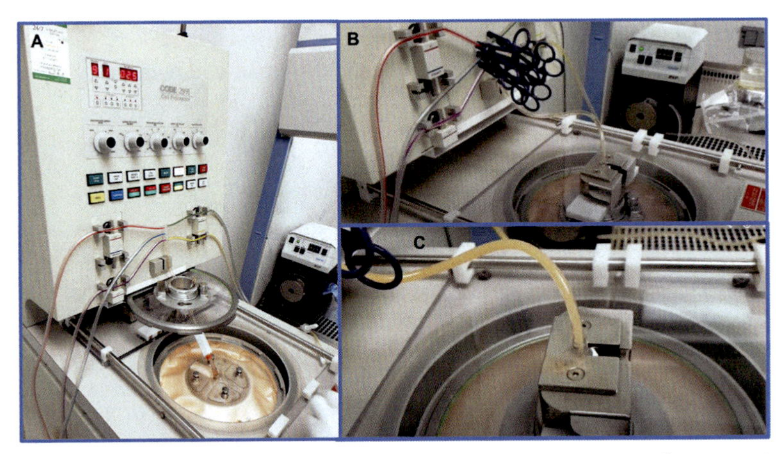

FIG. 3 (A) An image of the COBE 2991 cell processor that helps obtain pure islets based on density gradient separation. (B) Our group uses the continuous iodixanol (opti-prep) solutions for purification purposes. (C) Inflow of washed and recombined islet preparation into the COBE bag that contains the layered density gradients.

changes incrementally, rather than in large steps. Though they were originally used as a tool for analysis of islet and exocrine densities [100], continuous density gradients have been adopted as the primary gradients to separate islets and exocrine tissue for purification [101].

Continuous density gradients show improved recovery and decreased contamination [93,101]. Lake et al. first established the use of density gradients for large-scale islet purification using a discontinuous gradient of bovine serum albumin [102] that demonstrated similar results to other techniques used at the time and established use of the COBE 2991 Cell Processor as a viable method for islet purification [93,102]. Ficoll solutions were popularized over bovine serum albumin as density gradients to avoid the use of xenoprotein products [93]. Biocoll has been used as a viable density gradient but presents a level of toxicity to the islet cells [98]. Barbaro et al. have developed and utilized Biocoll solutions diluted with UW solution to produce high islet yield and viability while minimizing the toxicity of the Biocoll solution [98,103]. Iodixanol solutions (Opti-Prep) have been shown to possess antiinflammatory properties when compared to other density gradients, which improves islet recovery [104,105]. Balamurugan et al. have demonstrated the use of an analytical test gradient system to assess islet and exocrine density prior to purification to create a custom density gradient [99]. Using a custom density gradient rather than a fixed standard density gradient allows for adjustments to be made to the gradient range to maximize islet yield based on density variability [99].

Noguchi et al. suggested that purification using the COBE 2991 Cell Processor applies shear force within the COBE bag that results in mechanical damage to the isolated islets [93]. This results in potential loss of otherwise viable islets. Thus they utilize an alternative method of purification utilizing large cylindrical plastic bottles that lack narrow segments which would create shear force and do not require centrifugation [93,106].

Potential shear force is not the only factor that could contribute to islet loss during purification. Some islets remain embedded in exocrine tissue (mantled islets), especially in young donors [107,108]. These islets have a density closer to exocrine cells than fully isolated islets and therefore are often lost during the purification step [107,108]. IT success is often dependent on the total number of islets infused [109], and thus it is sometimes worthwhile to attempt rescue purification to recover mantled islets when standard purification results in low islet yields.

Several methods can be applied to rescue mantled islets. Golebiewska et al. suggested that postpurification gravity sedimentation procedures based on older methods could be utilized to isolate mantled islets after low-yield purification procedures [110]. Shimoda et al. demonstrated that gentle agitation using a high osmolality solution can separate mantled islets from their surrounding exocrine tissue [108]. Rescue gradient purification procedures can be performed when initial isolations show high islet yields but purification results in low islet yields. In this process the low-purity fractions are pooled, washed by centrifugation, resuspended in stock Ficoll, and then purified by a discontinuous gradient of Euroficoll solutions [107,109].

Alternative purification methods can be explored that might reduce islet loss or improve collection of mantled islets. Methods utilizing magnetic microspheres and magnetic fields for islet purification have been explored. Although few have seen continued use on a large scale, they could prove useful as secondary purification methods to recover islets from low-purity fractions [111–113]. Noguchi et al. employed a newer alternative to purification using the COBE 2991 Cell Processor under a modified method using 500 mL plastic bottles. In this method, digested tissue is purified by loading it into a 500 mL bottle of continuous density gradient [93,106,114]. The bottle methods result in high islet yield, reduced loss of islets indicated by larger average islet size, and an elimination of the shear forces present in the COBE 2991 Cell Processor that Noguchi et al. believed to cause mechanical damage to the islets [93,106,114].

After purification, the islets are cultured for 24–72 h at 20°C or 37°C (center dependent) in media supplemented with insulin, transferrin, and selenium. Noguchi et al., based on a recent literature survey, suggest that islet culturing at 22–26°C may be superior to culturing at 37°C [115]. As per the original Edmonton protocol, it was a prerequisite to infuse islets into the recipient within 4 h after islet isolation [116]. Since then, many centers have introduced the culturing of human islets prior to transplantation, indicating that it provides many benefits to clinical IT [117–119]. The culture period facilitates recipient conditioning and may allow for transplantation of a more immunologically quiescent graft [120]. Furthermore, other benefits include performing additional quality control testing of the isolated islets, initiating time-dependent immunosuppressive protocols, and permitting recipients the time to travel to transplant centers, even if they live far away [115]. Even in context of shipping the allo-islet product, a recent study suggested that allowing time for islet recovery post isolation, prior to shipping, yields less islet loss during shipment without decreasing islet function [121].

However, it is also well documented that isolated islets deteriorate rapidly in culture [118,122,123]. Islets are susceptible to multiple insults during DBD and organ procurement, organ preservation, isolation, and in the culturing process itself. One of the possible mechanisms that contributes to the reduction of islet mass during culture is the activation of intracellular cell death and damage-associated pathways. Studies have confirmed that after islet isolation, there is an increase in protein kinase (c-Jun and p38) activities that are triggered by cytokines such as tumor necrosis factor alpha and interleukin 1 beta [115]. Addition of antiapoptotic agents to culture media has been shown to prevent reduction in islet mass [124,125]. The supplementation of culture media with serum over its albumin component alone is generally considered better as serum contains many factors that are beneficial for islet viability and also because it neutralizes the endogenous pancreatic enzymes or exogenous enzymes left over from the isolation process [115]. However, because of drawbacks associated with fetal bovine serum and AB serum, many centers continue to supplement their media with human serum albumin. Once the islets have been cultured and before they are ready for transplantation, the purified islet preparation undergoes a mandatory and detailed quality control assessment for islet viability, purity, cell number, and insulin secretory response.

Islet release criteria

Islet potency is a measure of the probable success of the isolated product and encompasses islet quantity, viability, and insulin release functionality. The criteria that must be met before the isolation product can be released for transplantation include adequate dose (>5000 IEQ/kg of recipient body weight with a concentration of ≥20,000 IEQ/mL in a total settled tissue volume of <7 cc), purity (>30%), and sterility based on Gram stain [6,126]. Prior to release, visual inspection ensures that the recipient information is correctly matched to the product.

Islet quantity is assessed using dithizone staining with observation using light microscopy to estimate the islet concentration per milliliter of settled tissue [6,120,126]. Dithizone staining confirms that islets are present in the final product and allows for easy estimation of islet quantity as they bind to zinc ions present in the islet beta cells and color them red [120,127].

Islet viability is assessed using fluorescein diacetate/propidium iodide staining. Fluorescein diacetate, an inclusion dye, can permeate the plasma membrane of living cells and fluoresces green. Propidium iodide, an exclusion dye, is unable to penetrate living cells but easily penetrates dead or dying cells and fluoresces orange [120,128]. Viability requirements of ≥70%, indicated by green fluorescing islets, must be met for release [120,126].

A high-purity islet isolation product is assessed prior to culture for glucose-stimulated insulin release by enzyme-linked immunosorbent assay. These results are for information gathering purposes only [126], as they require several days to obtain and are therefore not reasonable for pretransplant product release [120].

The safety of the product intended to be transplanted to a patient is extremely important. An evaluation by visual inspection is performed to ensure that the final product has a typical yellow or amber color with visible aggregates in each product bag [126]. The presence of endotoxins in the islet isolation final product can ultimately lead to reduced graft function and even failure [129,130]. Endotoxin levels measured by limulus amoebocyte lysate analysis require a concentration of ≤5.0 EU/kg of recipient body weight to be accepted for release [120,126]. Product contamination is assessed by Gram staining for absence of microbiological organisms [120,126] as well as by microbiological analysis for a negative sterility result [120,126]. Since the sterility confirmation can take several days to complete, a preemptive plan is put in place to notify the investigators of any contamination, and subsequent steps should be taken to ensure patient health and safety in the event of contamination [126].

Lastly, to avoid unsafe levels of portal vein pressure during infusion, the total tissue volume of the product must be ≤15.0 mL in up to three product bags and ≤7.5 mL for a single product bag [126]. The need for multiple donor islet infusions is minimized by maximizing the infused islet mass matched to the ABO blood type of the recipient, which reduces risk of HLA-recipient sensitization.

Future perspectives

The division of the islet isolation process into separate steps provides multiple opportunities for improvements to the procedure, as the various cofactors during each stage can have a significant impact on the ultimate outcome of the islet preparation and transplantation. Improvements in donor availability, donor selection, and organ procurement and transportation, alongside innovations regarding preservation solutions and techniques, enzyme development, purification process, and culture techniques, can converge to maximize viable islet yields and improve transplantation outcomes.

Donor scarcity is a major factor in islet isolation, as successful transplantations often require isolations from multiple donors [1]. Although IT has been traditionally restricted to the use of DBD pancreases, recent evidence suggests that DCD pancreases are equally viable for islet isolation and transplantation [44,47]. Accepting DCD pancreases in addition to DBD pancreases could significantly widen the pool of pancreases available for isolation and transplantation.

Many improvements in donor selection have already been made using donor scoring systems that assess pancreatic viability prior to procurement [39,131,132]. Further refinement of these scoring systems and consideration of other variables, donor height [132] as an example, could continue to contribute to improved isolation outcomes.

Organ procurement of the pancreas is a critical step. Prolonged CIT [133] and poor oxygenation [134] negatively affect islet survival and function. Keeping CIT to a minimum is a standard practice for islet isolation and can only be improved by reducing that time, but innovations in organ oxygenation during preservation and transportation prior to islet isolation could help achieve positive outcomes. The addition of cytoprotective agents in the storage solution could also contribute to isolation success by limiting and reducing oxidative stress caused by poor oxygenation [52].

Maximizing pancreas digestion efficacy is dependent on tissue-dissociation enzyme blends; critical analysis of how tissue dissociation enzymes may act on known and novel tissue substrates present within the extracellular matrix for better tissue dissociation is equally important. This could facilitate the improvement of islet yields and also advance the potential cellular and molecular therapies associated with clinical IT [135–138]. Discovery and development of new enzymes and enzyme mixtures could drastically improve islet digestion and islet yield.

As indicated earlier, alternative purification methods can be explored and warrant systematic investigations that might reduce islet loss or improve collection of mantled islets. The islet culture steps can act as a double-edged sword to help mitigate the number of dying islets infused into the recipient, but these dying islets can create a toxic environment for the remaining islets in culture [139]. Utilization of a decellularized pancreas as a base for islet culture has been shown to be a promising potential alternative to traditional cell culture [140], though it brings its own limitations, requiring a separate pancreas as the scaffold for culture. Taking steps to improve all such limitations will help further establish IT as an effective mainstream clinical therapy for patients suffering from type 1 diabetes mellitus.

Acknowledgments

We extend our deepest gratitude to the courageous families who generously donated their loved one's organs and tissues for biomedical research. Such important research like this would not be possible without this selfless gift of hope. Thanks to our partner organ procurement organizations, Kentucky Organ Donor Affiliates (KODA) and LifeCenter, Ohio, for supporting these special families.

References

[1] Rickels MR, Robertson RP. Pancreatic islet transplantation in humans: recent progress and future directions. Endocr Rev 2019;40:631–68.

[2] Bellin MD, Dunn TB. Transplant strategies for type 1 diabetes: whole pancreas, islet and porcine beta cell therapies. Diabetologia 2020;63:2049–56.

[3] Alejandro R, Barton FB, Hering BJ, Wease S, Collaborative Islet Transplant Registry Investigators. 2008 update from the collaborative islet transplant registry. Transplantation 2008;86:1783–8.

[4] Johnson PR, Jones KE. Pancreatic islet transplantation. Semin Pediatr Surg 2012;21:272–80.

[5] Hering BJ, Clarke WR, Bridges ND, Eggerman TL, Alejandro R, Bellin MD, Chaloner K, Czarniecki CW, Goldstein JS, Hunsicker LG, Kaufman DB, Korsgren O, Larsen CP, Luo X, Markmann JF, Naji A, Oberholzer J, Posselt AM, Rickels MR, Ricordi C, Robien MA, Senior PA, Shapiro AM, Stock PG, Turgeon NA, Clinical Islet Transplantation Consortium. Phase 3 trial of transplantation of human islets in type 1 diabetes complicated by severe hypoglycemia. Diabetes Care 2016;39:1230–40.

[6] Gamble A, Pepper AR, Bruni A, Shapiro AMJ. The journey of islet cell transplantation and future development. Islets 2018;10:80–94.

[7] Tatum JA, Meneveau MO, Brayman KL. Single-donor islet transplantation in type 1 diabetes: patient selection and special considerations. Diab Metab Syndr Obes 2017;10:73–8.

[8] Lablanche S, Borot S, Wojtusciszyn A, Bayle F, Tetaz R, Badet L, Thivolet C, Morelon E, Frimat L, Penfornis A, Kessler L, Brault C, Colin C, Tauveron I, Bosco D, Berney T, Benhamou PY, Network G. Five-year metabolic, functional, and safety results of patients with type 1 diabetes transplanted with allogenic islets within the Swiss-French GRAGIL network. Diabetes Care 2015;38:1714–22.

[9] Hawthorne WJ. Factors related to successful clinical islet isolation. In: Transplantation, bioengineering, and regeneration of the endocrine pancreas. Elsevier; 2020. p. 485–502.

[10] Marfil-Garza BA, Shapiro AMJ, Kin T. Clinical islet transplantation: current progress and new frontiers. J Hepatobiliary Pancreat Sci 2021;28:243–54.

[11] Kin T. Islet isolation for clinical transplantation. Adv Exp Med Biol 2010;654:683–710.

[12] Balamurugan AN, Naziruddin B, Lockridge A, Tiwari M, Loganathan G, Takita M, Matsumoto S, Papas K, Trieger M, Rainis H, Kin T, Kay TW, Wease S, Messinger S, Ricordi C, Alejandro R, Markmann J, Kerr-Conti J, Rickels MR, Liu C, Zhang X, Witkowski P, Posselt A, Maffi P, Secchi A, Berney T, O'Connell PJ, Hering BJ, Barton FB. Islet product characteristics and factors related to successful human islet transplantation from the collaborative islet transplant registry (CITR) 1999-2010. Am J Transplant Off J Am Soc Transplant Am Soc Transplant Surg 2014;14:2595–606.

[13] Benhamou PY, Watt PC, Mullen Y, Ingles S, Watanabe Y, Nomura Y, Hober C, Miyamoto M, Kenmochi T, Passaro EP, et al. Human islet isolation in 104 consecutive cases. Factors affecting isolation success. Transplantation 1994;57:1804–10.

[14] Goto M, Eich TM, Felldin M, Foss A, Kallen R, Salmela K, Tibell A, Tufveson G, Fujimori K, Engkvist M, Korsgren O. Refinement of the automated method for human islet isolation and presentation of a closed system for in vitro islet culture. Transplantation 2004;78:1367–75.

[15] Ihm SH, Matsumoto I, Sawada T, Nakano M, Zhang HJ, Ansite JD, Sutherland DE, Hering BJ. Effect of donor age on function of isolated human islets. Diabetes 2006;55:1361–8.

[16] Lakey JR, Rajotte RV, Warnock GL, Kneteman NM. Human pancreas preservation prior to islet isolation. Cold ischemic tolerance. Transplantation 1995;59:689–94.

[17] Lakey JR, Warnock GL, Rajotte RV, Suarez-Alamazor ME, Ao Z, Shapiro AM, Kneteman NM. Variables in organ donors that affect the recovery of human islets of Langerhans. Transplantation 1996;61:1047–53.
[18] Matsumoto I, Sawada T, Nakano M, Sakai T, Liu B, Ansite JD, Zhang HJ, Kandaswamy R, Sutherland DE, Hering BJ. Improvement in islet yield from obese donors for human islet transplants. Transplantation 2004;78:880–5.
[19] Nano R, Clissi B, Melzi R, Calori G, Maffi P, Antonioli B, Marzorati S, Aldrighetti L, Freschi M, Grochowiecki T, Socci C, Secchi A, Di Carlo V, Bonifacio E, Bertuzzi F. Islet isolation for allotransplantation: variables associated with successful islet yield and graft function. Diabetologia 2005;48:906–12.
[20] Zeng Y, Torre MA, Karrison T, Thistlethwaite JR. The correlation between donor characteristics and the success of human islet isolation. Transplantation 1994;57:954–8.
[21] Loganathan G, Subhashree V, Narayanan S, Tweed B, Andrew Goedde M, Gunaratnam B, Tucker WW, Goli P, Mokshagundam S, McCarthy RC, Williams SK, Hughes MG, Balamurugan AN. Improved recovery of human islets from young donor pancreases utilizing increased protease dose to collagenase for digesting peri-islet extracellular matrix. Am J Transplant Off J Am Soc Transplant Am Soc Transplant Surg 2019;19:831–43.
[22] Niclauss N, Bosco D, Morel P, Demuylder-Mischler S, Brault C, Milliat-Guittard L, Colin C, Parnaud G, Muller YD, Giovannoni L, Meier R, Toso C, Badet L, Benhamou PY, Berney T. Influence of donor age on islet isolation and transplantation outcome. Transplantation 2011;91:360–6.
[23] Hering BJ, Kandaswamy R, Ansite JD, Eckman PM, Nakano M, Sawada T, Matsumoto I, Ihm SH, Zhang HJ, Parkey J, Hunter DW, Sutherland DE. Single-donor, marginal-dose islet transplantation in patients with type 1 diabetes. JAMA 2005;293:830–5.
[24] Kin T, Murdoch TB, Shapiro AM, Lakey JR. Estimation of pancreas weight from donor variables. Cell Transplant 2006;15:181–5.
[25] Balamurugan AN, Chang Y, Bertera S, Sands A, Shankar V, Trucco M, Bottino R. Suitability of human juvenile pancreatic islets for clinical use. Diabetologia 2006;49:1845–54.
[26] Socci C, Davalli AM, Vignali A, Pontiroli AE, Maffi P, Magistretti P, Gavazzi F, De Nittis P, Di Carlo V, Pozza G. A significant increase of islet yield by early injection of collagenase into the pancreatic duct of young donors. Transplantation 1993;55:661–3.
[27] de la Grandmaison GL, Clairand I, Durigon M. Organ weight in 684 adult autopsies: new tables for a Caucasoid population. Forensic Sci Int 2001;119:149–54.
[28] Saisho Y, Butler AE, Meier JJ, Monchamp T, Allen-Auerbach M, Rizza RA, Butler PC. Pancreas volumes in humans from birth to age one hundred taking into account sex, obesity, and presence of type-2 diabetes. Clin Anat 2007;20:933–42.
[29] Hanley SC, Paraskevas S, Rosenberg L. Donor and isolation variables predicting human islet isolation success. Transplantation 2008;85:950–5.
[30] Ponte GM, Pileggi A, Messinger S, Alejandro A, Ichii H, Baidal DA, Khan A, Ricordi C, Goss JA, Alejandro R. Toward maximizing the success rates of human islet isolation: influence of donor and isolation factors. Cell Transplant 2007;16:595–607.
[31] Dybala MP, Olehnik SK, Fowler JL, Golab K, Millis JM, Golebiewska J, Bachul P, Witkowski P, Hara M. Pancreatic beta cell/islet mass and body mass index. Islets 2019;11:1–9.
[32] Wang Y, Danielson KK, Ropski A, Harvat T, Barbaro B, Paushter D, Qi M, Oberholzer J. Systematic analysis of donor and isolation factor's impact on human islet yield and size distribution. Cell Transplant 2013;22:2323–33.
[33] Kaddis JS, Danobeitia JS, Niland JC, Stiller T, Fernandez LA. Multicenter analysis of novel and established variables associated with successful human islet isolation outcomes. Am J Transplant 2010;10:646–56.
[34] Sakuma Y, Ricordi C, Miki A, Yamamoto T, Pileggi A, Khan A, Alejandro R, Inverardi L, Ichii H. Factors that affect human islet isolation. Transplant Proc 2008;40:343–5.
[35] Ris F, Toso C, Veith FU, Majno P, Morel P, Oberholzer J. Are criteria for islet and pancreas donors sufficiently different to minimize competition? Am J Transplant Off J Am Soc Transplant Am Soc Transplant Surg 2004;4:763–6.
[36] Stegall MD, Dean PG, Sung R, Guidinger MK, McBride MA, Sommers C, Basadonna G, Stock PG, Leichtman AB. The rationale for the new deceased donor pancreas allocation schema. Transplantation 2007;83:1156–61.
[37] Koh A, Kin T, Imes S, Shapiro AM, Senior P. Islets isolated from donors with elevated HbA1c can be successfully transplanted. Transplantation 2008;86:1622–4.
[38] O'Gorman D, Kin T, Murdoch T, Richer B, McGhee-Wilson D, Ryan EA, Shapiro JA, Lakey JR. The standardization of pancreatic donors for islet isolations. Transplantation 2005;80:801–6.
[39] Wang LJ, Kin T, O'Gorman D, Shapiro AMJ, Naziruddin B, Takita M, Levy MF, Posselt AM, Szot GL, Savari O, Barbaro B, McGarrigle J, Yeh CC, Oberholzer J, Lei J, Chen T, Lian M, Markmann JF, Alvarez A, Linetsky E, Ricordi C, Balamurugan AN, Loganathan G, Wilhelm JJ, Hering BJ, Bottino R, Trucco M, Liu C, Min Z, Li Y, Naji A, Fernandez LA, Ziemelis M, Danobeitia JS, Millis JM, Witkowski P. A multicenter study: north American islet donor score in donor pancreas selection for human islet isolation for transplantation. Cell Transplant 2016;25:1515–23.
[40] Takita M, Matsumoto S, Noguchi H, Shimoda M, Chujo D, Sugimoto K, Itoh T, Lamont JP, Lara LF, Onaca N, Naziruddin B, Klintmalm GB, Levy MF. One hundred human pancreatic islet isolations at Baylor research institute. Proc (Baylor Univ Med Cent) 2010;23:341–8.
[41] Pepper AR, Gala-Lopez B, Kin T. Advances in clinical islet isolation. In: Islam MS, editor. Islets of Langerhans. Dordrecht: Springer; 2015.
[42] Leemkuil M, Leuvenink HGD, Pol RA. Pancreas transplantation from donors after circulatory death: an irrational reluctance? Curr Diab Rep 2019;19:129.
[43] Shahrestani S, Webster AC, Lam VW, Yuen L, Ryan B, Pleass HC, Hawthorne WJ. Outcomes from pancreatic transplantation in donation after cardiac death: a systematic review and meta-analysis. Transplantation 2017;101:122–30.
[44] Doppenberg JB, Nijhoff MF, Engelse MA, de Koning EJP. Clinical use of donation after circulatory death pancreas for islet transplantation. Am J Transplant 2021;21:3077–87.
[45] Lakey JR, Kneteman NM, Rajotte RV, Wu DC, Bigam D, Shapiro AM. Effect of core pancreas temperature during cadaveric procurement on human islet isolation and functional viability. Transplantation 2002;73:1106–10.

[46] Noguchi H. Pancreas procurement and preservation for islet transplantation: personal considerations. J Transplant 2011;2011, 783168.
[47] Saito T, Gotoh M, Satomi S, Uemoto S, Kenmochi T, Itoh T, Kuroda Y, Yasunami Y, Matsumoto S, Teraoka S. Islet transplantation using donors after cardiac death: report of the Japan islet transplantation registry. Transplantation 2010;90:740–7.
[48] Lakey JR, Burridge PW, Shapiro AM. Technical aspects of islet preparation and transplantation. Transpl Int 2003;16:613–32.
[49] Lee TC, Barshes NR, Brunicardi FC, Alejandro R, Ricordi C, Nguyen L, Goss JA. Procurement of the human pancreas for pancreatic islet transplantation. Transplantation 2004;78:481–3.
[50] Lakey JR, Tsujimura T, Shapiro AM, Kuroda Y. Preservation of the human pancreas before islet isolation using a two-layer (UW solution-perfluorochemical) cold storage method. Transplantation 2002;74:1809–11.
[51] Dalle Valle R, Capocasale E, Mazzoni MP, Busi N, Sianesi M. Pancreas procurement technique. Lessons learned from an initial experience. Acta Biomed 2006;77:152–6.
[52] Iwanaga Y, Sutherland DE, Harmon JV, Papas KK. Pancreas preservation for pancreas and islet transplantation. Curr Opin Organ Transplant 2008;13:445–51.
[53] Ridgway D, Manas D, Shaw J, White S. Preservation of the donor pancreas for whole pancreas and islet transplantation. Clin Transpl 2010;24:1–19.
[54] Fraker CA, Alejandro R, Ricordi C. Use of oxygenated perfluorocarbon toward making every pancreas count. Transplantation 2002;74:1811–2.
[55] Hering BJ, Matsumoto I, Sawada T, Nakano M, Sakai T, Kandaswamy R, Sutherland DE. Impact of two-layer pancreas preservation on islet isolation and transplantation. Transplantation 2002;74:1813–6.
[56] Tsujimura T, Kuroda Y, Kin T, Avila JG, Rajotte RV, Korbutt GS, Ryan EA, Shapiro AM, Lakey JR. Human islet transplantation from pancreases with prolonged cold ischemia using additional preservation by the two-layer (UW solution/perfluorochemical) cold-storage method. Transplantation 2002;74:1687–91.
[57] Kuroda Y, Kawamura T, Suzuki Y, Fujiwara H, Yamamoto K, Saitoh Y. A new, simple method for cold storage of the pancreas using perfluorochemical. Transplantation 1988;46:457–60.
[58] Morita A, Kuroda Y, Fujino Y, Tanioka Y, Ku Y, Saitoh Y. Assessment of pancreas graft viability preserved by a two-layer (University of Wisconsin solution/perfluorochemical) method after significant warm ischemia. Transplantation 1993;55:667–9.
[59] Ricordi C, Fraker C, Szust J, Al-Abdullah I, Poggioli R, Kirlew T, Khan A, Alejandro R. Improved human islet isolation outcome from marginal donors following addition of oxygenated perfluorocarbon to the cold-storage solution. Transplantation 2003;75:1524–7.
[60] Qin H, Matsumoto S, Klintmalm GB, De Vol EB. A meta-analysis for comparison of the two-layer and university of Wisconsin pancreas preservation methods in islet transplantation. Cell Transplant 2011;20:1127–37.
[61] Caballero-Corbalan J, Eich T, Lundgren T, Foss A, Felldin M, Kallen R, Salmela K, Tibell A, Tufveson G, Korsgren O, Brandhorst D. No beneficial effect of two-layer storage compared with UW-storage on human islet isolation and transplantation. Transplantation 2007;84:864–9.
[62] Kin T, Mirbolooki M, Salehi P, Tsukada M, O'Gorman D, Imes S, Ryan EA, Shapiro AM, Lakey JR. Islet isolation and transplantation outcomes of pancreas preserved with University of Wisconsin solution versus two-layer method using preoxygenated perfluorocarbon. Transplantation 2006;82:1286–90.
[63] Komatsu H, Kandeel F, Mullen Y. Impact of oxygen on pancreatic islet survival. Pancreas 2018;47:533–43.
[64] Scott 3rd WE, Weegman BP, Ferrer-Fabrega J, Stein SA, Anazawa T, Kirchner VA, Rizzari MD, Stone J, Matsumoto S, Hammer BE, Balamurugan AN, Kidder LS, Suszynski TM, Avgoustiniatos ES, Stone SG, Tempelman LA, Sutherland DE, Hering BJ, Papas KK. Pancreas oxygen persufflation increases ATP levels as shown by nuclear magnetic resonance. Transplant Proc 2010;42:2011–5.
[65] Kelly AC, Smith KE, Purvis WG, Min CG, Weber CS, Cooksey AM, Hasilo C, Paraskevas S, Suszynski TM, Weegman BP, Anderson MJ, Camacho LE, Harland RC, Loudovaris T, Jandova J, Molano DS, Price ND, Georgiev IG, Scott 3rd WE, Manas DMD, Shaw JAM, O'Gorman D, Kin T, McCarthy FM, Szot GL, Posselt AM, Stock PG, Karatzas T, Shapiro AMJ, Lynch RM, Limesand SW, Papas KK. Oxygen perfusion (Persufflation) of human pancreata enhances insulin secretion and attenuates islet proinflammatory signaling. Transplantation 2019;103:160–7.
[66] Nagata H, Matsumoto S, Okitsu T, Iwanaga Y, Noguchi H, Yonekawa Y, Kinukawa T, Shimizu T, Miyakawa S, Shiraki R, Hoshinaga K, Tanaka K. Procurement of the human pancreas for pancreatic islet transplantation from marginal cadaver donors. Transplantation 2006;82:327–31.
[67] Florack G, Sutherland DE, Heil J, Squifflet JP, Najarian JS. Preservation of canine segmental pancreatic autografts: cold storage versus pulsatile machine perfusion. J Surg Res 1983;34:493–504.
[68] Shimoda M, Itoh T, Sugimoto K, Iwahashi S, Takita M, Chujo D, Sorelle JA, Naziruddin B, Levy MF, Grayburn PA, Matsumoto S. Improvement of collagenase distribution with the ductal preservation for human islet isolation. Islets 2012;4:130–7.
[69] Barlow AD, Hosgood SA, Nicholson ML. Current state of pancreas preservation and implications for DCD pancreas transplantation. Transplantation 2013;95:1419–24.
[70] Dholakia S, Royston E, Sharples EJ, Sankaran V, Ploeg RJ, Friend PJ. Preserving and perfusing the allograft pancreas: past, present, and future. Transplant Rev (Orlando) 2018;32:127–31.
[71] Becker T, Ringe B, Nyibata M, Meyer zu Vilsendorf A, Schrem H, Luck R, Neipp M, Klempnauer J, Bektas H. Pancreas transplantation with histidine-tryptophan-ketoglutarate (HTK) solution and University of Wisconsin (UW) solution: is there a difference? JOP 2007;8:304–11.
[72] Potdar S, Malek S, Eghtesad B, Shapiro R, Basu A, Patel K, Broznick B, Fung J. Initial experience using histidine-tryptophan-ketoglutarate solution in clinical pancreas transplantation. Clin Transpl 2004;18:661–5.
[73] Stewart ZA, Cameron AM, Singer AL, Dagher NN, Montgomery RA, Segev DL. Histidine-tryptophan-ketoglutarate (HTK) is associated with reduced graft survival in pancreas transplantation. Am J Transplant Off J Am Soc Transplant Am Soc Transplant Surg 2009;9:217–21.
[74] Takita M, Itoh T, Shimoda M, Kanak MA, Shahbazov R, Kunnathodi F, Lawrence MC, Naziruddin B, Levy MF. Pancreatic ductal perfusion at organ procurement enhances islet yield in human islet isolation. Pancreas 2014;43:1249–55.

[75] Giraud S, Claire B, Eugene M, Debre P, Richard F, Barrou B. A new preservation solution increases islet yield and reduces graft immunogenicity in pancreatic islet transplantation. Transplantation 2007;83:1397–400.

[76] Balamurugan AN, Breite AG, Anazawa T, Loganathan G, Wilhelm JJ, Papas KK, Dwulet FE, McCarthy RC, Hering BJ. Successful human islet isolation and transplantation indicating the importance of class 1 collagenase and collagen degradation activity assay. Transplantation 2010;89:954–61.

[77] Balamurugan AN, Loganathan G, Bellin MD, Wilhelm JJ, Harmon J, Anazawa T, Soltani SM, Radosevich DM, Yuasa T, Tiwari M, Papas KK, McCarthy R, Sutherland DE, Hering BJ. A new enzyme mixture to increase the yield and transplant rate of autologous and allogeneic human islet products. Transplantation 2012;93:693–702.

[78] Balamurugan AN, Elder DA, Abu-El-Haija M, Nathan JD. Islet cell transplantation in children. Semin Pediatr Surg 2020;29, 150925.

[79] Brandhorst H, Friberg A, Nilsson B, Andersson HH, Felldin M, Foss A, Salmela K, Tibell A, Tufveson G, Korsgren O, Brandhorst D. Large-scale comparison of Liberase HI and collagenase NB1 utilized for human islet isolation. Cell Transplant 2010;19:3–8.

[80] Szot GL, Lee MR, Tavakol MM, Lang J, Dekovic F, Kerlan RK, Stock PG, Posselt AM. Successful clinical islet isolation using a GMP-manufactured collagenase and neutral protease. Transplantation 2009;88:753–6.

[81] Anazawa T, Balamurugan AN, Bellin M, Zhang HJ, Matsumoto S, Yonekawa Y, Tanaka T, Loganathan G, Papas KK, Beilman GJ, Hering BJ, Sutherland DE. Human islet isolation for autologous transplantation: comparison of yield and function using SERVA/Nordmark versus Roche enzymes. Am J Transplant Off J Am Soc Transplant Am Soc Transplant Surg 2009;9:2383–91.

[82] Tzakis AG, Ricordi C, Alejandro R, Zeng Y, Fung JJ, Todo S, Demetris AJ, Mintz DH, Starzl TE. Pancreatic islet transplantation after upper abdominal exenteration and liver replacement. Lancet 1990;336:402–5.

[83] Kin T, O'Gorman D, Zhai X, Pawlick R, Imes S, Senior P, Shapiro AM. Nonsimultaneous administration of pancreas dissociation enzymes during islet isolation. Transplantation 2009;87:1700–5.

[84] Shapiro AM, Ricordi C, Hering BJ, Auchincloss H, Lindblad R, Robertson RP, Secchi A, Brendel MD, Berney T, Brennan DC, Cagliero E, Alejandro R, Ryan EA, DiMercurio B, Morel P, Polonsky KS, Reems JA, Bretzel RG, Bertuzzi F, Froud T, Kandaswamy R, Sutherland DE, Eisenbarth G, Segal M, Preiksaitis J, Korbutt GS, Barton FB, Viviano L, Seyfert-Margolis V, Bluestone J, Lakey JR. International trial of the Edmonton protocol for islet transplantation. N Engl J Med 2006;355:1318–30.

[85] Balamurugan AN, Green ML, Breite AG, Loganathan G, Wilhelm JJ, Tweed B, Vargova L, Lockridge A, Kuriti M, Hughes MG, Williams SK, Hering BJ, Dwulet FE, McCarthy RC. Identifying effective enzyme activity targets for recombinant class I and class II collagenase for successful human islet isolation. Transplant Direct 2016;2, e54.

[86] Caballero-Corbalan J, Friberg AS, Brandhorst H, Nilsson B, Andersson HH, Felldin M, Foss A, Salmela K, Tibell A, Tufveson G, Korsgren O, Brandhorst D. Vitacyte collagenase HA: a novel enzyme blend for efficient human islet isolation. Transplantation 2009;88:1400–2.

[87] Qi M, Valiente L, McFadden B, Omori K, Bilbao S, Juan J, Rawson J, Scott S, Ferreri K, Mullen Y, El-Shahawy M, Dafoe D, Kandeel F, Al-Abdullah IH. The choice of enzyme for human pancreas digestion is a critical factor for increasing the success of islet isolation. Transplant Direct 2015;1.

[88] Loganathan G, Subhashree V, Breite AG, Tucker WW, Narayanan S, Dhanasekaran M, Mokshagundam S, Green ML, Hughes MG, Williams SK, Dwulet FE, McCarthy RC, Balamurugan AN. Beneficial effect of recombinant rC1rC2 collagenases on human islet function: efficacy of low-dose enzymes on pancreas digestion and yield. Am J Transplant Off J Am Soc Transplant Am Soc Transplant Surg 2018;18:478–85.

[89] Ricordi C, Alejandro R, Rilo HH, Carroll PB, Tzakis AG, Starzl TE, Mintz DH. Long-term in vivo function of human mantled islets obtained by incomplete pancreatic dissociation and purification. Transplant Proc 1995;27:3382.

[90] Ricordi C, Alejandro R, Zeng Y, Tzakis A, Casavilla A, Jaffe R, Mintz DH, Starzl TE. Human islet isolation and purification from pediatric-age donors. Transplant Proc 1991;23:783–4.

[91] Meier RP, Sert I, Morel P, Muller YD, Borot S, Badet L, Toso C, Bosco D, Berney T. Islet of Langerhans isolation from pediatric and juvenile donor pancreases. Transpl Int 2014;27:949–55.

[92] Loganathan G, Hughes MG, Szot GL, Smith KE, Hussain A, Collins DR, Green ML, Dwulet FE, Williams SK, Papas KK, McCarthy RC, Balamurugan AN. Low cost, enriched collagenase-purified protease enzyme mixtures successfully used for human islet isolation. OBM Transplant 2019;3:14.

[93] Noguchi H. Pancreatic islet purification from large mammals and humans using a COBE 2991 cell processor versus large plastic bottles. J Clin Med 2020;10:10.

[94] Wilhelm JJ, Bellin MD, Dunn TB, Balamurugan AN, Pruett TL, Radosevich DM, Chinnakotla S, Schwarzenberg SJ, Freeman ML, Hering BJ, Sutherland DE, Beilman GJ. Proposed thresholds for pancreatic tissue volume for safe intraportal islet autotransplantation after total pancreatectomy. Am J Transplant 2013;13:3183–91.

[95] Williams BM, Baldwin X, Vonderau JS, Hyslop WB, Desai CS. Portal flow dynamics after total pancreatectomy and autologous islet cell transplantation. Clin Transpl 2020;34, e14112.

[96] Eckhard M, Brandhorst D, Brandhorst H, Brendel MD, Bretzel RG. Optimization in osmolality and range of density of a continuous ficoll-sodium-diatrizoate gradient for isopycnic purification of isolated human islets. Transplant Proc 2004;36:2849–54.

[97] Noguchi H, Naziruddin B, Shimoda M, Fujita Y, Chujo D, Takita M, Peng H, Sugimoto K, Itoh T, Kobayashi N, Onaca N, Levy MF, Matsumoto S. Evaluation of osmolality of density gradient for human islet purification. Cell Transplant 2012;21:493–500.

[98] Barbaro B, Salehi P, Wang Y, Qi M, Gangemi A, Kuechle J, Hansen MA, Romagnoli T, Avila J, Benedetti E, Mage R, Oberholzer J. Improved human pancreatic islet purification with the refined UIC-UB density gradient. Transplantation 2007;84:1200–3.

[99] Anazawa T, Matsumoto S, Yonekawa Y, Loganathan G, Wilhelm JJ, Soltani SM, Papas KK, Sutherland DE, Hering BJ, Balamurugan AN. Prediction of pancreatic tissue densities by an analytical test gradient system before purification maximizes human islet recovery for islet autotransplantation/allotransplantation. Transplantation 2011;91:508–14.

[100] van Suylichem PT, Wolters GH, van Schilfgaarde R. The efficacy of density gradients for islet purification: a comparison of seven density gradients. Transpl Int 1990;3:156–61.
[101] Robertson GS, Chadwick DR, Contractor H, James RF, Bell PR, London NJ. The use of continuous density gradients for the assessment of islet and exocrine tissue densities and islet purification. Acta Diabetol 1993;30:175–80.
[102] Lake SP, Bassett PD, Larkins A, Revell J, Walczak K, Chamberlain J, Rumford GM, London NJ, Veitch PS, Bell PR, et al. Large-scale purification of human islets utilizing discontinuous albumin gradient on IBM 2991 cell separator. Diabetes 1989;38(Suppl 1):143–5.
[103] Huang GC, Zhao M, Jones P, Persaud S, Ramracheya R, Lobner K, Christie MR, Banga JP, Peakman M, Sirinivsan P, Rela M, Heaton N, Amiel S. The development of new density gradient media for purifying human islets and islet-quality assessments. Transplantation 2004;77:143–5.
[104] Mita A, Ricordi C, Messinger S, Miki A, Misawa R, Barker S, Molano RD, Haertter R, Khan A, Miyagawa S, Pileggi A, Inverardi L, Alejandro R, Hering BJ, Ichii H. Antiproinflammatory effects of iodixanol (OptiPrep)-based density gradient purification on human islet preparations. Cell Transplant 2010;19:1537–46.
[105] Noguchi H, Ikemoto T, Naziruddin B, Jackson A, Shimoda M, Fujita Y, Chujo D, Takita M, Kobayashi N, Onaca N, Levy MF, Matsumoto S. Iodixanol-controlled density gradient during islet purification improves recovery rate in human islet isolation. Transplantation 2009;87:1629–35.
[106] Shimoda M, Itoh T, Iwahashi S, Takita M, Sugimoto K, Kanak MA, Chujo D, Naziruddin B, Levy MF, Grayburn PA, Matsumoto S. An effective purification method using large bottles for human pancreatic islet isolation. Islets 2012;4:398–404.
[107] Miki A, Ricordi C, Messinger S, Yamamoto T, Mita A, Barker S, Haetter R, Khan A, Alejandro R, Ichii H. Toward improving human islet isolation from younger donors: rescue purification is efficient for trapped islets. Cell Transplant 2009;18:13–22.
[108] Shimoda M, Itoh T, Sugimoto K, Takita M, Chujo D, Iwahashi S, SoRelle JA, Naziruddin B, Levy MF, Grayburn PA, Matsumoto S. An effective method to release human islets from surrounding acinar cells with agitation in high osmolality solution. Transplant Proc 2011;43:3161–6.
[109] Ichii H, Pileggi A, Molano RD, Baidal DA, Khan A, Kuroda Y, Inverardi L, Goss JA, Alejandro R, Ricordi C. Rescue purification maximizes the use of human islet preparations for transplantation. Am J Transplant Off J Am Soc Transplant Am Soc Transplant Surg 2005;5:21–30.
[110] Golebiewska JE, Golab K, Gorycki T, Sledzinski M, Gulczynski J, Zygowska I, Wolnik B, Hoffmann M, Witkowski P, Ricordi C, Szurowska E, Sledzinski Z, Debska-Slizien A. "old school" islet purification based on the unit gravity sedimentation as a rescue technique for intraportal islet transplantation-A case report. Cell Transplant 2020;29, 963689720947098.
[111] Banerjee M, Otonkoski T. A simple two-step protocol for the purification of human pancreatic beta cells. Diabetologia 2009;52:621–5.
[112] Davies JE, James RF, London NJ, Robertson GS. Optimization of the magnetic field used for immunomagnetic islet purification. Transplantation 1995;59:767–71.
[113] Winoto-Morbach S, Leyhausen G, Schunke M, Ulrichs K, Muller-Ruchholtz W. Magnetic microspheres (MMS) coupled to selective lectins: a new tool for large-scale extraction and purification of human pancreatic islets. Transplant Proc 1989;21:2628–30.
[114] Miyagi-Shiohira C, Kobayashi N, Saitoh I, Watanabe M, Noguchi Y, Matsushita M, Noguchi H. The evaluation of islet purification methods that use large bottles to create a continuous density gradient. Cell Med 2017;9:45–51.
[115] Noguchi H, Miyagi-Shiohira C, Kurima K, Kobayashi N, Saitoh I, Watanabe M, Noguchi Y, Matsushita M. Islet culture/preservation before islet transplantation. Cell Med 2015;8:25–9.
[116] Shapiro AM, Lakey JR, Ryan EA, Korbutt GS, Toth E, Warnock GL, Kneteman NM, Rajotte RV. Islet transplantation in seven patients with type 1 diabetes mellitus using a glucocorticoid-free immunosuppressive regimen. N Engl J Med 2000;343:230–8.
[117] Froud T, Ricordi C, Baidal DA, Hafiz MM, Ponte G, Cure P, Pileggi A, Poggioli R, Ichii H, Khan A, Ferreira JV, Pugliese A, Esquenazi VV, Kenyon NS, Alejandro R. Islet transplantation in type 1 diabetes mellitus using cultured islets and steroid-free immunosuppression: Miami experience. Am J Transplant Off J Am Soc Transplant Am Soc Transplant Surg 2005;5:2037–46.
[118] Kin T, Senior P, O'Gorman D, Richer B, Salam A, Shapiro AM. Risk factors for islet loss during culture prior to transplantation. Transpl Int 2008;21:1029–35.
[119] Ryan EA, Paty BW, Senior PA, Bigam D, Alfadhli E, Kneteman NM, Lakey JR, Shapiro AM. Five-year follow-up after clinical islet transplantation. Diabetes 2005;54:2060–9.
[120] Yamamoto T, Horiguchi A, Ito M, Nagata H, Ichii H, Ricordi C, Miyakawa S. Quality control for clinical islet transplantation: organ procurement and preservation, the islet processing facility, isolation, and potency tests. J Hepato-Biliary-Pancreat Surg 2009;16:131–6.
[121] Olack BJ, Alexander M, Swanson CJ, Kilburn J, Corrales N, Flores A, Heng J, Arulmoli J, Omori K, Chlebeck PJ, Zitur L, Salgado M, Lakey JRT, Niland JC. Optimal time to ship human islets post tissue culture to maximize islet. Cell Transplant 2020;29, 963689720974582.
[122] Keymeulen B, Gillard P, Mathieu C, Movahedi B, Maleux G, Delvaux G, Ysebaert D, Roep B, Vandemeulebroucke E, Marichal M, In't Veld P, Bogdani M, Hendrieckx C, Gorus F, Ling Z, van Rood J, Pipeleers D. Correlation between beta cell mass and glycemic control in type 1 diabetic recipients of islet cell graft. Proc Natl Acad Sci U S A 2006;103:17444–9.
[123] Noguchi H, Yamada Y, Okitsu T, Iwanaga Y, Nagata H, Kobayashi N, Hayashi S, Matsumoto S. Secretory unit of islet in transplantation (SUIT) and engrafted islet rate (EIR) indexes are useful for evaluating single islet transplantation. Cell Transplant 2008;17:121–8.
[124] Noguchi H, Nakai Y, Ueda M, Masui Y, Futaki S, Kobayashi N, Hayashi S, Matsumoto S. Activation of c-Jun NH2-terminal kinase (JNK) pathway during islet transplantation and prevention of islet graft loss by intraportal injection of JNK inhibitor. Diabetologia 2007;50:612–9.
[125] Paraskevas S, Aikin R, Maysinger D, Lakey JR, Cavanagh TJ, Agapitos D, Wang R, Rosenberg L. Modulation of JNK and p38 stress activated protein kinases in isolated islets of Langerhans: insulin as an autocrine survival signal. Ann Surg 2001;233:124–33.
[126] Ricordi C, Goldstein JS, Balamurugan AN, Szot GL, Kin T, Liu C, Czarniecki CW, Barbaro B, Bridges ND, Cano J, Clarke WR, Eggerman TL, Hunsicker LG, Kaufman DB, Khan A, Lafontant DE, Linetsky E, Luo X, Markmann JF, Naji A, Korsgren O, Oberholzer J, Turgeon NA, Brandhorst D, Chen X, Friberg AS, Lei J, Wang LJ, Wilhelm JJ, Willits J, Zhang X, Hering BJ, Posselt AM, Stock PG, Shapiro AM, Chen X. National Institutes of health-sponsored clinical islet transplantation consortium phase 3 trial: manufacture of a complex cellular product at eight processing facilities. Diabetes 2016;65:3418–28.

[127] Latif ZA, Noel J, Alejandro R. A simple method of staining fresh and cultured islets. Transplantation 1988;45:827–30.

[128] NIH CIT Consortium Chemistry Manufacturing Controls Monitoring Committee, NIH CIT Consortium. Purified human pancreatic islet—viability estimation of islet using fluorescent dyes (FDA/PI): standard operating procedure of the NIH clinical islet transplantation consortium. CellR4 Repair Replace Regen Reprogram 2015;3.

[129] Berney T, Molano RD, Cattan P, Pileggi A, Vizzardelli C, Oliver R, Ricordi C, Inverardi L. Endotoxin-mediated delayed islet graft function is associated with increased intra-islet cytokine production and islet cell apoptosis. Transplantation 2001;71:125–32.

[130] Vargas F, Vives-Pi M, Somoza N, Armengol P, Alcalde L, Marti M, Costa M, Serradell L, Dominguez O, Fernandez-Llamazares J, Julian JF, Sanmarti A, Pujol-Borrell R. Endotoxin contamination may be responsible for the unexplained failure of human pancreatic islet transplantation. Transplantation 1998;65:722–7.

[131] Golebiewska JE, Bachul PJ, Wang LJ, Matosz S, Basto L, Kijek MR, Fillman N, Golab K, Tibudan M, Debska-Slizien A, Millis JM, Fung J, Witkowski P. Validation of a new north American islet donor score for donor pancreas selection and successful islet isolation in a medium-volume islet transplant center. Cell Transplant 2019;28:185–94.

[132] Wang LJ, Cochet O, Wang XJ, Krzystyniak A, Misawa R, Golab K, Tibudan M, Grose R, Savari O, Millis JM, Witkowski P. Donor height in combination with islet donor score improves pancreas donor selection for pancreatic islet isolation and transplantation. Transplant Proc 2014;46:1972–4.

[133] Wassmer CH, Perrier Q, Combescure C, Pernin N, Parnaud G, Cottet-Dumoulin D, Brioudes E, Bellofatto K, Lebreton F, Berishvili E, Lablanche S, Kessler L, Wojtusciszyn A, Buron F, Borot S, Bosco D, Berney T, Lavallard V. Impact of ischemia time on islet isolation success and posttransplantation outcomes: a retrospective study of 452 pancreas isolations. Am J Transplant 2021;21:1493–502.

[134] Papas KK, De Leon H, Suszynski TM, Johnson RC. Oxygenation strategies for encapsulated islet and beta cell transplants. Adv Drug Deliv Rev 2019;139:139–56.

[135] Jawahar AP, Narayanan S, Loganathan G, Pradeep J, Vitale GC, Jones CM, Hughes MG, Williams SK, Balamurugan AN. Ductal cell reprogramming to insulin-producing Beta-like cells as a potential Beta cell replacement source for chronic pancreatitis. Curr Stem Cell Res Ther 2019;14:65–74.

[136] Narayanan S, Loganathan G, Mokshagundam S, Hughes MG, Williams SK, Balamurugan AN. Endothelial cell regulation through epigenetic mechanisms: depicting parallels and its clinical application within an intra-islet microenvironment. Diabetes Res Clin Pract 2018;143:120–33.

[137] Chinnuswami R, Hussain A, Loganathan G, Narayanan S, Porter GD, Balamurugan AN. Porcine islet cell xenotransplantation. Xenotransplantation-Comprehensive Study; 2020.

[138] Narayanan S, Loganathan G, Hussain A, Williams SK, Balamurugan AN. Ductal cell reprograming to insulin-producing cells as a potential beta cell replacement source for islet auto-transplant recipients. In: Transplantation, bioengineering, and regeneration of the endocrine pancreas. Elsevier; 2020. p. 397–405.

[139] Shapiro AM, Pokrywczynska M, Ricordi C. Clinical pancreatic islet transplantation. Nat Rev Endocrinol 2017;13:268–77.

[140] Guruswamy Damodaran R, Vermette P. Decellularized pancreas as a native extracellular matrix scaffold for pancreatic islet seeding and culture. J Tissue Eng Regen Med 2018;12:1230–7.

Chapter 6

Auto islet isolation: Methods in removal and isolation from fibrosed and autolyzed pancreata

David Whaley[a], Kimia Damyar[a], Alicia Wells[a], Adam Good[a], Colleen Luong[a], Ivana Xu[a], Michael Alexander[a], Horacio Rilo[c], David Imagawa[a], and Jonathan R.T. Lakey[a,b]

[a]*Department of Surgery, University of California Irvine, Irvine, CA, United States,* [b]*Department of Biomedical Engineering, University of California Irvine, Irvine, CA, United States,* [c]*Department of Surgery, Zucker School of Medicine at Hofstra/Northwell, Hempstead, NY, United States*

Introduction

Autologous islet transplant

In contrast to allo- and xenotransplant procedures, an individual undergoing autologous islet transplant serves as both the donor and recipient of their own tissue. This method of grafting makes several guarantees, including donor-host tissue compatibility, organ availability, a reduction in disease transmission risk, and may also include the presence of fibrotic tissue and calcified pancreatic ductal tissue [1].

An individual is likely to undergo autologous islet transplant as a result of chronic pancreatitis and its intractable pain. In this treatment, the patient's otherwise fully functional pancreas is surgically removed, processed ex vivo enzymatically and mechanically, and the resulting purified islets are infused back into the same patient, most commonly via the hepatic portal vein (HPV) to restore pancreatic function [2]. While tissue compatibility, organ availability, and reduced disease transmission are attractive features of auto islet transplant for pancreatitis, the fibrotic tissue and duct obstructions inherent to pancreatitis introduce challenges to the processing stage of autotransplant. In conjunction with the utilization of an appropriate surgical approach, observation of proper facility and team management, and additional adjunctive measures during processing, autologous islet transplant can improve the survival rate and quality of life for those suffering from pancreatic complications [3].

Pancreatitis: Characterized by fibrotic tissue and duct obstruction

Pancreatitis is associated with both increasing rates of hospitalizations and rising ambulatory care visits, becoming the seventh most diagnosed digestive disease within the United States [4]. Despite being less common than the acute form of pancreatitis—with a prevalence of about 50 chronic cases per 100,000 pancreatitis cases annually—chronic pancreatitis presents a formidable challenge to patients, physicians, and the healthcare system due to its significant morbidity and mortality [5]. The most common etiology of chronic pancreatitis in the United States is due to excessive alcohol consumption but can also result from biliary duct obstruction by stones, tumors or pseudocysts, pancreas divisum, hypertriglyceridemia, and autoimmune or familial pancreatitis [6]. The clinical manifestations of chronic pancreatitis are extensive, with patients experiencing variable exocrine and endocrine insufficiency and pain that drastically reduces quality of life and is often debilitating [7,8]. Although the pathophysiology of the pain is unknown, one hypothesis suggests peri-pancreatic scarring induces an increase in ductal and interstitial pressure that the patient perceives as pain through a complex interaction with the central nervous system [9].

A major component of the pathophysiology of chronic pancreatitis involves the development of fibrosis or scar tissue [10]. Fibrosis is characterized by accumulation of extracellular matrix molecules and results in response to repetitive injury and inflammation, leading to disordered tissue architecture and organ failure. Myofibroblasts are generally thought to be the

Pancreas and Beta Cell Replacement. https://doi.org/10.1016/B978-0-12-824011-3.00012-6

key cell regulating tissue fibrosis and are believed to generate fibrotic tissue through the deposition of matrix and exertion of tensile force. Long-term inflammation and the development of fibrosis can cause irreversible damage to islet and acinar cells, altering normal endocrine and exocrine tasks [11]. Leading to severe complications such as improper digestion of nutrients, diabetes, and organ failure, fibrosis is a serious complication of chronic pancreatitis that poses many threats to the individual's health and well-being. The irreversible damage to islets and the tissue framework of the organ also presents a challenge for autotransplantation as the success rate is primarily based on the number of islets that remain following total pancreatectomy and laboratory processing.

Pancreatectomy

At present, there are a variety of pharmacologic, endoscopic, and surgical therapeutic options available to alleviate the pain accompanying chronic pancreatitis. Pharmacologic intervention is usually the first-line method of pain relief for pancreatitis. Commonly-prescribed drugs include: nonsteroidal antiinflammatory drugs (NSAIDs), acetaminophen, dextropropoxyphene, and prednisolone [7]. Due to the chronic and intense nature of the pain associated with untreated pancreatitis, patients often require more potent opioid analgesics for pain management.

To circumvent the use of strong narcotics and opioids, surgical interventions may be in the patient's best interest. Minor surgical procedures include pancreatic decompressions for large duct obstruction and partial pancreatic resections for localized pancreatic inflammation [12–15]. With a mortality rate of less than 2%, these operations are safe and have achieved sustained pain relief in 80%–90% of patients [16,17]. In some instances, however, a total pancreatectomy is the only viable treatment option available [18–20]. A total pancreatectomy is generally considered for individuals who have diffuse or focal disease that cannot be decompressed and for those who have exhausted more conservative surgical and medical treatments and continue to have intractable pain.

The total pancreatectomy procedure was first reported in the 1960s [21]; however, it was associated with significant morbidity and mortality [22]. Patients receiving a total pancreatectomy without islet transplant can suffer from significant long-term metabolic effects and are open to the complications associated with surgically induced diabetes resulting from total pancreatic endocrine and exocrine deficiency [23–25]. A study by Gall reported a 19.1% mortality at 6.5 years—with many late deaths caused by hypoglycemia—for 117 patients that underwent a total pancreatectomy [26].

Total pancreatectomy with islet autotransplantation (TP-IAT)

To reduce likelihood of complications associated with receiving a total pancreatectomy and preserve insulin independence, total pancreatectomy has been coupled with islet autotransplantation (TP-IAT) to preserve the insulin-producing beta cell mass while still reducing the pain associated with chronic pancreatitis [27]. TP-IAT has been shown to be safe overall while achieving a low rate of morbidity and mortality and achieving insulin independence in a good number of cases [28]. To achieve these results, a great deal of research was undertaken at both the initial stages of development and in recent years to improve the clinical outcomes for patients receiving this treatment.

In the 1960s, the initial success of pancreas transplantation was poor, with only one in 14 patients of the first recipients surviving 1-year posttransplant [29]. Around this same time, however, Lacy demonstrated the efficacy of intraductal enzymatic digestion for isolating pancreatic islet cells that could then be used for transplantation [30,31]. Performing the first reported islet cell implant into diabetic rats in 1972, he demonstrated that islet cells could be isolated and transplanted into an insulin-dependent organism to restore normoglycemia. In the 1970s, Najarian demonstrated that postsurgical diabetes resulting from a total pancreatectomy could be attenuated by isolating pancreatic islet cells from the pancreas of the patient and infusing them back through the portal vein [27]. The major steps of Najarian's procedure are depicted in Fig. 1 [32].

Following the initial success of TP-IAT in the 1970s, the next two decades were fraught with difficulty as investigators faced challenges involving surgical complications and suboptimal isolation and purification methods [33]. Patients undergoing TP-IAT during this early era sometimes received poorly purified islet infusions that led to life-threatening complications such as acute portal hypertension and disseminated intravascular coagulation (DIC) with subsequent diffuse hemorrhage [34]. Within a decade, however, the continual refinement of isolation and purification protocols culminated in 1988 with the development of the now-ubiquitous Ricordi method which automated the isolation of human pancreatic islets and significantly increased the islet cell yield that could be obtained [35].

Just over a decade later in 2000, the development of the allotransplant-based Edmonton protocol by James Shapiro, Edmond Ryan, Jonathan Lakey, and several others further galvanized interest in islet transplantation. Utilizing a novel immunosuppressive regimen of low-dose tacrolimus and daclizumab, sirolimus, and avoiding diabetogenic corticosteroids,

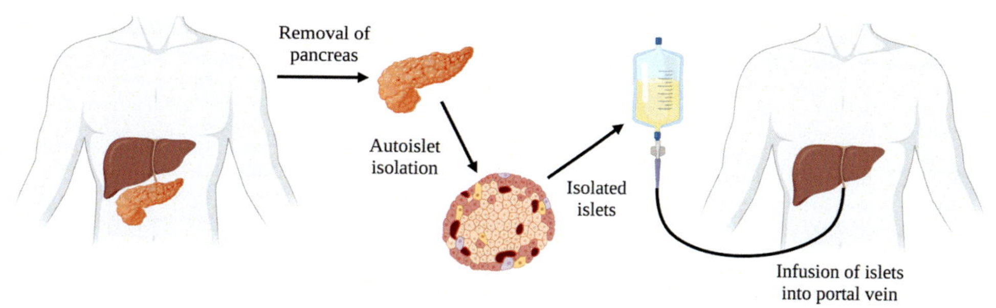

FIG. 1 Diagram of the major steps of an autologous islet transplant. *(Modified from Tucker T., Labadie K., Kopan C., Alexander M., Lakey J.R.T. Strategies in preventing diabetes after pancreatectomy using islet auto- and allo-transplantation. Int J Transplant Res Med 2017;3(1):1–8. https://doi.org/10.23937/2572-4045.1510029.)*

an initial trial for seven type 1 diabetics using the Edmonton protocol demonstrated complete insulin independence at 1 year [36]. In an international follow-up trial of the Edmonton Protocol with 36 patients at nine international sites, they demonstrated that the protocol could successfully restore glycemic stability in type 1 diabetics [37]. Two years postoperation, however, only 31% of patients remained insulin independent. Another important outcome of the study revealed that high volume centers had improved outcomes as compared to low volume centers with only 22% of cases at low volume centers achieving insulin independence at 1 year as compared to 67% of patients at high volume centers. This finding led to greater collaboration among institutions and inspired the National Institute of Health to sponsor a Clinical Islet Transplantation consortium aimed at developing a standard procedure that could be followed across different isolation centers [38].

In addition to investigating the relationship between the volume of centers and their patient outcomes, several studies have examined the factors that predict positive outcomes for TP-IAT [39–42]. While variables commonly measured include graft survival and function, narcotic use, persistent pain, and changes to quality of life, the study found that duration of narcotic use, islet yield, transplant tissue volume, and isolation and purification techniques were the most influential toward the final clinical outcome [41]. The study also identified the strongest independent risk factor for islet graft failure postoperatively as low islet isolation yield. Due to an increase in awareness and the ever-increasing viability of the procedure, the number of TP-IAT cases has steadily increased over the last few decades and is projected to increase further as greater developments and refinements are made over the coming years [43].

Pancreatitis: Importance of prompt surgical treatment

Patients in a study on TP-IAT from 2009 through 2013 at the University of Arizona Medical Center also had some success in attaining insulin independence in patients receiving TP-IAT [24]. Out of a total of 61 patients with chronic pancreatitis who underwent TP-IAT, 19% became insulin independent, 27% required <10 units of insulin, 23% 11–25 units, and 31% >25 units. Seventy-one percent of total patients no longer experienced pain 12 months after surgery and did not require further use of analgesics. Gruessner et al. argued that TP-IAT should be considered early on in treating chronic pancreatitis, and that the low postsurgery rate of insulin independence found in the patients was because the vast majority were in an advanced disease state and no longer had viable islets.

As disease duration increases, maximum recoverable islet yield decreases along with the likelihood of obtaining insulin independence following transplantation [44]. Sutherland demonstrated that islet yields <2500 IEQ/kg after a total pancreatectomy led to a proportional decrease in insulin independence. Further reinforcing dependence of autologous islet transplant on islet yield, Gruessner demonstrated initial islet yields >2500 IEQ/kg were more likely to confer insulin independence post-TP-IAT [24]. Such examples highlight the necessity of prompt surgical treatment to maximize the number of viable islets available for subsequent autotransplantation.

Pancreatitis and inherent challenges to successful autotransplantation

Due to the pathophysiological changes that result from chronic pancreatitis, complications may arise in purifying and obtaining islet cells through TP-IAT. The presence of fibrotic tissue in the pancreatic parenchyma can make the isolation of islets difficult due to the increased presence of extracellular matrix (ECM) [45,46]. In order to adequately digest the increased ECM content, fibrotic tissue requires higher doses and longer exposure to the digestive enzymes which can result in decreased islet viable yields and ultimately reduced autotransplant success. In addition, prior surgical manipulation of the pancreatic duct or the presence of extensive intraductal calcification can narrow pancreatic ducts and hinder the delivery of the enzymatic solution to the entirety of the pancreas. Without effective delivery of the enzymatic solution, maximum viable islet yields are reduced.

Total pancreatectomy with islet autotransplantation (TP-IAT): Outcomes

Ultimately, the goal of total pancreatectomy (TP) with islet autotransplantation (IAT) is to restore an individual's quality of life by alleviating pain and treating underlying cause of the acute symptoms of pancreatitis. In addition, the procedure must aim to counter chronic diabetic complications after transplant and restore glycemic control [47]. According to the Clinical Islet Transplant Registry (CITR), there have been more than 827 autologous islet transplantations with 819 total recipients since 1999 [43]. In that time, TP-IAT outcomes have shown a strong dependence on islet yield (IEQ/kg) on treatment outcome [20,48,49]. In fact, one study of 581 individuals that underwent TP-IAT revealed lower islet yield to be the most significant contributor to graft failure. A strong dose-response relationship was identified between the lowest yield category (<2000 IEQ/kg) and the highest yield category (≥5000 IEQ/kg) in the study cohort. In addition, the lowest yield category was 25 times more likely to experience graft failure [39]. Methods to ensure the best possible outcomes for autologous transplant recipients are further discussed in this chapter and include surgical approaches, facility and team management, as well as adjunctive procedures to deal with the fibrotic and calcified tissues associated with pancreatitis.

Clinical islet transplant facility requirements reduce disease transmission, increase safety, standardize procedures, and maximize patient outcomes

Introduction

Islet transplantation to treat chronic pancreatitis and type 1 diabetes has been performed by a select group of institutions around the world for nearly four decades. Over time, regulatory requirements governing tissue transplantation have evolved into a highly regulated and standardized set of protocols. These "Tissue Rules" provide guidance on human cell, tissue, and tissue-based product (HCT/P) manufacturing, products of which include autologous islet cells. Overall, regulations help prevent the transmission of infectious diseases and contamination of human products that could result during improper recovery, handling and processing of cells and tissues. The rules also standardize operations within and across institutions and ultimately increase safety as well as procedural outcomes.

Regulation of autologous islets

The Tissue Rules, which govern autologous islets in the United States, are listed with the Food and Drug Administration (FDA) under Section 21 of the Federal Code of Regulations (CFR); Section 1271 defines human cells, tissues, and cellular and tissue-based products (HCT/P), including islets as "*articles containing or consisting of human cells or tissues that are intended for implantation, transplantation, infusion, or transfer to a human recipient*" [50]. These rules provide a detailed explanation of what constitutes safe manufacturing and should be adhered to at every step when handling auto islets. Autologous islets are regulated solely under the Tissue Rules (Section 361 of the FDA's Public Health Service Act) since the cells are recovered from and reimplanted into the same individual during the *same surgical procedure*. Briefly, the Tissue Rules stipulate that each clinical auto islet lab has policies that address Current Good Tissue Practice (cGTP), establishment and maintenance of a quality program, personnel, procedures, facilities, environmental controls, equipment, supplies, recovery, processing and processing controls, labeling, storage, receipt, predistribution shipment, and distribution of an HCT/P, records, tracking, and complaint files.

Requirements and regulations to prepare islets for transplant outside of the United States have comparable regulatory bodies. In Canada, the regulatory framework for islet cell transplant is under the purview of the Health Products and Food Branch [51]. In Europe, islets are regulated by the EMEA (EU Regulatory Body) [52]. In Japan, the Japan Ministry of Health, Labor and Welfare, under the Pharmaceuticals and Medical Device Agency [53], govern clinical islet transplantation.

It is crucial for each clinical islet lab to understand the federal, state, and local regulations to which they must report and to work closely with their healthcare institutions' leadership and regulatory bodies to ensure buy-in and agreement on the standards to which the clinical auto islet lab must adhere. Ultimately, compliance with these regulations is necessary to prevent the contamination of the clinical islet prep during the manufacturing process, thereby maximizing tissue yields and transplant outcomes.

Islet quality assessment

To assess the quality of autologous islets throughout the transplant procedure, several assays are commonly used. Viability is assessed using fluorescein diacetate (FDA), calcein AM, or SYTO Green stain in combination with a propidium iodide (PI), ethidium bromide (EtBr) counter stain. Commonly, a FDA/PI, Calcein AM/PI, or SYTO Green/EtBr combination is used for this purpose [54–56]. Ideally, islet viability should be >80% to proceed with islet processing and autologous transplant, although fibrotic tissue sometimes precludes this threshold from being met. As such, islet preparations with viabilities below this value are often deemed acceptable for autotransplant. In addition to viability measurements, cell counting is also a required component of quality assessment. Due to high concentrations of zinc within pancreatic islets relative to other tissues, the zinc chelating agent dithizone (DTZ) is commonly utilized for islet counting during processing [57]. A minimum islet count of 2000 IEQ/kg must be obtained for an auto islet transplant to confer any degree of insulin independence. However, greater IEQ/kg yields (>5000 IEQ/kg) are correlated with increased duration of insulin independence posttransplant [20,58]. Finally, potential bacterial contamination is assessed via Gram staining and endotoxin detection just prior to autotransplant. A table of testing parameters is summarized in Table 1.

The vulnerability of islet cells to contamination

Islets are particularly vulnerable to contamination for several reasons. First, excised pancreata typically come from hospital operating rooms where aseptic procedures are not under the same strict control as a clean room due to technical and other environmental limits. Second, auto islet transplant patients often endure multiple pancreatic interventions involving manipulation of the main pancreatic duct before the surgery [59]. Those interventions can transfer bacterial flora from the intestinal tract into the drainage system of the pancreas and represent a common source of microbial contamination of excised pancreas transport solutions. It is therefore essential to implement a robust microbiology surveillance program that monitors samples of pancreatic transport solutions taken from critical time points during transport to and from the operating room [59]. Finally, some of the larger laboratory equipment—incubators, centrifuges, refrigerators, and microscopes—used in islet cell manufacturing are not generally designed to undergo automated decontamination cycle, thereby increasing the potential risk of contamination [59].

Use of a clean room to manufacture islet cells

Given these factors, sterility in islet cell processing represents a challenge from a good tissue practice perspective. It is for these reasons that most clinical auto islet programs process pancreas in a clean room facility, making the control of contaminants easier to manage. While not required by the FDA for auto islet transplant programs, the use of a clean room to process tissues helps meet FDA-mandated environmental controls since these facilities are constructed specifically to prevent the possible contamination of human tissues and cells [59].

Specific equipment needs for an auto islet lab

While the use of a clean room is not a requirement by the FDA, there is some basic equipment that a clinical auto islet lab needs to manufacture cells for transplant. To digest and isolate islets, a digestion chamber, COBE cell purification

TABLE 1 Summary of quality assessment assays during autologous islet transplant.

Measurement	Assay	Tolerance	Sample and timing
Viability	Dye Exclusion (FDA/PI)	≥70%	Islets postisolation
Cell Count/Identity	DTZ	5000–20,000 (IEQ/kg) first transplant. 3000–20,000 (IEQ/kg) second transplant.	
Endotoxin	Endotoxin assay	<5 EU/kg	Supernatant of islet suspension in transplant media immediately prior to transplant
Bacterial Type	Gram stain	No organisms detected within limits of assay	

Modified from Papas KK, Suszynski TM., Colton CK. Islet assessment for transplantation. Curr Opin Organ Transplant 2009;14(6):674–82. https://doi.org/10.1097/MOT.0b013e328332a489.

unit, self-contained clinical biohazard vacuum and disposal system, and a dedicated enzyme chiller are required. For islet maintenance and culture, an auto islet program should have at least three Class 2 biological safety cabinets (BSCs). The BSCs also provide an engineered environment for cell preparation, counting, sampling, and microbiological surveillance. Additional equipment include incubators; refrigerated centrifuges; low-temperature incubators; compressed oxygen, nitrogen, and carbon dioxide delivery capabilities; as well as liquid nitrogen delivery systems to support freezers. Ultrapure water generators, water baths, −80°C, −20°C freezers, autoclaves, tabletop centrifuges, and an extensive chemical library facilitate manufacture and storage of reagents and enzymes used in autologous islet processing and transplant.

Procedures

Clinical auto islet labs must develop and maintain standard operating procedures (SOPs) that cover core cGTP requirements. The SOPs fulfill the dual purpose of describing in detail the processes and methods to manufacture islets for transplant as well as the procedures implemented to prevent the risk of introducing, transmitting, or spreading contaminants or diseases to the islet prep. An overview of an SOP's required sections and the contents they should address can be found in Table 2.

Standard operating procedures should be revised as needed and reviewed on a periodic basis, following the guidelines laid forth in the clinical islet lab's Quality Program. SOPs should always be available to personnel in the area where the work is performed and available if an audit was conducted.

Supplies and reagents

Supplies and reagents used in the islet manufacturing process must be purchased from validated vendors and include documentation of quality, sterility, and recordable supply chain records to ensure their ability to conform to appropriate good manufacturing practice.

TABLE 2 Standard operating procedure requirements.

SOP section	Content
Header	The document should have a title, document number, and version. The header should make apparent the activity covered under the SOP.
Revision History	Record the changes made to a procedure and justification of why the procedure was created.
Author(s)	Key roles include the *Author, Reviewer, Approver*, and *Quality Reviewer/Approver*.
Approval	The approval is generally found on the cover page. Each healthcare institution is different, and the SOPs may require more than one approval signatures.
Purpose	The purpose should summarize the intent of the SOP in one or two sentences. It should be detailed enough so users can recognize what the document covers.
Scope	The scope should define to whom or what the particular set of procedures applies.
References	The SOP should include references and related documents that are needed in addition to the SOP to understand and execute the procedure described in the text.
Definitions	The SOP should define terms that may be unfamiliar to users and spell out acronyms and abbreviations used in the document.
Roles and Responsibilities	Define the personnel responsible for performing activities written in the procedure.
Procedure	This is the critically important section of the SOP. The FDA wants to ensure that standards meet compliance expectations. It is also equally important to remember that the FDA and other regulatory agencies will hold clinical islet labs to the standards outlined in the written procedures, evaluating whether the standards and practices described have been maintained.
Appendices/Attachments	These documents are often blank worksheets and flow charts that serve as an aid in explaining the procedures.

Modified from Rilo LR, Cercone R, Lakey JRT. Requirements for clinical islet laboratories. In: Transplantation, bioengineering, and regeneration of the endocrine pancreas, vol. 2. Elsevier; 2019. p. 51–66. https://doi.org/10.1016/B978-0-12-814831-0.00003-8.

Outcome analysis

The clinical islet lab team should review the isolation data of all autologous transplant recipients after each isolation and conduct more extensive reviews on a quarterly basis. Data from these reviews should be presented at the healthcare institution's Quality Assurance meetings on an annual basis at a minimum.

Auto islet transplantation

Autologous transplant: From patient to processing

Basic surgical procedure to remove pancreas

Prior to digestion and processing of autologous islets, surgical removal of the pancreas is necessary. Removal can be performed as a near total or total pancreatectomy. Near total pancreatectomy involves the removal of the pancreas while leaving a small rim along the duodenal C-loop, common bile duct, with pancreaticoduodenal arteries intact. Conversely, total pancreatectomy entails the removal of the entire pancreas, spleen, duodenum, and the distal common bile duct. To reduce warm ischemia time to the islet cells, blood supply to the pancreas is preserved throughout the mobilization and resection process [3].

Total pancreatectomy procedure

Initially, the total pancreatectomy operation begins with the opening of the abdominal cavity by midline or bilateral subcostal incision. Upon necessary adhesion lysis, the duodenum and pancreatic head are mobilized by Kocher maneuver such that there is clear sight of the left renal vein and the superior mesenteric artery [60]. Then, the outer peritoneum of the portal triad is opened, and the gastroduodenal artery is dissected and looped. After the division of short gastric vessels and mobilization of the spleen are performed [60], the distal portion of the pancreas is mobilized at the level of the superior mesenteric vein, preserved, and then processed for islet cell harvest [3]. After the tissue lining the splenic artery and veins are dissected and looped, the duodenum is transected, and the stomach is positioned such that the head and body of the pancreas are visible. The superior mesenteric vein is dissected allowing for the detachment of the pancreatic neck from the portal vein. Succeeding the transection of the bile duct and the dissection of the superior mesenteric vein, small tributaries to the portal vein are divided to remove the pancreatic uncinate process from the portal vein. The procured pancreas is then precannulated in the surgical room in anticipation of the subsequent enzyme perfusion and placed into a preservation solution for transfer [60]. In addition to precannulation, an intravenous insulin drip is set to conserve blood glucose levels, prevent acute metabolic decomposition, and to produce favorable environmental conditions for returning islet cells. Gastrointestinal reconstruction can be achieved by several methods, including but not limited to side-to-side two-layer gastrojejunostomy or end-to-side duodenojejunostomy. In addition, the restoration of bile duct continuity may involve end-to-side hepaticojejunostomy proximal to gastrojejunostomy [3].

Pancreas transfer: UW and HTK solutions

After pancreatectomy, the organ must be transported to a processing location. Whether the final destination is directly adjacent to the surgical suite or at an offsite location, excised pancreata are placed in a preservation medium, commonly either University of Wisconsin (UW) or alternatively Histidine-Tryptophan-Ketoglutarate (HTK) solution. Currently, UW solution is considered the gold standard for organ preservation; however, HTK produces similar results at a reduced cost, and both are suitable for autologous pancreas procedures [61,62].

UW solution is constituted of lactobionate, raffinose (osmotic maintenance), hydroethyl starch (colloid carrier), a phosphate (physiologic buffer), potassium, and oxygen radical scavengers. HTK solution includes histidine (physiologic buffer), tryptophan (cell membrane stabilization), ketoglutarate (anaerobic metabolism substrate), mannitol (osmotic maintenance), and an oxygen free radical scavenger [63]. Both UW and HTK are low in sodium, but HTK is also low in potassium. HTK is a crystalloid fluid in contrast to UW, which is a colloid fluid [64]. The low sodium concentration (15 mM) in HTK solution serves to reduce excessive sodium and water influx during ischemia and is safe for entry into circulation [63]. HTK has a lower viscosity and can therefore be better suited for flushing the organs compared to UW solution [62]. A low flow rate is used with HTK, with a greater total volume of solution used, to reach equilibrium [64].

While UW solution is considered the gold standard for pancreas islet isolations, studies have concluded no statistically significant differences in islet size, size distribution, digestion times, digestion rates, AIC, and IEQ/g between UW or HTK solutions. However, with increased ischemia time, both solutions decrease in efficacy [65]. Additionally, studies have indicated no significant difference in pancreatic graft survival rates at 1 month, 3 months, 6 months, and 1 year when using UW or HTK solutions [61,64]. UW and HTK solutions also yield similar results in conditions of short cold ischemia times (mean ischemic time < 10hours) and low flush volumes (3–4 L) [63].

Pancreas transfer: Temperature

In addition to solution type, an appropriate transfer solution temperature also needs to be selected to maintain the highest possible islet viability postpancreatectomy. Chilled UW/HTK solution (4°C) is optimal as higher temperatures are associated with decreased postprocessing islet viability, yields, and cell diameters [66]. Although solution temperatures as high as 18°C offer some islet protection during transport, chilled solutions enhance their survival further by slowing cellular metabolism and subsequent buildup of harmful metabolic by-products [67].

Digestion of fibrotic pancreata

Autologous tissue processing: Enzymes and modifications of enzymatic parameters for fibrotic pancreas

Recurrent pancreatitis is generally associated with a progressively fibrotic pancreas. Hence, there is greater deposition of extracellular matrix materials resulting in an overabundance of calcified tissue. Because this tissue is difficult to digest enzymatically, parameters can be altered to optimize islet yield [45]. GMP-grade enzyme blends consisting of a collagenase and neutral protease are customarily used to digest fibrotic pancreata for autotransplant. Currently, manufacturers who supply enzymes suitable for this type of work in humans include VitaCyte and SERVA/Roche, although other manufacturers like Nordmark Biochemicals are working to develop GMP-grade enzymes as well. At University of California Irvine, our team conducts autologous pancreas transplant processing with a VitaCyte enzyme blend consisting of GMP-grade Collagenase HA and BP Protease [68]. Collagenase HA itself is composed of 60% w/w Class 1 (C1) and 40% w/w Class 2 (C2) collagenases from *Clostridium histolyticum*, and BP Protease provides a *Bacillus thermoproteolyticus*-derived neutral protease (NP) [69]. Collagenase with an increased C1:C2 ratio is used since it has been shown to increase viable islet yields compared to lower ratio collagenases [69]. Of note, researchers have reported islet yields > 200,000 IEQ in 90% of autologous islet isolations when VitaCyte BP Protease is substituted with SERVA neutral protease [70]. Many factors influence autologous transplant outcomes, and the enzyme blend that suits the specific needs of each islet transplant program should be utilized.

Lyophilized enzymes are reconstituted in an ice cold, Ca^{2+}-containing buffer solution with 10 U/mL heparin for at least 30 minutes prior to their addition to the pancreas perfusion solution [69]. The University of California Irvine autologous islet program also reconstitutes an additional vial of neutral protease at this time to be introduced after initial perfusion depending on the extent of fibrosis. The primary collagenase and neutral protease enzymes are added to approximately 350 mL (up to 450 mL for a large pancreas) of a cold Ca^{2+}-containing perfusion solution which is then dispersed throughout the pancreas manually for 10 minutes, and at 37°C for 4 minutes [69,71]. After manual perfusion, the pancreas is taken immediately to a Ricordi chamber in a sterile container for processing. If, after 10–15 minutes, no free islets are observed in the chamber, the additional previously reconstituted vial of neutral protease is added. This adjunctive step is the result of undocumented observations that an increased NP:C1/C2 ratio increases islet yield from excessively fibrotic pancreata and has been used successfully in several pancreas isolations at the University of California Irvine. Caution should be exercised when increasing neutral protease concentrations since excessively high concentrations are associated with decreased islet yield and viability [72,73].

Modification of traditional mechanical digestion methods

Another approach to improve processing of fibrotic pancreas tissue involves modification of mechanical digestion. Standard digestion of human pancreas uses the digestion chamber, which is manually or automatically shaken at different speeds and uses marbles to "mill" the pancreas in enzyme digestion solution [35,74,75]. The shaking speed and circulating fluid flow speed are adjusted based on the status of the digested tissue and appearance of free islets.

Due to the presence of calcification, however, it may be useful to mince a fibrotic pancreas into smaller pieces prior to placement in the Ricordi chamber. An automated tissue processor, such as Retsch Grindomix (Verder Scientific), reduces the organ to 1–2 mm pieces, thereby increasing the surface area accessible to the enzyme for digestion. This technique has been used to scale up isolation of pig islets with better yields compared to the standard protocol which calls for the pancreas to be chopped with scissors [76]. The tissue processor can alternatively be used as a precursor to noncollagenase-based islet isolation, such as using the selective osmotic shock technique. This technique has a strong requirement for mechanical disruption of the connective tissue, thereby obviating the need for collagenase, which the tissue processor provides [77].

Improving pancreatic duct patency with calculi removal

Two clinical techniques that may influence future islet autotransplantation for pancreatitis are extracorporeal shock wave lithotripsy (ESWL) and endoscopic retrograde cholangiopancreatography (ERCP). Both have demonstrated efficacy in the removal of obstructive pancreatic calculi. ESWL is used to fragment large pancreatic stones (the calculi inside pancreatic ducts) and was first used in 1987 [78]. It has also been used for pancreatic calculi that coexist with pseudocyst [79]. Compared to endoscopic lithotripsy, ESWL is considered safe, effective, and is preferred in both randomized clinical trials [80,81] and in a metaanalysis [82]. ESWL can result in complete clearance and decompression of pancreatic duct in up to 74% and 86% of patients, respectively [80]. Small calculi can be removed using ERCP followed by sphincterotomy and extraction. ERCP and ESWL have been shown to be a safe treatment for acute [83] and chronic pancreatitis [84]. Large calculi require ESWL that is optionally followed by ERCP [85]. A potential treatment regimen for chronic pancreatitis would then be to use ESWL (and optionally ERCP) to remove as much obstruction of the pancreatic ducts by the calculi, followed by total or partial pancreatectomy and islet isolation. This would provide better access to the enzyme solution for distension and digestion, and thus improving the efficacy of the enzymatic activity and may reduce the need for extended digestion time or increased enzyme dose. Adding these adjunct treatments, however, would increase duration of hospitalization postsurgery and add additional costs.

Autologous transplant: From processing to patient

Auto islet transport

Immediate packaging of processed autologous islets for transplant is important to avoid excessive delay and operating time for patients. When packing the islets for transportation, additional care should be taken to maintain sterility. Depending on the tissue volume, the islets can be transferred in multiple transplant bags. For packed tissue volumes greater than 10 mL, the pellet must be divided evenly and transferred into two 200 mL transplant bags. The bags must be labeled with patient identification and FDA-required labeling. Each tissue load is then resuspended in 100 mL transplant media and requires an additional 100 mL media as a rinse solution. In patients with no known allergy to ciprofloxacin, 0.4 mL of Cipro (1% = 10,000 μg/mL) is added per volume of rinse media [86].

An intravenous fluid bag can be used for transportation of islet cell suspension [87,88]. Gas-permeable bags have also been designed with the ability to maintain islet mass, potency, and sterility for islet shipment. They are made from the same material used for the storage of platelet products and delivery under aseptic conditions as well as provide an environment that allows PO_2 levels to remain constant during cell delivery, thereby maintaining islet quality. Thus utilization of gas-permeable bags to contain processed islet cells can be beneficial for auto islet transplantation procedures as well as transportation to remote research institutions [89,90].

A limiting factor to the widespread application of islet autotransplantation (IAT) immediately after partial or complete pancreatic resection is that very few facilities have the technology and expertise required for isolation and processing of islets suitable for human transplantation. To overcome this issue, there have been reports indicating the possibility of distance processing for both allo- and autotransplantation of islets. Rabkin et al. demonstrated the feasibility of processing and preparation of islets in a clinical facility located 1500 miles away. Following total or partial pancreatectomy in five patients with a history of chronic pancreatitis (CP), the resected pancreata were preserved in ViaSpan solution and transferred to a clinical islet processing facility by a commercial airliner and returned [91]. A case study has also reported the feasibility of transcontinental shipping of the pancreas following pancreatectomy. The islets were processed in a laboratory half a continent away from the operative site and returned to the original site 16 hours after pancreas resection. Follow-up examination indicated that the patient became normoglycemic and insulin independent at 10 months postsurgery. Total pancreatectomy (TP) with islet autotransplantation (AIT) can remain a viable treatment for qualified patients even when the islets are processed at a remote clinical islet facility. Moreover, remote processing can offer this therapy to patients without the need to travel [92].

Surgical transplantation of autologous islets

The portal venous system of the liver is the current standard site for clinical islet transplantation. Pancreatic islet cell infusion is commonly performed in two ways: via insertion of a catheter into the portal vein or mesenteric venous tributaries for engraftment into the liver. Infusion of a low-volume endotoxin-free islet cell suspension into the portal vein with intravenous heparin injection is implemented to reduce the rate of complications such as acute portal hypertension and thrombosis. Generally, 35 U/kg of heparin is administered intravenously; in addition, 35 U/kg of heparin combined with islets is given during infusion into the portal vein. Thus a patient receives a total dose of 70 U/kg of heparin [93,94]. Monitoring portal venous pressure during islet infusion is particularly important since changes greater than 25 cm H_2O in the portal pressure can significantly increase the risk of portal vein thrombosis [95]. After transplantation, recipients' arterial and central venous pressure is also measured.

Transplantation site: Risk vs benefit

The transplantation site is a critical aspect of any type of islet transplant. The intrahepatic portal duct has remained the gold standard as it represents a minimally invasive and highly efficacious site for clinical islet transplantation with high blood exposure and the ability to secrete insulin from the graft at near physiological levels directly to the liver [96]. For all the benefits that the liver offers, there are risks and limitations associated with intraportal islet infusion. Portal vein thrombosis is one of the major complications believed to be related to the final volume of infused islets [36,93]. Recently, this concept has been under inspection, and there are debates over the most effective preparation approach regarding the use of purified versus unpurified islets. Unpurified islet preparation provides a large mass of islets for transplantation, which increases the chance of improved metabolic performance, but it also increases the risk of thrombosis [97]. Utilization of purified islets, on the other hand, reduces tissue volume and thus the risk of portal vein thrombosis; however, it also reduces the mass and quality of infused islets, as the gradients used during the purification process are toxic to the cells [98]. New evidence indicates that the release of tissue factors during the transplant procedure may induce a hypercoagulable state. Therefore blockade of tissue factors has been proposed as a new approach to reduce the risk of thrombosis [99]. Other reported complications associated with intraportal transplantation include portal hypertension, hepatic infarction, liver failure, and loss of islet mass due to the instant blood-mediated inflammatory reaction (IBMIR) that can occur when islets encounter the recipient's blood [100,101].

To address the risks and limitations of intraportal islet infusion, alternative transplant sites have been investigated. Stice et al. reported the efficacy of a dual autotransplant site approach in four patients with intraoperative issues that prevented the complete intraportal infusion of autologous islets. In this procedure, after partial intraportal infusion of islets, an omental pouch was created to transplant the remaining islet graft into the recipients. Follow-up examinations indicated that all patients had a decrease in exogenous insulin requirements. Furthermore, at 3 months posttransplant, there was no significant difference in glycemic control or graft function between the combined site recipients and their matched controls that only received intraportal islet transplant [102]. Finding an optimal transplantation site is a highly active area of research for insulin-secreting cell replacement therapies. Currently, there is limited clinical evidence to indicate superior safety and efficacy of the alternates to the intrahepatic portal duct transplant. Further randomized clinical trials and long-term follow-up studies are required to determine if optimal engraftment and long-term glycemic control can be achieved within other investigated body tissues [96].

Autologous transplant: Factors affecting patient outcomes

Pancreatic islet mass

The outcome of IAT is mostly associated with islet viability and graft yield as suggested by the 2017 Clinical Islet Transplant Registry (CITR) data, which indicated that approximately 90% of patients that have undergone TP-IAT had graft function but only one-third became insulin independent [43,103]. It is suggested that improved graft function is associated with transplantation of more islet equivalents per kilogram body weight (IEQ/kg) with a minimum of 3000 IEQ/kg [49]. Our study of 22 patients with a median age of 36 that have undergone total or partial pancreatectomy from 2000 to 2003 indicated that 41% of patients became insulin independent with all the patients received more than 3000 IEQ/kg, which compares favorably with other larger studies [3,49,104–107]. A similar finding was reported in the study of 45 autologous islet recipients with the median age of 39. It was found that the patients who became insulin independent had received a significantly higher mean islet mass (6635 IEQ/kg) compared to the insulin-dependent group who received a mean islet mass of 3799 IEQ/kg [48]. Moreover, one of the largest studies on patients that have undergone TP-IAT demonstrated that after 3 years, 72% of patients who received more than 5000 IEQ/kg remained insulin independent while 22% of those who received 2500–5000 IEQ/kg and only 12% that received less than 2500 IEQ/kg were insulin free [20]. Therefore a higher mass of islet graft is correlated with improved glycemic control and IAT outcome.

Tissue volume

When a large volume of unpurified islets is obtained from a resected pancreas, cells can be purified to reduce exocrine tissue contamination and tissue mass which reduce the risk of developing complications such as portal vein thrombosis, bleeding, and elevated portal pressure [95]. Considering that islet mass is the prominent factor in determining posttransplant diabetes outcome, the decision to purify islets is critical and demands an evaluation of both short-term complication risks and long-term metabolic outcomes. Therefore it is crucial to determine how much dispersed pancreatic tissue volume is safe for TP-IAT to minimize the risks and ensure an islet yield with improved clinical outcomes [95]. Ultimately, the decision to purify islets, as well as the extent of purification, should be made by qualified personnel and in consultation with the patient's surgeon [86]. As a general rule, auto islet purification is performed when the tissue volume is large for transplantation. Wilhelm et al. have recommended transplanting no more than 0.25 cc/kg body weight of tissue volume during intraportal infusion [41,95].

Age and disease duration

Patient age at the time of transplant and duration of CP can also affect islet isolation results, graft function, and pain outcomes after TP-IAT. Bellin et al. investigated the effect of age and disease duration on TP-IAT in patients at ages 3–56 who suffered from CP. Regarding isolation results, islet yield has been suggested to decrease with increasing age and longer duration of CP, with a 13% and 22% decrease in obtainable IEQ/kg for every 5 years of age and disease duration. In terms of diabetes outcomes and glycemic control, a shorter duration of the disease and a younger age at the time of transplant were associated with full graft function at 1 year and 2 years posttransplant. All patients with full graft function were younger than 21 years and had less than 17 years of CP. Additionally, the patients who became insulin independent at 1-year follow-up were younger and had a shorter disease duration [108]. Autograft attrition after transplant, which can reduce islet function over time, is also associated with age at the time of transplant. According to the 2017 CITR report, patients who underwent TP-IAT at 35 years or older had low insulin retention 5 years posttransplant. However, more than 70% of patients aged 18–35 years, who received islet doses of ≥275,000 IEQ/kg, had remained insulin independent for 5 years after transplant [43,103]. Lastly, persistent use of opioids, although infrequent at 1 year (15% of patients), posttransplant was found to increase with age and disease duration. Overall, TP-IAT outcomes are negatively impacted by older age and prolonged CP [108].

Postoperative complications

IAT requires major surgery and involves risks that may impede a patient's quality of life posttransplant. The recipients may also suffer from postoperative complications, which can be fatal [20,94]. In a study of patients who underwent IAT from 2000 to 2003, 68% of the recipients experienced postoperative complications, most of which were minor. Major postoperative complications of TP-IAT include acute respiratory distress syndrome, intraabdominal abscess, and pulmonary embolism, and minor complications include delayed gastric emptying, urinary tract infection, pneumonia, wound and line infection, hyperglycemia, hematoma, and cellulitis [3]. One report on patients that have undergone TP-IAT indicated that 15.9% of patients required relaparotomy due to the following major complications: bleeding (9.5%), anastomotic leak (4.2%), gastrointestinal distress (4.7%), and intraabdominal infection (1.9%). Additionally, these patients also require lifelong pancreatic enzyme replacement therapy (PERT) after transplant due to complete pancreatic exocrine insufficiency [20,94].

Quality of life after autologous transplant

Despite the risks and postoperative complications associated with TP-IAT, studies have shown an overall positive outcome on patients' quality of life. According to a Sutherland et al. study on more than 400 patients, the rate of survival at 5 years postoperation was 89% in adults and 98% in children suggesting a minimal risk of death after TP-IAT. Moreover, IAT represents an efficacious therapy with meaningful islet function in most patients and substantial graft function in two-thirds of recipients with insulin independence achieved in one-quarter of adults and half of children [20]. In addition to its favorable effect on glycemic control, TP-IAT has been shown to substantially ameliorate chronic pain, one of the most common complications of CP. The procedure has been shown to significantly reduce abdominal pain, enhance the quality of life from decreased disability and pain, and improve depression and anxiety associated with the chronic pain prior to TP-IAT.

References

[1] Dykewicz CA. Summary of the guidelines for preventing opportunistic infections among hematopoietic stem cell transplant recipients. Clin Infect Dis 2001;33(2):139–44. https://doi.org/10.1086/321805.

[2] Maffi P, Balzano G, Ponzoni M, Nano R, Sordi V, Melzi R, Mercalli A, Scavini M, Esposito A, Peccatori J, Cantarelli E, Messina C, Bernardi M, Del Maschio A, Staudacher C, Doglioni C, Ciceri F, Secchi A, Piemonti L. Autologous pancreatic islet transplantation in human bone marrow. Diabetes 2013;62(10):3523–31. https://doi.org/10.2337/db13-0465.
[3] Rodriguez Rilo HL, Ahmad SA, D'Alessio D, Iwanaga Y, Kim J, Choe KA, Moulton JS, Martin J, Pennington LJ, Soldano DA, Biliter J, Martin SP, Ulrich CD, Somogyi L, Welge J, Matthews JB, Lowy AM. Total pancreatectomy and autologous islet cell transplantation as a means to treat severe chronic pancreatitis. J Gastrointest Surg 2003;7(8):978–89. Elsevier Inc https://doi.org/10.1016/j.gassur.2003.09.008.
[4] Everhart JE, Ruhl CE. Burden of digestive diseases in the United States part III: liver, biliary tract, and pancreas. Gastroenterology 2009;136(4):1134–44. https://doi.org/10.1053/j.gastro.2009.02.038.
[5] Yadav D, Timmons L, Benson JT, Dierkhising RA, Chari ST. Incidence, prevalence, and survival of chronic pancreatitis: a population-based study. Am J Gastroenterol 2011;106(12):2192–9. https://doi.org/10.1038/ajg.2011.328.
[6] Coté GA, Yadav D, Slivka A, Hawes RH, Anderson MA, Burton FR, Brand RE, Banks PA, Lewis MD, Disario JA, Gardner TB, Gelrud A, Amann ST, Baillie J, Money ME, O'Connell M, Whitcomb DC, Sherman S. Alcohol and smoking as risk factors in an epidemiology study of patients with chronic pancreatitis. Clin Gastroenterol Hepatol 2011;9(3):266–73. https://doi.org/10.1016/j.cgh.2010.10.015.
[7] Andrén-Sandberg A, Hoem D, Gislason H. Pain management in chronic pancreatitis. Eur J Gastroenterol Hepatol 2002;14(9):957–70. https://doi.org/10.1097/00042737-200209000-00006.
[8] Ohayon MM, Schatzberg AF. Chronic pain and major depressive disorder in the general population. J Psychiatr Res 2010;44(7):454–61. https://doi.org/10.1016/j.jpsychires.2009.10.013.
[9] Morgan K, Owczarski SM, Borckardt J, Madan A, Nishimura M, Adams DB. Pain control and quality of life after pancreatectomy with islet autotransplantation for chronic pancreatitis. J Gastrointest Surg 2012;16(1):129–34. https://doi.org/10.1007/s11605-011-1744-y.
[10] Greenhalgh SN, Iredale JP, Henderson NC. Origins of fibrosis: pericytes take centre stage. F1000Prime Rep 2013;5. https://doi.org/10.12703/P5-37.
[11] Hamada S, Masamune A, Shimosegawa T. Pancreatic fibrosis. Pancreapedia: Exocrine Pancreas Knowledge Base; 2015.
[12] Bliss LA, Yang CJ, Eskander MF, De Geus SWL, Callery MP, Kent TS, Moser AJ, Freedman SD, Tseng JF. Surgical management of chronic pancreatitis: current utilization in the United States. HPB 2015;17(9):804–10. Blackwell Publishing Ltd https://doi.org/10.1111/hpb.12459.
[13] Frey CF, Amikura K. Local resection of the head of the pancreas combined with longitudinal pancreaticojejunostomy in the management of patients with chronic pancreatitis. Ann Surg 1994;220(4):492–507. Lippincott Williams and Wilkins https://doi.org/10.1097/00000658-199410000-00008.
[14] Frey CF, Jeffrey Smith G. Description and rationale of a new operation for chronic pancreatitis. Pancreas 1987;2(6):701–7. https://doi.org/10.1097/00006676-198711000-00014.
[15] Puestow CB, Gillesby WJ. Retrograde surgical drainage of pancreas for chronic relapsing pancreatitis. AMA Arch Surg 1958;76(6):898–907. https://doi.org/10.1001/archsurg.1958.01280240056009.
[16] Cahen DL, Gouma DJ, Nio Y, Rauws EAJ, Boermeester MA, Busch OR, Stoker J, Laméris JS, Dijkgraaf MGW, Huibregtse K, Bruno MJ. Endoscopic versus surgical drainage of the pancreatic duct in chronic pancreatitis. N Engl J Med 2007;356(7):676–84. https://doi.org/10.1056/NEJMoa060610.
[17] Holmberg JT, Isaksson G, Ihse I. Long term results of pancreaticojejunostomy in chronic pancreatitis. Surg Gynecol Obstet 1985;160(4):339–46.
[18] Gachago C, Draganov PV. Pain management in chronic pancreatitis. World J Gastroenterol 2008;14(20):3137–48. https://doi.org/10.3748/wjg.14.3137.
[19] Schmulewitz N. Total pancreatectomy with autologous islet cell transplantation in children: making a difference. Clin Gastroenterol Hepatol 2011;9(9):725–6. https://doi.org/10.1016/j.cgh.2011.05.023.
[20] Sutherland DER, Radosevich DM, Bellin MD, Hering BJ, Beilman GJ, Dunn TB, Chinnakotla S, Vickers SM, Bland B, Balamurugan AN, Freeman ML, Pruett TL. Total pancreatectomy and islet autotransplantation for chronic pancreatitis. J Am Coll Surg 2012;214(4):409–24. https://doi.org/10.1016/j.jamcollsurg.2011.12.040.
[21] Warren KW, Poulantzas JK, Kune GA. Life after total pancreatectomy for chronic pancreatitis: clinical study of eight cases. Ann Surg 1966;164(5):830–4. https://doi.org/10.1097/00000658-196611000-00006.
[22] Braasch JW, Vito L, Nugent FW. Total pancreatectomy for end-stage chronic pancreatitis. Ann Surg 1978;188(3):317–22. https://doi.org/10.1097/00000658-197809000-00006.
[23] Berney T, Rudisuhli T, Oberholzer J, Caulfield A, Morel P. Long-term metabolic results after pancreatic resection for severe chronic pancreatitis. Arch Surg 2000;135(9):1106–11. https://doi.org/10.1001/archsurg.135.9.1106.
[24] Gruessner RWG, Cercone R, Galvani C, Rana A, Porubsky M, Gruessner AC, Rilo H. Results of open and robot-assisted pancreatectomies with autologous islet transplantations: treating chronic pancreatitis and preventing surgically induced diabetes. Transplant Proc 2014;46(6):1978–9. Elsevier USA https://doi.org/10.1016/j.transproceed.2014.06.005.
[25] Iino S. Japanese clinical statistical data of patients with hepatitis B. Nippon rinsho Jpn J Clin Med 1992;50:602–12.
[26] Gall FP, Mühe E, Gebhardt C. Results of partial and total pancreaticoduodenectomy in 117 patients with chronic pancreatitis. World J Surg 1981;5(2):269–73. https://doi.org/10.1007/BF01658311.
[27] Najarian JS, Sutherland DER, Matas AJ, Goetz FC. Human islet autotransplantation following pancreatectomy. Transplant Proc 1979;11(1):336–40.
[28] Fazlalizadeh R, Moghadamyeghaneh Z, Demirjian AN, Imagawa DK, Foster CE, Lakey JR, Stamos MJ, Ichii H. Total pancreatectomy and islet autotransplantation: a decade nationwide analysis. World J Transplant 2016;233. https://doi.org/10.5500/wjt.v6.i1.233.
[29] Sutherland DE, Goetz FC, Najarian JS. Review of world's experience with pancreas and islet transplantation and results of intraperitoneal segmental pancreas transplantation from related and cadaver donors at Minnesota. Transplant Proc 1981;13(1):291–7.
[30] Ballinger WF, Lacy PE. Transplantation of intact pancreatic islets in rats. Surgery 1972;72(2):175–86.

[31] Lacy PE, Kostianovsky M. Method for the isolation of intact islets of Langerhans from the rat pancreas. Diabetes 1967;16(1):35–9. https://doi.org/10.2337/diab.16.1.35.
[32] Tucker T, Labadie K, Kopan C, Alexander M, Lakey JRT. Strategies in preventing diabetes after pancreatectomy using islet auto- and allo-transplantation. Int J Transplant Res Med 2017;3(1):1–8. https://doi.org/10.23937/2572-4045.1510029.
[33] Agarwal A, Brayman KL. Update on islet cell transplantation for type 1 diabetes. Semin Interv Radiol 2012;29(2):90–8. https://doi.org/10.1055/s-0032-1312569.
[34] Mittal VK, Toledo-Pereyra LH, Sharma M, Ramaswamy K, Puri VK, Cortez JA, Gordon D. Acute portal hypertension and disseminated intravascular coagulation following pancreatic islet autotransplantation after subtotal pancreatectomy. Transplantation 1981;31(4):302–6. https://doi.org/10.1097/00007890-198104000-00014.
[35] Ricordi C, Lacy PE, Finke EH, Olack BJ, Scharp DW. Automated method for isolation of human pancreatic islets. Diabetes 1988;37(4):413–20. https://doi.org/10.2337/diab.37.4.413.
[36] Shapiro AJ, Lakey JR, Ryan EA, Korbutt GS, Toth E, Warnock GL, Kneteman NM, Rajotte RV. Islet transplantation in seven patients with type 1 diabetes mellitus using a glucocorticoid-free immunosuppressive regimen. N Engl J Med 2000;230–8. https://doi.org/10.1056/NEJM200007273430401.
[37] Shapiro AMJ, Ricordi C, Hering BJ, Auchincloss H, Lindblad R, Robertson RP, Secchi A, Brendel MD, Berney T, Brennan DC, Cagliero E, Alejandro R, Ryan EA, DiMercurio B, Morel P, Polonsky KS, Reems JA, Bretzel RG, Bertuzzi F, et al. International trial of the Edmonton protocol for islet transplantation. N Engl J Med 2006;355(13):1318–30. https://doi.org/10.1056/NEJMoa061267.
[38] Ricordi C, Goldstein JS, Balamurugan AN, Szot GL, Kin T, Liu C, Czarniecki CW, Barbaro B, Bridges ND, Cano J, Clarke WR, Eggerman TL, Hunsicker LG, Kaufman DB, Khan A, Lafontant DE, Linetsky E, Luo X, Markmann JF, et al. National institutes of health-sponsored clinical islet transplantation consortium phase 3 trial: manufacture of a complex cellular product at eight processing facilities. Diabetes 2016;65(11):3418–28. https://doi.org/10.2337/db16-0234.
[39] Chinnakotla S, Beilman GJ, Dunn TB, Bellin MD, Freeman ML, Radosevich DM, Arain M, Amateau SK, Mallery JS, Schwarzenberg SJ, Clavel A, Wilhelm J, Robertson RP, Berry L, Cook M, Hering BJ, Sutherland DER, Pruett TL. Factors predicting outcomes after a total pancreatectomy and islet autotransplantation lessons learned from over 500 cases. Ann Surg 2015;262(4):610–22. https://doi.org/10.1097/SLA.0000000000001453.
[40] Harris H. Systematic review of total pancreatectomy and islet autotransplantation for chronic pancreatitis. Br J Surg 2012;99(6):761–7. https://doi.org/10.1002/bjs.8747.
[41] Matsumoto S, Takita M, Shimoda M, Sugimoto K, Itoh T, Chujo D, SoRelle JA, Tamura Y, Rahman AM, Onaca N, Naziruddin B, Levy MF. Impact of tissue volume and purification on clinical autologous islet transplantation for the treatment of chronic pancreatitis. Cell Transplant 2012;21(4):625–32. https://doi.org/10.3727/096368911X623899.
[42] Naziruddin B, Matsumoto S, Noguchi H, Takita M, Shimoda M, Fujita Y, Chujo D, Tate C, Onaca N, Lamont J, Kobayashi N, Levy MF. Improved pancreatic islet isolation outcome in autologous transplantation for chronic pancreatitis. Cell Transplant 2012;21(2–3):553–8. https://doi.org/10.3727/096368911X605475.
[43] Clinical Islet Transplant Registry. Inaugural report on autologous islet transplantation; 2017.
[44] Takita M, Lara LF, Naziruddin B, Shahbazov R, Lawrence MC, Kim PT, Onaca N, Burdick JS, Levy MF. Effect of the duration of chronic pancreatitis on pancreas islet yield and metabolic outcome following islet autotransplantation. J Gastrointest Surg 2015;19(7):1236–46. https://doi.org/10.1007/s11605-015-2828-x.
[45] Dhanasekaran M, Loganathan G, Narayanan S, Tucker W, Jawahar AP, Patel A, et al. Technical modifications to improve islet yield from chronic pancreatitis pancreas (CPP) for islet auto-transplantation (IAT). J Clin Exp Transplant 2017;2(1). https://doi.org/10.4172/2475-7640.1000112.
[46] Michihiro M, Takashi K, Naotake A, Kazunori O, Taihei I, Ikuko M, Takehide A. A review of autologous islet transplantation. Cell Med 2013;59–62. https://doi.org/10.3727/215517913X666558.
[47] Walsh RM, Saavedra JRA, Lentz G, Guerron AD, Scheman J, Stevens T, Trucco M, Bottino R, Hatipoglu B. Improved quality of life following total pancreatectomy and auto-islet transplantation for chronic pancreatitis. J Gastrointest Surg 2012;16(8):1469–77. https://doi.org/10.1007/s11605-012-1914-6.
[48] Ahmad SA, Lowy AM, Wray CJ, D'Alessio D, Choe KA, James LE, Gelrud A, Matthews JB, Rilo HLR. Factors associated with insulin and narcotic independence after islet autotransplantation in patients with severe chronic pancreatitis. J Am Coll Surg 2005;201(5):680–7. https://doi.org/10.1016/j.jamcollsurg.2005.06.268.
[49] International Islet Transplant Registry Newsletter. 2001.
[50] Food and Drug Administration. Human cells, tissues, and cellular and tissue-based products. Food and Drug Administration 2017.
[51] Anon. Safety of human cells, tissues and organs for transplantation regulations. Health Products and Food Branch; 2015.
[52] European Medicines Agency. Scientific recommendation on classification of advanced therapy medicinal products. vol. 1394. European Medicines Agency; 2012. https://www.ema.europa.eu/en/human-regulatory/marketing-authorisation/advanced-therapies/advanced-therapy-classification.
[53] Labour and Welfare Ministry of Health. Pharmaceutical and Medical Devices Agency. 2018; https://www.pmda.go.jp/english/index.html. Accessed July 16, 2018.
[54] Barnett MJ, McGhee-Wilson D, Shapiro AMJ, Lakey JRT. Variation in human islet viability based on different membrane integrity stains. Cell Transplant 2004;13(5):481–8. https://doi.org/10.3727/000000004783983701.
[55] Boyd V, Cholewa O, Papas KK. Limitation in the use of fluorescein diacetate/propidium iodide (FDA/PI) and cell permeable nucleic acid stains for viability measurements of isolated islets of langerhans. Curr Trends Biotechnol Pharm 2008;2:66–84.
[56] Lau H, Corrales N, Alexander M, Mohammadi MR, Li S, Smink AM, de Vos P, Lakey JRT. Necrostatin-1 supplementation enhances young porcine islet maturation and in vitro function. Xenotransplantation 2020;27(1). https://doi.org/10.1111/xen.12555.

[57] Khiatah B, Qi M, Wu Y, Chen KT, Perez R, Valiente L, Omori K, Isenberg JS, Kandeel F, Yee JK, Al-Abdullah IH. Pancreatic human islets and insulin-producing cells derived from embryonic stem cells are rapidly identified by a newly developed Dithizone. Sci Rep 2019;9(1). https://doi.org/10.1038/s41598-019-45678-y.

[58] Papas KK, Suszynski TM, Colton CK. Islet assessment for transplantation. Curr Opin Organ Transplant 2009;14(6):674–82. https://doi.org/10.1097/MOT.0b013e328332a489.

[59] Rilo LR, Cercone R, Lakey JRT. Requirements for clinical islet laboratories. In: Transplantation, bioengineering, and regeneration of the endocrine pancreas, vol. 2. Elsevier; 2019. p. 51–66. https://doi.org/10.1016/B978-0-12-814831-0.00003-8.

[60] Chinnakotla S, Bellin MD, Schwarzenberg SJ, Radosevich DM, Cook M, Dunn TB, Beilman GJ, Freeman ML, Balamurugan AN, Wilhelm J, Bland B, Jimenez-Vega JM, Hering BJ, Vickers SM, Pruett TL, Sutherland DER. Total pancreatectomy and islet autotransplantation in children for chronic pancreatitis: indication, surgical techniques, postoperative management, and long-term outcomes. Ann Surg 2014;260(1):56–64. https://doi.org/10.1097/SLA.0000000000000569.

[61] Becker T, Ringe B, Nyibata M, Meyer zu Vilsendorf A, Schrem H, Lück R, Neipp M, Klempnauer J, Bektas H. Pancreas transplantation with histidine-tryptophan-ketoglutarate (HTK) solution and University of Wisconsin (UW) solution: is there a difference? J Pancreas 2007;8(3):304–11. http://www.joplink.net/prev/200705/200705_09.pdf.

[62] Caballero-Corbalán J, Brandhorst H, Malm H, Felldin M, Foss A, Salmela K, Tibell A, Tufveson G, Korsgren O, Brandhorst D. Using HTK for prolonged pancreas preservation prior to human islet isolation. J Surg Res 2012;175(1):163–8. https://doi.org/10.1016/j.jss.2011.03.012.

[63] Fridell JA, Mangus RS, Powelson JA. Histidine-tryptophan-ketoglutarate for pancreas allograft preservation: the Indiana university experience. Am J Transplant 2010;10(5):1284–9. https://doi.org/10.1111/j.1600-6143.2010.03095.x.

[64] Schneeberger S, Biebl M, Steurer W, Hesse UJ, Troisi R, Langrehr JM, Schareck W, Mark W, Margreiter R, Königsrainer A. A prospective randomized multicenter trial comparing histidine-tryptophane-ketoglutarate versus University of Wisconsin perfusion solution in clinical pancreas transplantation. Transpl Int 2009;22(2):217–24. https://doi.org/10.1111/j.1432-2277.2008.00773.x.

[65] Wang Y, Danielson KK, Ropski A, Harvat T, Barbaro B, Paushter D, Qi M, Oberholzer J. Systematic analysis of donor and isolation factor's impact on human islet yield and size distribution. Cell Transplant 2013;22(12):2323–33. https://doi.org/10.3727/096368912X662417.

[66] Noguchi H, Naziruddin B, Jackson A, Shimoda M, Ikemoto T, Fujita Y, Chujo D, Takita M, Kobayashi N, Onaca N, Levy MF, Matsumoto S. Low-temperature preservation of isolated islets is superior to conventional islet culture before islet transplantation. Transplantation 2010;89(1):47–54. https://doi.org/10.1097/TP.0b013e3181be3bf2.

[67] Obermaier R, Drognitz O, Benz S, Hopt UT, Pisarski P. Pancreatic ischemia/reperfusion injury: impact of different preservation temperatures. Pancreas 2008;37(3):328–32. https://doi.org/10.1097/MPA.0b013e31816d9283.

[68] Lakey JRT, Burridge PW, Shapiro AMJ. Technical aspects of islet preparation and transplantation. Transpl Int 2003;16(9):613–32. https://doi.org/10.1111/j.1432-2277.2003.tb00361.x.

[69] Balamurugan AN, Loganathan G, Bellin MD, Wilhelm JJ, Harmon J, Anazawa T, Soltani SM, Radosevich DM, Yuasa T, Tiwari M, Papas KK, McCarthy R, Sutherland DER, Hering BJ. A new enzyme mixture to increase the yield and transplant rate of autologous and allogeneic human islet products. Transplantation 2012;93(7):693–702. https://doi.org/10.1097/TP.0b013e318247281b.

[70] Balamurugan AN, Breite AG, Anazawa T, Loganathan G, Wilhelm JJ, Papas KK, Dwulet FE, McCarthy RC, Hering BJ. Successful human islet isolation and transplantation indicating the importance of class 1 collagenase and collagen degradation activity assay. Transplantation 2010;89(8):954–61. https://doi.org/10.1097/TP.0b013e3181d21e9a.

[71] Shapiro AJ, Ricordi C. Islet cell transplantation procedure and surgical technique. In: Kirk AD, Knechtle SJ, Larsen CP, Madsen JC, Pearson TC, Webber SA, editors. Textbook of Organ Transplantation. Wiley Online; 2014.

[72] Tsukada M, Saito T, Ise K, Kenjo A, Kimura T, Satoh Y, Saito T, Anazawa T, Oshibe I, Suzuki S, Hashimoto Y, Gotoh M. A model to evaluate toxic factors influencing islets during collagenase digestion: the role of serine protease inhibitor in the protection of islets. Cell Transplant 2012;21(2–3):473–82. https://doi.org/10.3727/096368911X605385.

[73] Wolters GHJ, Vos-Scheperkeuter GH, van Deijnen JHM, van Schilfgaarde R. An analysis of the role of collagenase and protease in the enzymatic dissociation of the rat pancreas for islet isolation. Diabetologia 1992;35(8):735–42. https://doi.org/10.1007/BF00429093.

[74] Paget M, Murray H, Bailey CJ, Downing R. Human islet isolation: semi-automated and manual methods. Diab Vasc Dis Res 2007;4(1):7–12. https://doi.org/10.3132/dvdr.2007.010.

[75] Ricordi C. Islet transplantation: a brave new world. Diabetes 2003;52(7):1595–603. American Diabetes Association Inc https://doi.org/10.2337/diabetes.52.7.1595.

[76] Ellis C, Lyon JG, Korbutt GS. Optimization and scale-up isolation and culture of neonatal porcine islets: potential for clinical application. Cell Transplant 2016;25(3):539–47. https://doi.org/10.3727/096368915X689451.

[77] Thompson EM, Sollinger JL, Opara EC, Adin CA. Selective osmotic shock for islet isolation in the cadaveric canine pancreas. Cell Transplant 2018;27(3):542–50. https://doi.org/10.1177/0963689717752947.

[78] Sauerbruch T, Holl J, Sackmann M, Werner R, Wotzka R, Paumgartner G. Disintegration of a pancreatic duct stone with extracorporeal shock waves in a patient with chronic pancreatitis. Endoscopy 1987;19(5):207–8. https://doi.org/10.1055/s-2007-1018284.

[79] Li BR, Liao Z, Du TT, Ye B, Chen H, Ji TT, et al. Extracorporeal shock wave lithotripsy is a safe and effective treatment for pancreatic stones coexisting with pancreatic pseudocysts. Gastrointest Endosc 2016;84(1):69–78. https://doi.org/10.1016/j.gie.2015.10.026.

[80] Costamagna G, Gabbrielli A, Mutignani M, Perri V, Pandolfi M, Boscaini M, Crucitti F. Extracorporeal shock wave lithotripsy of pancreatic stones in chronic pancreatitis: immediate and medium-term results. Gastrointest Endosc 1997;46(3):231–6. https://doi.org/10.1016/S0016-5107(97)70092-0.

[81] Dumonceau JM, Costamagna G, Tringali A, Vahedi K, Delhaye M, Hittelet A, Spera G, Giostra E, Mutignani M, De Maertelaer V, Devière J. Treatment for painful calcified chronic pancreatitis: extracorporeal shock wave lithotripsy versus endoscopic treatment: a randomised controlled trial. Gut 2007;56(4):545–52. https://doi.org/10.1136/gut.2006.096883.

[82] Guda NM, Partington S, Freeman ML. Extracorporeal shock wave lithotripsy in the management of chronic calcific pancreatitis: a meta-analysis. J Pancreas 2005;6(1):6–12. http://www.joplink.net/prev/200501/05.html.
[83] Neoptolemos JP, Carr-Locke DL, London NJ, Bailey IA, James D, Fossard DP. Controlled trial of urgent endoscopic retrograde cholangiopancreatography and endoscopic sphincterotomy versus conservative treatment for acute pancreatitis due to gallstones. Lancet 1988;2(8618):979–83. https://doi.org/10.1016/s0140-6736(88)90740-4.
[84] Troendle DM, Fishman DS, Barth BA, Giefer MJ, Lin TK, Liu QY, Abu-El-Haija M, Bellin MD, Durie PR, Freedman SD, Gariepy C, Gonska T, Heyman MB, Himes R, Husain SZ, Kumar S, Lowe ME, Morinville VD, Ooi CY, et al. Therapeutic endoscopic retrograde cholangiopancreatography in pediatric patients with acute recurrent and chronic pancreatitis. Pancreas 2017;46(6):764–9. https://doi.org/10.1097/MPA.0000000000000848.
[85] Tandan M, Talukdar R, Reddy DN. Management of pancreatic calculi: an update. Gut Liver 2016;10(6):873–80. https://doi.org/10.5009/gnl15555.
[86] Balamurugan AN, Loganathan G, Lockridge A, Soltani SM, Wilhelm JJ, Beilman GJ, Hering BJ, Sutherland DER. Islet isolation from pancreatitis pancreas for islet autotransplantation. In: Islets of Langerhans. 2nd ed. Springer Netherlands; 2015. p. 1199–227. https://doi.org/10.1007/978-94-007-6686-0_48.
[87] Ryan EA, Lakey JRT, Paty BW, Imes S, Korbutt GS, Kneteman NM, Bigam D, Rajotte RV, James Shapiro AM. Successful islet transplantation: continued insulin reserve provides long-term glycemic control. Diabetes 2002;51(7):2148–57. https://doi.org/10.2337/diabetes.51.7.2148.
[88] Ryan EA, Paty BW, Senior PA, Bigam D, Alfadhli E, Kneteman NM, et al. Five-year follow-up after clinical islet transplantation. Diabetes 2005;2060–9. https://doi.org/10.2337/diabetes.54.7.2060.
[89] Ichii H, Sakuma Y, Pileggi A, Fraker C, Alvarez A, Montelongo J, Szust J, Khan A, Inverardi L, Naziruddin B, Levy MF, Klintmalm GB, Goss JA, Alejandro R, Ricordi C. Shipment of human islets for transplantation. Am J Transplant 2007;7(4):1010–20. https://doi.org/10.1111/j.1600-6143.2006.01687.x.
[90] Lemoli RM, Tafuri A, Strife A, Andreeff M, Clarkson BD, Gulati SC. Proliferation of human hematopoietic progenitors in long-term bone marrow cultures in gas-permeable plastic bags is enhanced by colony-stimulating factors. Exp Hematol 1992;20(5):569–75.
[91] Rabkin JM, Olyaei AJ, Orloff SL, Geisler SM, Wahoff DC, Hering BJ, Sutherland DER. Distant processing of pancreas islets for autotransplantation following total pancreatectomy. Am J Surg 1999;177(5):423–7. https://doi.org/10.1016/S0002-9610(99)00078-1.
[92] Rabkin JM, Leone JP, Sutherland DER, Ahman A, Reed M, Papalois BE, Wahoff DC. Transcontinental shipping of pancreatic islets for autotransplantation after total pancreatectomy. Pancreas 1997;15(4):416–9. https://doi.org/10.1097/00006676-199711000-00013.
[93] Casey JJ, Lakey JRT, Ryan EA, Paty BW, Owen R, O'Kelly K, Nanji S, Rajotte RV, Korbutt GS, Bigam D, Kneteman NN, Shapiro AMJ. Portal venous pressure changes after sequential clinical islet transplantation. Transplantation 2002;74(7):913–5. https://doi.org/10.1097/00007890-200210150-00002.
[94] Kesseli SJ, Smith KA, Gardner TB. Total pancreatectomy with islet autologous transplantation: the cure for chronic pancreatitis? Clin Transl Gastroenterol 2015;6(1). https://doi.org/10.1038/ctg.2015.2.
[95] Wilhelm JJ, Bellin MD, Dunn TB, Balamurugan AN, Pruett TL, Radosevich DM, Chinnakotla S, Schwarzenberg SJ, Freeman ML, Hering BJ, Sutherland DER, Beilman GJ. Proposed thresholds for pancreatic tissue volume for safe intraportal islet autotransplantation after total pancreatectomy. Am J Transplant 2013;13(12):3183–91. https://doi.org/10.1111/ajt.12482.
[96] Pellegrini S. Alternative transplantation sites for islet transplantation. In: Transplantation, bioengineering, and regeneration of the endocrine pancreas, vol. 1. Elsevier; 2019. p. 833–47. https://doi.org/10.1016/B978-0-12-814833-4.00065-4.
[97] James Shapiro AM, Lakey JRT, Rajotte RV, Warnock GL, Friedlich MS, Jewell LD, Kneteman NM. Portal vein thrombosis after transplantation of partially purified pancreatic islets in a combined human liver/islet allograft. Transplantation 1995;59(7):1060–3. https://doi.org/10.1097/00007890-199504150-00027.
[98] London NJ, Contractor H, Robertson G, Chadwick D, Bell P, James RF. Human islet purification: what next? Diabetes Nutr Metab 1992;5.
[99] Moberg L, Johansson H, Lukinius A, Berne C, Foss A, Källen R, Østraat Ø, Salmela K, Tibell A, Tufveson G, Elgue G, Nilsson Ekdahl K, Korsgren O, Nilsson B. Production of tissue factor by pancreatic islet cells as a trigger of detrimental thrombotic reactions in clinical islet transplantation. Lancet 2002;360(9350):2039–45. https://doi.org/10.1016/S0140-6736(02)12020-4.
[100] Naziruddin B, Iwahashi S, Kanak MA, Takita M, Itoh T, Levy MF. Evidence for instant blood-mediated inflammatory reaction in clinical autologous islet transplantation. Am J Transplant 2014;14(2):428–37. https://doi.org/10.1111/ajt.12558.
[101] Walsh TJ, Eggleston JC, Cameron JL. Portal hypertension, hepatic infarction, and liver failure complicating pancreatic islet autotransplantation. Surgery 1982;91(4):485–7.
[102] Stice MJ, Dunn TB, Bellin MD, Skube ME, Beilman GJ. Omental pouch technique for combined site islet autotransplantation following total pancreatectomy. Cell Transplant 2018;27(10):1561–8. https://doi.org/10.1177/0963689718798627.
[103] Rodriguez S, Alexander M, Lakey JR. Pancreatic islet transplantation: A surgical approach to type 1 diabetes treatment. Springer Science and Business Media LLC; 2020. p. 655–64. https://doi.org/10.1007/978-3-030-53370-0_48.
[104] International Islet Transplant Registry Newsletter. Number 8. 1995.
[105] Sutherland DER, Gores PF, Farney AC, Wahoff DC, Matas AJ, Dunn DL, Gruessner RWG, Najarian JS. Evolution of kidney, pancreas, and islet transplantation for patients with diabetes at the University of Minnesota. Am J Surg 1993;166(5):456–91. https://doi.org/10.1016/S0002-9610(05)81142-0.
[106] White SA, Davies JE, Pollard C, Swift SM, Clayton HA, Sutton CD, Weymss-Holden S, Musto PP, Berry DP, Dennison AR. Pancreas resection and islet autotransplantation for end-stage chronic pancreatitis. Ann Surg 2001;233(3):423–31. https://doi.org/10.1097/00000658-200103000-00018.
[107] White SA, Pollard C, Davies JE, Sutton CD, Hales CN, Dennison AR. Temporal relationship of insulin, intact proinsulin and split proinsulin after islet autotransplantation. Transplant Proc 2001;33(1–2):680. https://doi.org/10.1016/S0041-1345(00)02199-0.
[108] Bellin MD, Prokhoda P, Hodges JS, Schwarzenberg SJ, Freeman ML, Dunn TB, Wilhelm JJ, Pruett TL, Kirchner VA, Beilman GJ, Chinnakotla S. Age and disease duration impact outcomes of total pancreatectomy and islet autotransplant for PRSS1 hereditary pancreatitis. Pancreas 2018;47(4):466–70. https://doi.org/10.1097/MPA.0000000000001028.

Chapter 7

Oxygenation of the pancreas

Amy C. Kelly[a], Thomas M. Suszynski[b], and Klearchos K. Papas[a]
[a]*University of Arizona, Department of Surgery, Institute of Cellular Transplantation, Tucson, AZ, United States,* [b]*University of Minnesota, Medical Center and M Health, Departments of Surgery (Plastic Surgery) and Orthopedic Surgery, Minneapolis, MN, United States*

Introduction

Organ transplantation is increasingly used for the treatment of a variety of organ failure and end-stage organ disease. Despite advances in procurement and preservation, the logistics of transporting donor organs between facilities and the need to coordinate operating times, especially during the middle of the night, continue to pose challenges with how transplant surgical care delivery is done. Furthermore, the demand for donated organs that qualify for transplantation far exceeds the supply. There is a shortage of ideal organs undergoing donation following brain death (DBD). Thus criteria for organ donation have been expanded to include older and higher risk extended criteria donors (ECD) and donors after circulatory or cardiac death (DCD) [1–3]. While these expanded criteria have increased the number of available organs for transplant, there are concerns about the quality and function that may have consequential impact on posttransplant outcomes. The other major challenge in organ transplantation is the unavoidable period of cold ischemic time (CIT), when the commonly employed static cold storage (SCS) preservation is used, ischemia during these steps has been identified as a key detrimental factor for kidney [4–6], liver [7], heart [8], small bowel [9], and pancreas for transplant [10]. For pancreas preservation for islet isolation and transplantation, ischemia during preservation has compounding detrimental effects on islet viability and function and likely predisposes islets to further damage during the islet isolation process. It is widely accepted that improving the oxygenation of preserved pancreata may improve islet quality and may thus reduce the number of donors needed per patient, expand the donor pool, or increase the allowable preservation time. Oxygenation of the pancreas during preservation prior to pancreas and islet transplantation is the focus of the present chapter and the reasons behind the susceptibility of islets to ischemia and low oxygen conditions along with the need for adequate oxygenation will be discussed in detail in this chapter. Current clinical and research methodologies for pancreas preservation in pancreas and islet transplantation will also be discussed with special emphasis on pancreas oxygenation.

Pancreas transplantation

Human pancreas transplantation is an available treatment option for certain patients with severe type 1 diabetes (T1D). The procedure can provide normoglycemia without the need for exogenous insulin and in a way that better approximates the physiological glucose control mechanism. Pancreas transplants can be conducted as single procedures or more often, performed in combination with kidney transplants (simultaneous pancreas-kidney, SPK) as a standard therapeutic option for patients with kidney failure [11].

Between 2004 and 2008 in the United States, the most common pancreas transplant category was a SPK (73%) compared to pancreas transplants alone accounting for only 9% of all pancreas transplants in diabetic patients [12]. Overall, graft survival rates have progressively improved for all pancreas transplant categories one year after transplantation (79%–90%) [12]. In US cases, pancreas graft survival rates in SPK transplant category went from 75% in 1988 to 85% in 2001 and have been maintained greater than 85% since. In whole pancreas alone cases, graft survival has improved from around 50% to greater than 75% across the same time frame [13]. The improvements were attributed to less technical and immunological failure rates. A recent retrospective study at a single center from 2005 to 2018 showed a whole pancreas graft survival rate of 85% at 3 months and revealed that when a pancreas graft failed early there was no significant impact on patient mortality [14]. Despite these improvements, the need for major surgery, lifetime immunosuppression, and anticoagulation therapy in some patients are limiting the use of pancreas transplantation as a therapeutic option for more T1D patients [11,15,16].

Pancreas and Beta Cell Replacement. https://doi.org/10.1016/B978-0-12-824011-3.00010-2

Pancreas transplantation as a therapy has been complemented by the continued development of intrahepatic islet allotransplantation—using islets obtained from deceased donor pancreata and a steroid-free immunosuppressive drug regimen (the Edmonton Protocol) [17–19]. This procedure has gained attention and is currently implemented for a number of qualified T1D patients in a variety of countries around the world and on a clinical research basis in the United States [20,21]. Over the past decade, islet allotransplantation has continued to evolve with substantial improvements in terms of outcomes [22,23]. In 5- and 10-year follow-up observational studies, islet transplantation using the Edmonton protocol demonstrates marked metabolic improvement at 10 years with hypoglycemic awareness in the majority of patients and graft survival rates >80% [23,24]. The efficacy of islet transplantation compared to whole pancreas transplant is ongoing. An NIH-sponsored multicenter phase 3 clinical trial has shown that transplantation of purified human islets in T1D with severe hypoglycemia resulted in 87% of patients with glycemic control and restoration of hypoglycemic awareness one year following transplant [25], with parallel studies reporting quality of life improvements up to 2 years following islet transplant [20]. Recently, a Biologics Licensing Application has been filed in the United States by Cellular Transplantation (CellTrans Inc, https://www.celltransinc.com/) for intraportal islet transplantation. If approved, islet transplantation could be reimbursed by insurance in the United States in the near future [26] (FDA briefing document).

Specific challenges in pancreas preservation for islet transplantation

When transplanting organs such as the kidney, heart, and liver, the procurement and transplant team can coordinate directly, and often this is done locally. Pancreas procurement for islet cell transplantation poses additional challenges. Due to the expense and complexity of the islet isolation process, very few centers are capable of producing clinical-grade islet cell preparations. Therefore the pancreas must be preserved and transported to appropriate facilities where trained personnel are capable of isolating islets [27]. Moreover, patients receiving islet transplantation may require islets from more than one pancreas donor to achieve the same short-term outcomes compared to whole pancreas transplant [17,22,28]. Additionally, to achieve long-term independence from insulin, patients receiving islet transplantation often need more than one procedure [28], which exacerbates the cost, the logistics of transport to the isolation facility, and the shortage of suitable organs. Together, transportation considerations and the need for multiple donors lead to more organs needing to be preserved for longer, often more than 12 h.

The pancreas is a technically challenging organ to procure, preserve, and ultimately transplant. The challenge of procuring a pancreas arises from the fact that most pancreas donors are also going to donate a liver allograft, and the vasculature must be shared between these two organs. Additionally, unlike a kidney or liver for example, there are several inflow and outflow vessels that make adequate and complete flushing more uncertain. This surgery is executed with a unique technique along with liver *en bloc* [29]. Since there are many steps involved in the procurement of the pancreas, there is ample opportunity for the organ to experience ischemia and possibly reperfusion injury. These vulnerabilities are exacerbated by the difficulty of the procedures to drain and flush the organ that increase the risk of edema.

Hypoxia in islets: Structural, biochemical, and molecular considerations

Islets of Langerhans are highly vascularized multicellular spheroidal cell clusters ranging from 50 to 400 μm in diameter containing a variety of cell types that include insulin-producing β-cells (the most abundant cell type), other endocrine cells, immune cells, and autonomic nerves [30–32]. The extensive microvascular network within islets supports function, which is tightly regulated by autocrine, paracrine, and endocrine signaling [33]. While there is variability in the cellular composition and spatial organization of islets between species, β-cells are in close contact with capillaries branching from the afferent arteriole and often exhibit polarity in hormone secretion [34,35]. Within the islet, parasympathetic cholinergic axons are sparse and sympathetic nerve endings are responsible for innervating vascular smooth muscle cells [32]. The regulation of islet vascular tone by sympathetic control suggests that local blood flow and oxygen supply play a key role in hormone secretion regulation. Isolated islet function becomes affected at pO_2 levels below 38 mm Hg, emphasizing the limits of oxygen transfer by diffusion alone to the islet interior when the natural intra-islet vasculature is collapsed or not connected to the blood supply [31,36–39].

The critical need for adequate oxygen delivery is highlighted by the fact that islet cells, particularly the insulin-producing β-cells, are particularly sensitive to hypoxia. The metabolic features characteristic of β-cells that enable the coupling of glucose to adenosine triphosphate (ATP) production also make them functionally susceptible to hypoxia. First, transport of glucose into β-cells is not restricted and, once in the cell, is phosphorylated by hexokinase and glucokinase for entry into glycolysis [40,41]. β-cells predominantly express glucokinase, which has a low affinity for glucose and virtually no inhibitory feedback, which facilitates high glycolytic flux to generate pyruvate for mitochondrial respiration [42]. This is demonstrated by studies that show as many as 90% of oxidized carbons in purified β-cell mitochondria are derived from glucose [43,44]. Pyruvate preferentially enters the mitochondria because the β-cell has limited ability to shunt glucose to the pentose-phosphate pathway or produce lactate through lactate dehydrogenase (LDHA). With little capacity to produce lactate, β-cells cannot function anaerobically and the inability to produce ATP under conditions of oxygen deprivation results in cell death.

In addition to the inherent metabolic vulnerabilities, islets and β-cells have oxygen-sensing pathways that can further modulate their function. Like many cell types, islet cells exhibit a regulated hypoxia-mediated response through the activity of the transcription factor hypoxia-inducible factor-1α (HIF-1α) [45]. Mechanistically, hypoxia stabilizes HIF-1α which increases a hypoxic-response gene set, including LDHA, which promotes glycolysis and lactate production and impairs the glucose-sensing function in islets [46,47]. β-cell overexpression of LDHα, the enzyme that would enable energy production under hypoxia, disrupts GSIS [46,48,49] Therefore β-cell function (or glucose-coupled insulin secretion) is more vulnerable to hypoxia than β-cell/islet viability. This limits the use of antiapoptotic agents or other approaches to abate cell death to minimize the effects of hypoxia on viability because the effect of hypoxia on islet function may still severely impair outcomes [39]. In fact, experiments in which targeted delivery of ATP to β-cells was able to minimize the effect of hypoxia on viability were unable to block the negative effect of hypoxia on GSIS [50].

Interestingly, there is growing evidence that even if β-cells survive a hypoxic insult, they may exhibit persistently impaired function, even after reoxygenation [50–53]. This is consistent with emerging literature on islets, which suggests that even brief exposure to hypoxia may be sufficient to impair GSIS in the long term via initiation of a persistent hypoxia-induced genetic signature [39,46,48,53–55]. Work by Cantley et al. has shown similar findings in isolated human islets prepared for transplantation. Islets exposed to 6h of hypoxia demonstrated HIF-1α activation and expression of the characteristic hypoxia-response gene signature, such as elevated LDHA [51]. Interestingly, this was maintained posttransplant indicating hypoxic preexposure to an islet graft, such as would occur during pancreas preservation, drives a HIF-1α-dependent switch to glycolysis with subsequent poor glucose-coupled insulin secretion. These findings are complemented by a study by Smith et al. demonstrating ischemia in isolated human islets is damaging even with 8–12h exposure [53]. Both studies found that islets exposed to ischemia had significant transcriptional changes and many of the genes were associated with increased inflammatory and hypoxia-response signaling, and decreased nutrient transport and metabolism [53]. Moreover, exposure to hypoxia appears to induce a proinflammatory signature in the islets, which may further trigger the immune system if transplanted [53].

Fundamental concepts of organ oxygenation

Currently, preservation of solid organs is achieved through static or dynamic approaches. Static cold storage (SCS) in chilled preservation solution remains the primary preservation method. Dynamic preservation methods involve intravascular perfusion of either liquid or gas media. There is an array of preservation solutions optimized to minimize osmotic stress on the organ that can be used to flush, bathe, or perfuse a donor organ. Gas mixtures containing oxygen can be used statically in a hyperbaric chamber or can be perfused intravascularly into an organ, a process referred to as persufflation (PSF). These alternative preservation methods to improve oxygenation of the pancreas are discussed and summarized in Table 1.

SCS is by far the most utilized preservation method for transplant organs because it is low cost, simple, and reasonably effective for use with many organs. This type of hypothermic preservation relies on low temperature to reduce the organ's demands for oxygen and thus ATP. The reduction in metabolic demands therefore reduces tissue OCR following Arrhenius kinetics (e.g., a decrease from 37°C to 4–8°C corresponds to an OCR decline of ~10-fold) [56] and helps reduce hypoxic damage. However, since viable cells and tissue still consume oxygen even at cold preservation temperatures (4–8°C) as oxygen diffuses into the bulk of an organ from its surface, it will be consumed. Consequently, the cells located within the core of the organ may not be adequately oxygenated. The depth of oxygen "penetration" into tissue can be calculated using mathematical modeling and considering both oxygen diffusion and consumption. Small rodent organs, as those used in preclinical studies, may be adequately oxygenated during SCS by virtue of their small dimensions (on the order of millimeters). However, large human organs are inadequately oxygenated during SCS in lieu of larger dimensions (centimeters) [56]. The two-layer method (TLM) has been advocated as an improved SCS wherein high oxygen-carrying perfluorocarbon (PFC) is used to enhance the surface oxygenation of organs. In the TLM, the hydrophobic PFC with a greater specific gravity and the aqueous preservation solution form two separate layers, with the organ suspended between these two layers. The PFC is oxygenated to provide a "reservoir" for oxygen. However, as with standard SCS, the TLM suffers from the same passive oxygen delivery limitations. If assuming an organ is a cylinder of oxygen-consuming tissue and the preservation temperature is 4–8°C, calculations suggest that the oxygen penetration depth is only 1mm and any deeper, the pO_2 would be virtually zero [56]. Consequently, a solid organ that has dimensions beyond 1–2mm (which accounts for all human-sized organs) must be preserved by delivering oxygen into the bulk of the organ by some form of direct perfusion that obviates the limitation of surface diffusion alone.

TABLE 1 Summary comparison of pancreas preservation methods.

Preservation method	Advantages	Disadvantages
Two-layer method	• Relatively simple • Highly portable • Improves oxygenation of small organs	• Oxygenation has limited surface penetration depth and not effective with larger human-sized organs • PFCs add additional cost to static cold storage
Machine perfusion hypothermic	• Flushes waste products during preservation • Rapidly cools organ to reduce oxygen demand • May reduce unwanted enzymatic digestion of pancreas • Established track record with other organs for transplant	• Causes parenchymal edema and potential reperfusion injury • Requires specialized equipment to continuously oxygenate liquid perfusate
Machine perfusion normothermic	• Flushes waste products during preservation • Can be used to assess pancreas function ex vivo at physiological temperatures	• Relies on complex devices that replace or reoxygenate blood in closed circulation ex vivo • Causes parenchymal edema
Persufflation	• Does not significantly pressurize the organ so less likely to cause edema • May enable improved oxygen delivery through the smallest vessels	• Does not flush out waste products during preservation • Requires specialized equipment like electrolyzer to generate continuous oxygen gas delivery

Commonly used and alternative pancreas preservation methods to improve oxygenation

Static cold storage is the most commonly used clinical pancreas preservation method

SCS remains the most common method to preserve organs for transplant because it is low cost, simple to implement, and has a long track record with successful transplantation outcomes. SCS relies on low temperature to reduce the organ demand for oxygen and ATP. The variety of solutions available have been refined over decades, with successful formulations containing some osmotic controlling substances such as lactobionate, raffinose, or hydroxyethyl starch—reviewed by Kenmochi et al. [57] and Wojtusciszyn et al. [58]. The University of Wisconsin (UW) solution is currently the standard preservation solution used for abdominal organ transplantation at most centers. UW solution was initially developed for pancreas preservation and is characterized as an intracellular colloidal solution and has a high potassium/sodium ratio [59]. A leading alternative is Celsior, a crystalloid solution with a low potassium/sodium ratio and low viscosity. Both Celsior and UW solutions have been associated with successful transplantations without any safety concerns [60]. A more recent analysis of the preservation solutions on the outcomes of subsequent islet isolations in humans revealed no significant difference in islet viability or function between Celsior and UW but did note better yields with UW. Furthermore, the effectiveness of transplanted islets following preservation with Celsior, UW, and Institut George Lopez 1 (IGL-1) showed similar results between the preservation solutions [58,59,61].

The long history of SCS has taught us that hypothermia during preservation is critical to enabling transplantation. However, conventional SCS does not address the oxygenation problem. The important role of oxygen was demonstrated by Hackl et al. showing preoxygenation of different preservation solutions improved the quality of pancreas and islets [62].

Alternative preservation methods to improve oxygenation of the pancreas

Two-layer method

TLM for pancreas preservation was a coordinated effort to enhance pancreas oxygenation compared to SCS during preservation. The basis of TLM is to preserve by using conventional aqueous cold storage solution in combination with hydrophobic high oxygen-carrying liquids called PFCs. SCS solutions are aqueous and PFCs hydrophobic, and consequently they are immiscible. Since PFCs have a higher specific gravity, SCS solutions and PFC separate to form two layers. The pancreas, which has a specific gravity higher than aqueous SCS solution but lower than PFC, then suspends at the two-layer interface. The rationale for using the TLM is to both cool as well as better oxygenate the pancreas.

However, the efficacy of TLM in preserving human-sized pancreata has been debated and remains unclear. Pancreata preserved with TLM have been associated with improved islet yields in rats [63], dogs [64], pigs [65–67], and in humans [68–70]. While most studies compare TLM to SCS, Noguchi et al. reported that the islet yield from pancreata preserved with the TLM using a modified preservation solution was significantly higher when compared with the TLM using traditional SCS solution [67]. Interestingly, Noguchi et al. found no significant difference in ATP content of the pancreas or in islet viability, in vitro or in vivo function between the two preservation methods.

However, there was evidence to suggest that TLM improves rodent pancreas preservation [71–73]. In a study by Tanaka et al, TLM was used to preserve rat pancreata after exposure to warm ischemia, mimicking the ischemia and preservation time expected with human DCD donors. The authors found 3 h of TLM followed by conventional SCS improved pancreas ATP levels and subsequent islet yields. However, this was in rodent rather than larger human organs.

Based on these positive findings, TLM was justified for use with human pancreata. However, the use of the rat pancreas model to justify clinical application of TLM remains problematic [74]. No different from conventional SCS, a pancreas preserved with TLM depends solely on oxygen diffusion from the organ surface. Theoretical calculations show that even at 4°C and in the absence of pO_2 gradients in preservation media equilibrated with pure oxygen ($pO_2 = 760$ mm Hg at the medium/organ interface), oxygen penetration depth into the organ is only about 1 mm [56,74]. Direct oxygen partial pressure (pO_2) measurements carried out in 10-mm-thick porcine pancreatic segments preserved with the TLM at 8°C confirmed the theoretical calculations and have shown that pO_2 in their core was virtually zero [56]. Since the rat pancreas is just 1–4 mm thick, TLM may be an acceptable method for organ preservation, providing adequate oxygenation. However, a human-sized pancreas may be centimeters thick and TLM is unlikely to significantly enhance organ oxygenation compared with conventional SCS. This is consistent with reports in human organs that demonstrate no measurable improvement with TLM compared to SCS [68,70]. Some larger retrospective studies have reported that TLM does not significantly improve isolation outcomes [75,76], highlighting the need for continued development of pancreas preservation protocols.

Machine perfusion

Previously in this chapter we discussed that oxygen availability during preservation by SCS and by the TLM relies solely on passive oxygen and nutrient diffusion from the organ surfaces and this is insufficient given the size of a human pancreas for transplant. Machine perfusion is a method developed to help circumvent surface oxygen and nutrient diffusion limitations by delivery of either nonphysiological fluid or blood product directly through the vasculature of an organ. Perfusion may be performed at hypothermic temperatures (4–8°C) with aqueous solutions optimized for preservation or it may be performed at normothermic temperatures (37°C) with blood. An important underlying concept is that because the perfusion circuit is closed, the perfusate needs to be maintained at adequate oxygenation, no matter which fluid or which temperature is used. For example, warm blood perfused into an organ would be depleted of oxygen within seconds, so it needs to be continuously reoxygenated to be effective. Machine perfusion as a preservation technique relies on the premise that devices can be designed to replace or reoxygenate blood in circulation of an organ ex vivo or with fluids optimized for hypothermia to abate ischemic damage [77].

While machine perfusion has evolved, it was observed that immediate intravascular flushing helped to better cool the organ for hypothermic preservation [78]. The first perfusions were chilled solutions hung above the surgical field and leveraged static gravity-directed column pressure to deliver solution to the organ. Currently, following cannulation of inflow artery, hand-driven syringe perfusion is generally used to flush the organ shortly after procurement. The first use of machine perfusion was in hypothermic conditions to preserve canine hearts for up to 12 h [79].

Hypothermic

The rationale behind the use of hypothermic machine perfusion (HMP) for organ preservation is that by reducing the temperature, cellular metabolism will also decrease, thus reducing the metabolic rate, which also means reducing the rate of accumulation of toxic metabolic by-products. Importantly, HMP has been widely successful in preservation of the kidney for transplantation, underscoring that the risk of edema can be managed. However, unlike the kidney, the pancreas does not have a thick capsule and under the pressure of liquid perfusion the pancreas swells significantly and this may cause irreversible tissue injury.

Earlier work to determine if HMP could be used to preserve larger pancreata without damage showed promising results, with some reports of the edema helping to reduce unwanted enzymatic digestion [77,80]. Leeser et al. used HMP to preserve human pancreas for islet isolation and described partial perfusion (4 h) following longer duration of SCS resulted in better yields and insulin secretion of isolated islets when compared to shorter duration of SCS alone [81].

Recent studies support the use of HMP to improve the quality of pancreata allocated for transplant. Several groups utilizing HMP in donor human pancreata have shown the perfusion can be conducted with no edema in either pancreas or duodenum throughout the 6-h perfusion period and with no obvious signs of apoptosis or formation of reactive oxygen species [82,83]. HMP also improved adenosine triphosphate levels without adverse effects compared with SCS [82,83]. Furthermore, islet cell proportions are normal [82] and isolated islets function to transplant standards.

A clinical feasibility study in 2021 compared DBD pancreata preserved by SCS to normally discarded DCD pancreata preserved by 6h of HMP. There was no edema or apoptosis in the pancreas following HMP and the isolated islets were as viable (>90%) between HMP preservation and optimal organs preserved by SCS [84]. This study demonstrated that 6h of HMP can safely increase cold preservation time and may salvage marginal human donor pancreata for use in clinical transplantation.

Normothermic

Apart from SCS, normothermic regional perfusion (NRP) is the only other preservation technique that has been used clinically in pancreas transplantation. NRP is based on extracorporeal membrane oxygenation technology for use in DCD donors [85]. Cannulas are placed in the iliac artery and vein and the donor's own blood is then warmed, oxygenated, and recirculated into the abdominal organs to increase oxygenation in order to minimize injury from warm ischemia [86]. This strategy has shown some success in liver transplantation and variable results in kidney transplantation following DCD organ retrieval [87,88]. Some groups have attempted to apply NRP to clinical unsuitable human donor organs as a means of salvage, but reportedly it led to additional tissue injury [89]. A major disadvantage of normothermic perfusion is that the machinery involved in recirculating the blood is cumbersome and necessitates anticoagulation.

In addition to possible benefits associated with improved preservation, perfusion offers some unique opportunities to assess organ quality prior to islet isolation or transplantation. Several studies outline methods to assess organ quality during perfusion. In the 1990s, Kenmochi et al. correlated brief duration of HMP flow rate in pancreata with their transplantation outcomes [90,91]. Another attractive opportunity offered by machine perfusion involves the assessment of tissue viability by whole-organ oxygen consumption rate measurement [92] or the noninvasive assessment of ATP content by 31P-NMR spectroscopy [54,93]. This is especially true in normothermic perfusion preservation because the temperature is more physiological and thus the kinetics better reflect reality and enable the measurement of glucose-stimulated insulin secretion.

Oxygen persufflation

Persufflation (PSF) is a method of delivering humidified gaseous oxygen directly into the organ through its native vasculature. PSF, like machine liquid perfusion, enhances oxygen availability by circumventing the limitation in passive surface oxygen diffusion by direct delivery of oxygen into the bulk of the organ. However, gaseous PSF does not significantly pressurize the organ and is much less likely to cause edema compared to machine liquid perfusion. Furthermore, it may be that smaller vessels are better perfused with the lower viscosity gaseous medium. The drawback of PSF is that it does not flush out waste products during preservation as occurs in liquid perfusion.

PSF was discovered unexpectedly in 1902 by Rudolf Magnus while perfusing an isolated cat heart with defibrinated blood [94]. This event prompted a series of more extensive studies designed to elucidate the utility of PSF in preserving cardiac function and later, a group at McGill University demonstrated PSF could preserve spinal reflexes in frogs and cardiac muscle contractions in rabbits [95,96]. This work described the significant improvements of PSF over liquid perfusion, citing the lack of edema and the improved oxygenation. The authors even replaced oxygen gas with nitrogen gas to illustrate how anoxia eliminated these reflex activities.

Although PSF was first discovered using hearts, PSF research has since been predominantly with livers and kidneys [7,8,94,97]. In general, there are two PSF approaches that have been investigated in depth and are differentiated by whether vein or artery is cannulated: anterograde (gas enters from the arterial system and exits through the venous system) and retrograde (gas enters from the venous system and exits through needle punctures created in the organ surface). For pancreas to be properly preserved for islet isolation, the ductal network must remain intact for subsequent enzymatic distention. Fig. 1 depicts a schematic of persufflation of human pancreas. Please note that while being persufflated the pancreas is stored (bathed) in standard preservation aqueous solutions.

In pilot studies using porcine pancreata, PSF for 24h improved whole pancreas ATP content as measured by ^{31}P-NMR spectroscopy compared to the TLM [93]. Representative images of a well persufflated pancreas (Fig. 2A) and poorly persufflated pancreas (Fig. 2B) are shown with dark regions representing oxygen gas in the vasculature. The improved ATP content in human pancreas is visualized in the spectra images shown in Fig. 2C–E. In porcine pancreas preserved by TLM (Fig. 2D) there was virtually no ATP detected compared to the rodent pancreas preserved by TLM (Fig. 2C), reflecting the

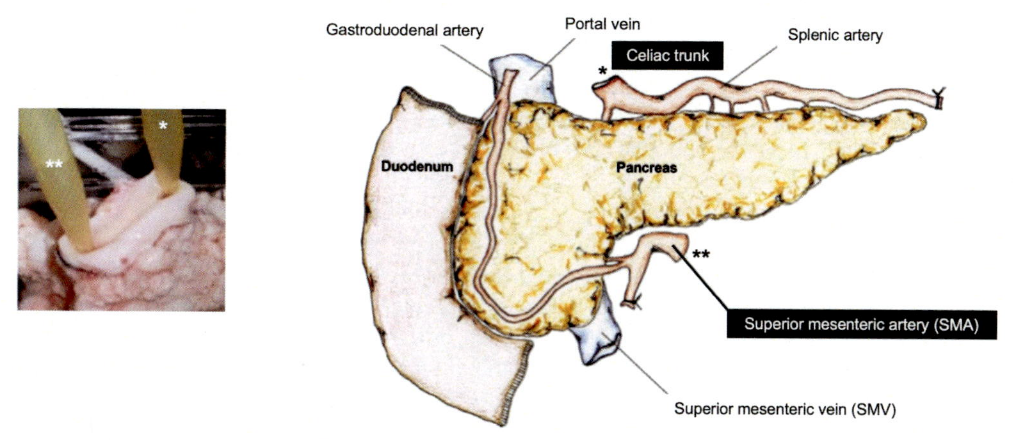

FIG. 1 On the left, the aorta is divided longitudinally and the celiac trunk (*) and the superior mesenteric artery (**) are identified and cannulated. On the right, the schematic drawing highlights the relevant vascular anatomy of the human pancreas in relation to the cannulated vessels (marked * for celiac trunk and ** for the superior mesenteric artery). *(Illustration created by coauthor TMS.)*

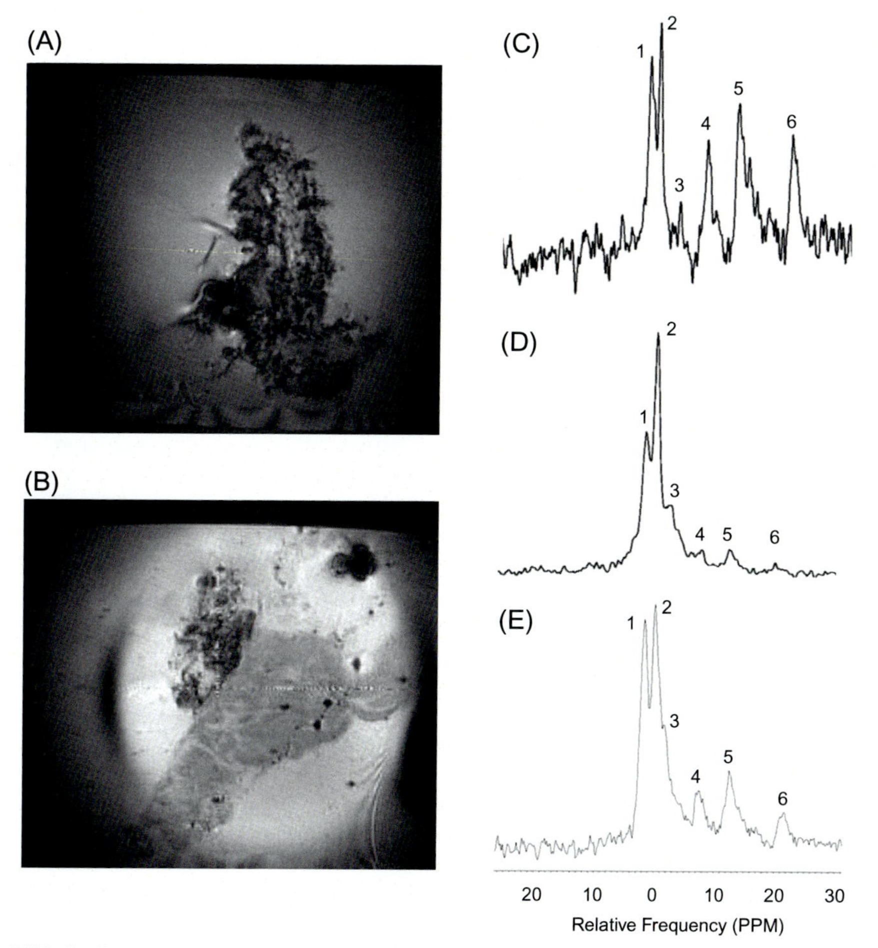

FIG. 2 Oxygenation by PSF. Gradient echo MRI of (A) a well persufflated pancreas with gas filling the vasculature indicated by dark regions and (B) a pancreas with poor persufflation stemming from a large arterial gas leak. The improved status of pancreatic ATP content with PSF can be seen in the spectra images in (C)–(E). ^{31}P-NMR spectra acquired from (C) a rat pancreas preserved by the two-layer method (TLM); (D) a porcine pancreas preserved by TLM; and (E) a persufflated human pancreas. Peaks are labeled accordingly (1) phosphomonoester; (2) inorganic phosphate; (3) phosphodiester; (4) γ-ATP; (5) α-ATP; and (6) β-ATP. *(Reproduced from Scott WE, Weegman BP, Ferrer-Fabrega J, Stein SA, Anazawa T, Kirchner VA, Rizzari MD, Stone J, Matsumoto S, Hammer BE, Balamurugan AN, Kidder LS, Suszynski TM, Avgoustiniatos ES, Stone SG, Tempelman LA, Sutherland DER, Hering BJ, Papas K K. Pancreas oxygen persufflation increases ATP levels as shown by nuclear magnetic resonance. Transplant Proc 2010; 42(6): 2011–2015. https://doi.org/10.1016/j.transproceed.2010.05.091.)*

inability of TLM to adequately preserve larger organs. In Fig. 2E, Scott et al. demonstrate that PSF of human pancreata increased ATP:Pi levels similar to those observed from the rat model preserved by TLM. Furthermore, the persufflated pancreatic tissue exhibited distended capillaries and significantly less cell death relative to regions not exposed to PSF or to tissues preserved with TLM [10].

In addition to improvements to whole pancreas, the beneficial effects of PSF can be seen in isolated human islets. In our work, a series of human islet preparations from deceased donors isolated after preservation by SCS alone ($n=11$) were compared with preparations of pancreata preserved by SCS followed by a period of oxygen PSF (SCS/PSF) ($n=13$) [52]. Pancreas preservation by SCS/PSF compared with SCS alone reduced inflammatory responses and promoted metabolic pathways in human islets. PSF improved glucose-stimulated insulin secretion and these effects persisted for at least 2 days in culture after islet isolation. Interestingly, this work demonstrated that these positive effects were diminished the longer the human pancreas remained in SCS postprocurement and prior to the onset of PSF, underscoring that initiating PSF as soon as possible postprocurement may be important [52]. The likely mechanism for these improvements is reduced cold ischemia, which as discussed has a demonstrable impact on β-cell function [38,53].

Albeit limited, the benefits of PSF for pancreas preservation appear positive and it is unclear why PSF has not been pursued with more vigor. It may be that a stigma accompanies the notion of introducing gas into the vasculature and, consequently, has made clinicians skeptical of the potential utility of PSF [94]. It must be noted that perhaps pancreas PSF prior to islet isolation, which requires enzymatic disruption of the organ and thus is accompanied by no risk of gas embolus since the organ is not reperfused in vivo, may continue to provide a suitable clinical model to establish the role of PSF in organ preservation. There have been clinical trials in liver and kidney transplantation using PSF as a method of organ preservation, however these have been limited [7,97].

Another ongoing limitation of PSF may be the logistics and safety concerns of transporting pressurized oxygen in gas cylinders, especially during air travel. However, technological advancements in portable devices that are able to generate in situ oxygen from water (via electrolysis) eliminate the need to have pressurized gas cylinders [98,99]. These types of devices have been used to generate breathable oxygen in deployed submarines [100]. These technologies could enable PSF to become more widely used as a preservation method for organ transplantation.

Conclusion

Oxygenation is critical to pancreas preservation. Active oxygen delivery using the native vasculature is necessary to adequately oxygenate human-sized organs during the preservation period. Preservation methodologies that enable active intravascular oxygenation such as HMP and PSF may enable the salvage and clinical utilization of ECD pancreata. Additionally, since pancreatic islets are particularly vulnerable to ischemia, active intravascular oxygenation during the preservation period may be critical to enhance isolated islet survival, to improve function, and to reduce the likelihood of rejection.

References

[1] Krieger NR, Odorico JS, Heisey DM, D'Alessandro AM, Knechtle SJ, Pirsch JD, Sollinger HW. Underutilization of pancreas donors. Transplantation 2003;75(8):1271–6. https://doi.org/10.1097/01.TP.0000061603.95572.BF.

[2] Muthusamy ASR, Mumford L, Hudson A, Fuggle SV, Friend PJ. Pancreas transplantation from donors after circulatory death from the United Kingdom. Am J Transplant 2012;12(8):2150–6. https://doi.org/10.1111/j.1600-6143.2012.04075.x.

[3] Neidlinger NA, Odorico JS, Sollinger HW, Fernandez LA. Can "extreme" pancreas donors expand the donor pool? Curr Opin Organ Transplant 2008;13(1):67–71. https://doi.org/10.1097/MOT.0b013e3282f44a51.

[4] Rolles K, Foreman J, Pegg DE. Preservation of ischemically injured canine kidneys by retrograde oxygen persufflation. Transplantation 1984;38(2):102–6. https://doi.org/10.1097/00007890-198408000-00002.

[5] Rolles K, Foreman J, Pegg DE. A pilot clinical study of retrograde oxygen persufflation in renal preservation. Transplantation 1989;48(2):339–42.

[6] Treckmann JW, Paul A, Saad S, Hoffmann J, Waldmann K-H, Broelsch CE, Nagelschmidt M. Primary organ function of warm ischaemically damaged porcine kidneys after retrograde oxygen persufflation. Nephrol Dial Transplant 2006;21(7):1803–8. https://doi.org/10.1093/ndt/gfl066.

[7] Treckmann J, Minor T, Saad S, Ozcelik A, Malagó M, Broelsch CE, Paul A. Retrograde oxygen persufflation preservation of human livers: a pilot study. Liver Transpl 2008;14(3):358–64. https://doi.org/10.1002/lt.21373.

[8] Kuhn-Régnier F, Fischer JH, Jeschkeit S. Coronary oxygen persufflation for long-term myocardial protection. Thorac Cardiovasc Surg 1998;46(Suppl 2):308–12. https://doi.org/10.1055/s-2007-1013091.

[9] Minor T, Klauke H, Isselhard W. Improved preservation of the small bowel by luminal gas oxygenation: energetic status during ischemia and functional integrity upon reperfusion. Transplant Proc 1997;29(7):2994–6. https://doi.org/10.1016/s0041-1345(97)00757-4.

[10] Scott WE, O'Brien TD, Ferrer-Fabrega J, Avgoustiniatos ES, Weegman BP, Anazawa T, Matsumoto S, Kirchner VA, Rizzari MD, Murtaugh MP, Suszynski TM, Aasheim T, Kidder LS, Hammer BE, Stone SG, Tempelman LA, Sutherland DER, Hering BJ, Papas KK. Persufflation improves pancreas preservation when compared with the two-layer method. Transplant Proc 2010;42(6):2016–9. https://doi.org/10.1016/j.transproceed.2010.05.092.

[11] Gruessner AC, Sutherland DER. Pancreas transplant outcomes for United States (US) and non-US cases as reported to the United Network for Organ Sharing (UNOS) and the International Pancreas Transplant Registry (IPTR) as of June 2004. Clin Transpl 2005;19(4):433–55. https://doi.org/10.1111/j.1399-0012.2005.00378.x.

[12] Gruessner AC, Sutherland DER. Pancreas transplant outcomes for United States (US) cases as reported to the United Network for Organ Sharing (UNOS) and the International Pancreas Transplant Registry (IPTR). Clin Transpl 2008;45–56.

[13] Gruessner AC, Sutherland DER. Pancreas transplant outcomes for United States (US) and non-US cases as reported to the United Network for Organ Sharing (UNOS) and the International Pancreas Transplant Registry (IPTR) as of October 2002. Clin Transpl 2002;41–77.

[14] Lehner LJ, Öllinger R, Globke B, Naik MG, Budde K, Pratschke J, Eckardt K-U, Kahl A, Zhang K, Halleck F. Impact of early pancreatic graft loss on outcome after simultaneous pancreas-kidney transplantation (SPKT)-A landmark analysis. J Clin Med 2021;10(15):3237. https://doi.org/10.3390/jcm10153237.

[15] Gruessner RWG, Sutherland DER, Kandaswamy R, Gruessner AC. Over 500 solitary pancreas transplants in nonuremic patients with brittle diabetes mellitus. Transplantation 2008;85(1):42–7. https://doi.org/10.1097/01.tp.0000296820.46978.3f.

[16] Stites E, Kennealey P, Wiseman AC. Current status of pancreas transplantation. Curr Opin Nephrol Hypertens 2016;25(6):563–9. https://doi.org/10.1097/MNH.0000000000000264.

[17] Brennan DC, Kopetskie HA, Sayre PH, Alejandro R, Cagliero E, Shapiro AMJ, Goldstein JS, DesMarais MR, Booher S, Bianchine PJ. Long-term follow-up of the Edmonton protocol of islet transplantation in the United States. Am J Transplant 2016;16(2):509–17. https://doi.org/10.1111/ajt.13458.

[18] Shapiro AMJ, Ricordi C, Hering BJ, Auchincloss H, Lindblad R, Robertson RP, Secchi A, Brendel MD, Berney T, Brennan DC, Cagliero E, Alejandro R, Ryan EA, DiMercurio B, Morel P, Polonsky KS, Reems J-A, Bretzel RG, Bertuzzi F, Lakey JRT. International trial of the Edmonton protocol for islet transplantation. N Engl J Med 2006;355(13):1318–30. https://doi.org/10.1056/NEJMoa061267.

[19] Shapiro AM, Lakey JR, Ryan EA, Korbutt GS, Toth E, Warnock GL, Kneteman NM, Rajotte RV. Islet transplantation in seven patients with type 1 diabetes mellitus using a glucocorticoid-free immunosuppressive regimen. N Engl J Med 2000;343(4):230–8. https://doi.org/10.1056/NEJM200007273430401.

[20] Foster ED, Bridges ND, Feurer ID, Eggerman TL, Hunsicker LG, Alejandro R, Clinical Islet Transplantation Consortium. Improved health-related quality of life in a phase 3 islet transplantation trial in type 1 diabetes complicated by severe hypoglycemia. Diabetes Care 2018;41(5):1001–8. https://doi.org/10.2337/dc17-1779.

[21] Rickels MR, Stock PG, de Koning EJP, Piemonti L, Pratschke J, Alejandro R, Bellin MD, Berney T, Choudhary P, Johnson PR, Kandaswamy R, Kay TWH, Keymeulen B, Kudva YC, Latres E, Langer RM, Lehmann R, Ludwig B, Markmann JF, White S. Defining outcomes for β-cell replacement therapy in the treatment of diabetes: a consensus report on the Igls criteria from the IPITA/EPITA opinion leaders workshop. Transplantation 2018;102(9):1479–86. https://doi.org/10.1097/TP.0000000000002158.

[22] Ryan EA, Lakey JR, Rajotte RV, Korbutt GS, Kin T, Imes S, Rabinovitch A, Elliott JF, Bigam D, Kneteman NM, Warnock GL, Larsen I, Shapiro AM. Clinical outcomes and insulin secretion after islet transplantation with the Edmonton protocol. Diabetes 2001;50(4):710–9. https://doi.org/10.2337/diabetes.50.4.710.

[23] Ryan EA, Paty BW, Senior PA, Bigam D, Alfadhli E, Kneteman NM, Lakey JRT, Shapiro AMJ. Five-year follow-up after clinical islet transplantation. Diabetes 2005;54(7):2060–9. https://doi.org/10.2337/diabetes.54.7.2060.

[24] Vantyghem M-C, Chetboun M, Gmyr V, Jannin A, Espiard S, Le Mapihan K, Raverdy V, Delalleau N, Machuron F, Hubert T, Frimat M, Van Belle E, Hazzan M, Pigny P, Noel C, Caiazzo R, Kerr-Conte J, Pattou F, Members of the Spanish Back Pain Research Network Task Force for the Improvement of Inter-Disciplinary Management of Spinal Metastasis. Ten-year outcome of islet alone or islet after kidney transplantation in type 1 diabetes: a prospective parallel-arm cohort study. Diabetes Care 2019;42(11):2042–9. https://doi.org/10.2337/dc19-0401.

[25] Hering BJ, Clarke WR, Bridges ND, Eggerman TL, Alejandro R, Bellin MD, Chaloner K, Czarniecki CW, Goldstein JS, Hunsicker LG, Kaufman DB, Korsgren O, Larsen CP, Luo X, Markmann JF, Naji A, Oberholzer J, Posselt AM, Rickels MR, Clinical Islet Transplantation Consortium. Phase 3 trial of transplantation of human islets in type 1 diabetes complicated by severe hypoglycemia. Diabetes Care 2016;39(7):1230–40. https://doi.org/10.2337/dc15-1988.

[26] Download.pdf. (n.d.). Retrieved November 12, 2021, from https://www.fda.gov/media/147524/download.

[27] Shapiro AMJ. State of the art of clinical islet transplantation and novel protocols of immunosuppression. Curr Diab Rep 2011;11(5):345–54. https://doi.org/10.1007/s11892-011-0217-8.

[28] Shapiro AMJ. Strategies toward single-donor islets of Langerhans transplantation. Curr Opin Organ Transplant 2011;16(6):627–31. https://doi.org/10.1097/MOT.0b013e32834cfb84.

[29] Brockmann JG, Vaidya A, Reddy S, Friend PJ. Retrieval of abdominal organs for transplantation. Br J Surg 2006;93(2):133–46. https://doi.org/10.1002/bjs.5228.

[30] Arrojo e Drigo R, Ali Y, Diez J, Srinivasan DK, Berggren P-O, Boehm BO. New insights into the architecture of the islet of Langerhans: a focused cross-species assessment. Diabetologia 2015;58(10):2218–28. https://doi.org/10.1007/s00125-015-3699-0.

[31] Patel SN, Ishahak M, Chaimov D, Velraj A, LaShoto D, Hagan DW, Buchwald P, Phelps EA, Agarwal A, Stabler CL. Organoid microphysiological system preserves pancreatic islet function within 3D matrix. Sci Adv 2021;7(7). https://doi.org/10.1126/sciadv.aba5515, eaba5515.

[32] Rodriguez-Diaz R, Abdulreda MH, Formoso AL, Gans I, Ricordi C, Berggren P-O, Caicedo A. innervation patterns of autonomic axons in the human endocrine pancreas. Cell Metab 2011;14(1):45–54. https://doi.org/10.1016/j.cmet.2011.05.008.

[33] Aamodt KI, Powers AC. Signals in the pancreatic islet microenvironment influence β-cell proliferation. Diabetes Obes Metab 2017;19(Suppl 1):124–36. https://doi.org/10.1111/dom.13031.

[34] Bonner-Weir S. Morphological evidence for pancreatic polarity of beta-cell within islets of Langerhans. Diabetes 1988;37(5):616–21. https://doi.org/10.2337/diab.37.5.616.

[35] Cortizo A, Espinal J, Hammonds P. Vectorial insulin secretion by pancreatic beta-cells. FEBS Lett 1990;272(1–2):137–40. https://doi.org/10.1016/0014-5793(90)80467-w.

[36] Avgoustiniatos ES, Colton CK. Effect of external oxygen mass transfer resistances on viability of immunoisolated tissue. Ann N Y Acad Sci 1997;831:145–67. https://doi.org/10.1111/j.1749-6632.1997.tb52192.x.

[37] Cotton CK. Engineering challenges in cell-encapsulation technology. Trends Biotechnol 1996;14(5):158–62. https://doi.org/10.1016/0167-7799(96)10021-4.

[38] Dionne KE, Colton CK, Yarmush ML. Effect of hypoxia on insulin secretion by isolated rat and canine islets of Langerhans. Diabetes 1993;42(1):12–21. https://doi.org/10.2337/diab.42.1.12.

[39] Papas KK, De Leon H, Suszynski TM, Johnson RC. Oxygenation strategies for encapsulated islet and beta cell transplants. Adv Drug Deliv Rev 2019;139:139–56. https://doi.org/10.1016/j.addr.2019.05.002.

[40] Matschinsky FM. Banting lecture 1995. A lesson in metabolic regulation inspired by the glucokinase glucose sensor paradigm. Diabetes 1996;45(2):223–41. https://doi.org/10.2337/diab.45.2.223.

[41] Thorens B, Mueckler M. Glucose transporters in the 21st century. Am J Physiol Endocrinol Metab 2010;298(2):E141–5. https://doi.org/10.1152/ajpendo.00712.2009.

[42] Berman HK, Newgard CB. Fundamental metabolic differences between hepatocytes and islet beta-cells revealed by glucokinase overexpression. Biochemistry 1998;37(13):4543–52. https://doi.org/10.1021/bi9726133.

[43] Schuit FC. Is GLUT2 required for glucose sensing? Diabetologia 1997;40(1):104–11. https://doi.org/10.1007/s001250050651.

[44] Schuit FC, De Vos A, Moens K, Quartier E, Heimberg H. Glucose-induced B-cell recruitment and the expression of hexokinase isoenzymes. Adv Exp Med Biol 1997;426:259–66. https://doi.org/10.1007/978-1-4899-1819-2_36.

[45] Lendahl U, Lee KL, Yang H, Poellinger L. Generating specificity and diversity in the transcriptional response to hypoxia. Nat Rev Genet 2009;10(12):821–32. https://doi.org/10.1038/nrg2665.

[46] Cantley J, Grey ST, Maxwell PH, Withers DJ. The hypoxia response pathway and β-cell function. Diabetes Obes Metab 2010;12(Suppl 2):159–67. https://doi.org/10.1111/j.1463-1326.2010.01276.x.

[47] Cantley J, Selman C, Shukla D, Abramov AY, Forstreuter F, Esteban MA, Claret M, Lingard SJ, Clements M, Harten SK, Asare-Anane H, Batterham RL, Herrera PL, Persaud SJ, Duchen MR, Maxwell PH, Withers DJ. Deletion of the von Hippel-Lindau gene in pancreatic beta cells impairs glucose homeostasis in mice. J Clin Invest 2009;119(1):125–35. https://doi.org/10.1172/JCI26934.

[48] Alcazar O, Tiedge M, Lenzen S. Importance of lactate dehydrogenase for the regulation of glycolytic flux and insulin secretion in insulin-producing cells. Biochem J 2000;352(Pt 2):373–80.

[49] Ishihara H, Wang H, Drewes LR, Wollheim CB. Overexpression of monocarboxylate transporter and lactate dehydrogenase alters insulin secretory responses to pyruvate and lactate in beta cells. J Clin Invest 1999;104(11):1621–9. https://doi.org/10.1172/JCI7515.

[50] Atchison N, Swindlehurst G, Papas KK, Tsapatsis M, Kokkoli E. Maintenance of ischemic β cell viability through delivery of lipids and ATP by targeted liposomes. Biomater Sci 2014;2(4):548–59. https://doi.org/10.1039/C3BM60094G.

[51] Cantley J, Walters SN, Jung M-H, Weinberg A, Cowley MJ, Whitworth TP, Kaplan W, Hawthorne WJ, O'Connell PJ, Weir G, Grey ST. A preexistent hypoxic gene signature predicts impaired islet graft function and glucose homeostasis. Cell Transplant 2013;22(11):2147–59. https://doi.org/10.3727/096368912X658728.

[52] Kelly AC, Smith KE, Purvis WG, Min CG, Weber CS, Cooksey AM, Hasilo C, Paraskevas S, Suszynski TM, Weegman BP, Anderson MJ, Camacho LE, Harland RC, Loudovaris T, Jandova J, Molano DS, Price ND, Georgiev IG, Scott WE, Papas KK. Oxygen perfusion (persufflation) of human pancreata enhances insulin secretion and attenuates islet proinflammatory signaling. Transplantation 2019;103(1):160–7. https://doi.org/10.1097/TP.0000000000002400.

[53] Smith KE, Kelly AC, Min CG, Weber CS, McCarthy FM, Steyn LV, Badarinarayana V, Stanton JB, Kitzmann JP, Strop P, Gruessner AC, Lynch RM, Limesand SW, Papas KK. Acute ischemia induced by high-density culture increases cytokine expression and diminishes the function and viability of highly purified human islets of Langerhans. Transplantation 2017;101(11):2705–12. https://doi.org/10.1097/TP.0000000000001714.

[54] Papas KK, Colton CK, Gounarides JS, Roos ES, Jarema MAC, Shapiro MJ, Cheng LL, Cline GW, Shulman GI, Wu H, Bonner-Weir S, Weir GC. NMR spectroscopy in β cell engineering and islet transplantation. Ann N Y Acad Sci 2001;944(1):96–119. https://doi.org/10.1111/j.1749-6632.2001.tb03826.x.

[55] Papas KK, Long RC, Sambanis A, Constantinidis I. Development of a bioartificial pancreas: I. long-term propagation and basal and induced secretion from entrapped betaTC3 cell cultures. Biotechnol Bioeng 1999;66(4):219–30.

[56] Papas KK, Hering BJ, Guenther L, Gunther L, Rappel MJ, Colton CK, Avgoustiniatos ES. Pancreas oxygenation is limited during preservation with the two-layer method. Transplant Proc 2005;37(8):3501–4. https://doi.org/10.1016/j.transproceed.2005.09.085.

[57] Kenmochi T, Miyamoto M, Sasaki H, Une S, Nakagawa Y, Moldovan S, Benhamou PY, Brunicardi FC, Tanaka H, Mullen Y. LAP-1 cold preservation solution for isolation of high-quality human pancreatic islets. Pancreas 1998;17(4):367–77. https://doi.org/10.1097/00006676-199811000-00007.

[58] Wojtusciszyn A, Bosco D, Morel P, Baertschiger R, Armanet M, Kempf MC, Badet L, Toso C, Berney T. A comparison of cold storage solutions for pancreas preservation prior to islet isolation. Transplant Proc 2005;37(8):3396–7. https://doi.org/10.1016/j.transproceed.2005.09.019.

[59] Niclauss N, Wojtusciszyn A, Morel P, Demuylder-Mischler S, Brault C, Parnaud G, Ris F, Bosco D, Badet L, Benhamou P-Y, Berney T. Comparative impact on islet isolation and transplant outcome of the preservation solutions Institut Georges Lopez-1, University of Wisconsin, and Celsior. Transplantation 2012;93(7):703–8. https://doi.org/10.1097/TP.0b013e3182476cc8.

[60] Boggi U, Vistoli F, Del Chiaro M, Signori S, Croce C, Pietrabissa A, Berchiolli R, Marchetti P, Del Prato S, Mosca F. Pancreas preservation with University of Wisconsin and Celsior solutions: a single-center, prospective, randomized pilot study. Transplantation 2004;77(8):1186–90. https://doi.org/10.1097/01.tp.0000120535.89925.ca.

[61] Hubert T, Gmyr V, Arnalsteen L, Jany T, Triponez F, Caiazzo R, Vandewalle B, Vantyghem M-C, Kerr-Conte J, Pattou F. Influence of preservation solution on human islet isolation outcome. Transplantation 2007;83(3):270–6. https://doi.org/10.1097/01.tp.0000251723.97483.16.

[62] Hackl F, Stiegler P, Stadlbauer V, Schaffellner S, Iberer F, Matzi V, Maier A, Klemen H, Smolle-Jüttner FM, Tscheliessnigg K. Preoxygenation of different preservation solutions for porcine pancreas preservation. Transplant Proc 2010;42(5):1621–3. https://doi.org/10.1016/j.transproceed.2010.02.071.

[63] Matsuda T, Suzuki Y, Tanioka Y, Toyama H, Kakinoki K, Hiraoka K, Fujino Y, Kuroda Y. Pancreas preservation by the 2-layer cold storage method before islet isolation protects isolated islets against apoptosis through the mitochondrial pathway. Surgery 2003;134(3):437–45. https://doi.org/10.1067/s0039-6060(03)00165-x.

[64] Fujino Y. Two-layer cold storage method for pancreas and islet cell transplantation. World J Gastroenterol 2010;16(26):3235–8. https://doi.org/10.3748/wjg.v16.i26.3235.

[65] Brandhorst D, Iken M, Brendel MD, Bretzel RG, Brandhorst H. Adaption of neutral protease activity for islet isolation from the long-term two-layer method-stored pig pancreas. Transplant Proc 2005;37(1):458–9. https://doi.org/10.1016/j.transproceed.2004.12.035.

[66] Brandhorst D, Iken M, Brendel MD, Bretzel RG, Brandhorst H. Successful pancreas preservation by a perfluorocarbon-based one-layer method for subsequent pig islet isolation. Transplantation 2005;79(4):433–7. https://doi.org/10.1097/01.tp.0000151765.96118.1b.

[67] Noguchi H, Ueda M, Nakai Y, Iwanaga Y, Okitsu T, Nagata H, Yonekawa Y, Kobayashi N, Nakamura T, Wada H, Matsumoto S. Modified two-layer preservation method (M-Kyoto/PFC) improves islet yields in islet isolation. Am J Transplant 2006;6(3):496–504. https://doi.org/10.1111/j.1600-6143.2006.01223.x.

[68] Matsumoto S, Zhang G, Qualley S, Clever J, Tombrello Y, Strong DM, Reems JA. The effect of two-layer (University of Wisconsin solution/perfluorochemical) preservation method on clinical grade pancreata prior to islet isolation and transplantation. Transplant Proc 2004;36(4):1037–9. https://doi.org/10.1016/j.transproceed.2004.04.012.

[69] Ramachandran S, Desai NM, Goers TA, Benshoff N, Olack B, Shenoy S, Jendrisak MD, Chapman WC, Mohanakumar T. Improved islet yields from pancreas preserved in perflurocarbon is via inhibition of apoptosis mediated by mitochondrial pathway. Am J Transplant 2006;6(7):1696–703. https://doi.org/10.1111/j.1600-6143.2006.01368.x.

[70] Zhang G, Matsumoto S, Newman H, Strong DM, Robertson RP, Reems J-A. Improve islet yields and quality when clinical grade pancreata are preserved by the two-layer method. Cell Tissue Bank 2006;7(3):195–201. https://doi.org/10.1007/s10561-006-0002-0.

[71] Gioviale MC, Damiano G, Puleio R, Bellavia M, Cassata G, Palumbo VD, Spinelli G, Altomare R, Barone R, Cacciabaudo F, Buscemi G, Lo Monte AI. Histologic effects of University of Wisconsin two-layer method preservation of rat pancreas. Transplant Proc 2013;45(5):1723–8. https://doi.org/10.1016/j.transproceed.2013.02.047.

[72] Takahashi T, Tanioka Y, Matsuda T, Toyama H, Kakinoki K, Li S, Hiraoka K, Fijino Y, Suzuki Y, Kuroda Y. Impact of the two-layer method on the quality of isolated pancreatic islets. Hepato-Gastroenterology 2006;53(68):179–82.

[73] Tanaka T, Suzuki Y, Tanioka Y, Sakai T, Kakinoki K, Goto T, Li S, Yoshikawa T, Matsumoto I, Fujino Y, Kuroda Y. Possibility of islet transplantation from a nonheartbeating donor pancreas resuscitated by the two-layer method. Transplantation 2005;80(6):738–42. https://doi.org/10.1097/01.tp.0000174136.70282.5a.

[74] Avgoustiniatos ES, Hering BJ, Papas KK. The rat pancreas is not an appropriate model for testing the preservation of the human pancreas with the two-layer method. Transplantation 2006;81(10):1471–2. https://doi.org/10.1097/01.tp.0000215389.64186.3f.

[75] Caballero-Corbalán J, Eich T, Lundgren T, Foss A, Felldin M, Källen R, Salmela K, Tibell A, Tufveson G, Korsgren O, Brandhorst D. No beneficial effect of two-layer storage compared with UW-storage on human islet isolation and transplantation. Transplantation 2007;84(7):864–9. https://doi.org/10.1097/01.tp.0000284584.60600.ab.

[76] Kin T, Mirbolooki M, Salehi P, Tsukada M, O'Gorman D, Imes S, Ryan EA, Shapiro AMJ, Lakey JRT. Islet isolation and transplantation outcomes of pancreas preserved with University of Wisconsin solution versus two-layer method using preoxygenated perfluorocarbon. Transplantation 2006;82(10):1286–90. https://doi.org/10.1097/01.tp.0000244347.61060.af.

[77] Taylor MJ, Baicu SC. Current state of hypothermic machine perfusion preservation of organs: the clinical perspective. Cryobiology 2010;60(3 Suppl):S20–35. https://doi.org/10.1016/j.cryobiol.2009.10.006.

[78] Weegman BP, Suszynski TM, Scott WE, Ferrer Fábrega J, Avgoustiniatos ES, Anazawa T, O'Brien TD, Rizzari MD, Karatzas T, Jie T, Sutherland DER, Hering BJ, Papas KK. Temperature profiles of different cooling methods in porcine pancreas procurement. Xenotransplantation 2014;21(6):574–81. https://doi.org/10.1111/xen.12114.

[79] Calhoon JH, Bunegin L, Gelineau JF, Felger MC, Naples JJ, Miller OL, Sako EY. Twelve-hour canine heart preservation with a simple, portable hypothermic organ perfusion device. Ann Thorac Surg 1996;62(1):91–3. https://doi.org/10.1016/0003-4975(96)00272-x.

[80] Taylor MJ, Baicu S, Leman B, Greene E, Vazquez A, Brassil J. Twenty-four hour hypothermic machine perfusion preservation of porcine pancreas facilitates processing for islet isolation. Transplant Proc 2008;40(2):480–2. https://doi.org/10.1016/j.transproceed.2008.01.004.

[81] Leeser DB, Bingaman AW, Poliakova L, Shi Q, Gage F, Bartlett ST, Farney AC. Pulsatile pump perfusion of pancreata before human islet cell isolation. Transplant Proc 2004;36(4):1050–1. https://doi.org/10.1016/j.transproceed.2004.04.041.

[82] Branchereau J, Renaudin K, Kervella D, Bernadet S, Karam G, Blancho G, Cantarovich D. Hypothermic pulsatile perfusion of human pancreas: preliminary technical feasibility study based on histology. Cryobiology 2018;85:56–62. https://doi.org/10.1016/j.cryobiol.2018.10.002.

[83] Leemkuil M, Lier G, Engelse MA, Ploeg RJ, de Koning EJP, t' Hart NA, Krikke C, Leuvenink HGD. Hypothermic oxygenated machine perfusion of the human donor pancreas. Transplant Direct 2018;4(10). https://doi.org/10.1097/TXD.0000000000000829, e388.

[84] Doppenberg JB, Leemkuil M, Engelse MA, Krikke C, de Koning EJP, Leuvenink HGD. Hypothermic oxygenated machine perfusion of the human pancreas for clinical islet isolation: a prospective feasibility study. Transpl Int 2021;34(8):1397–407. https://doi.org/10.1111/tri.13927.

[85] Prudhomme T, Mulvey JF, Young LAJ, Mesnard B, Lo Faro ML, Ogbemudia AE, Dengu F, Friend PJ, Ploeg R, Hunter JP, Branchereau J. Ischemia-reperfusion injuries assessment during pancreas preservation. Int J Mol Sci 2021;22(10):5172. https://doi.org/10.3390/ijms22105172.
[86] Rojas-Peña A, Sall LE, Gravel MT, Cooley EG, Pelletier SJ, Bartlett RH, Punch JD. Donation after circulatory determination of death: the University of Michigan experience with extracorporeal support. Transplantation 2014;98(3):328–34. https://doi.org/10.1097/TP.0000000000000070.
[87] Antoine C, Savoye E, Gaudez F, Cheisson G, Badet L, Videcoq M, Legeai C, Bastien O, Barrou B, National Steering Committee of Donation After Circulatory Death. Kidney transplant from uncontrolled donation after circulatory death: contribution of normothermic regional perfusion. Transplantation 2020;104(1):130–6. https://doi.org/10.1097/TP.0000000000002753.
[88] Oniscu GC, Randle LV, Muiesan P, Butler AJ, Currie IS, Perera MTPR, Forsythe JL, Watson CJE. In situ normothermic regional perfusion for controlled donation after circulatory death—The United Kingdom experience. Am J Transplant 2014;14(12):2846–54. https://doi.org/10.1111/ajt.12927.
[89] Branchereau J, Hunter J, Friend P, Ploeg R. Pancreas preservation: clinical practice and future developments. Curr Opin Organ Transplant 2020;25(4):329–35. https://doi.org/10.1097/MOT.0000000000000784.
[90] Asano T, Kenmochi T, Isono K. Organ preservation. Nihon Geka Gakkai Zasshi 1996;97(11):958–63.
[91] Kenmochi T, Asano T, Nakagouri T, Enomoto K, Isono K, Horie H. Prediction of viability of ischemically damaged canine pancreatic grafts by tissue flow rate with machine perfusion. Transplantation 1992;53(4):745–50. https://doi.org/10.1097/00007890-199204000-00007.
[92] Bunegin L, Tolstykh GP, Gelineau JF, Cosimi AB, Anderson LM. Oxygen consumption during oxygenated hypothermic perfusion as a measure of donor organ viability. ASAIO J 2013;59(4):427–32. https://doi.org/10.1097/MAT.0b013e318292e865.
[93] Scott WE, Weegman BP, Ferrer-Fabrega J, Stein SA, Anazawa T, Kirchner VA, Rizzari MD, Stone J, Matsumoto S, Hammer BE, Balamurugan AN, Kidder LS, Suszynski TM, Avgoustiniatos ES, Stone SG, Tempelman LA, Sutherland DER, Hering BJ, Papas KK. Pancreas oxygen persufflation increases ATP levels as shown by nuclear magnetic resonance. Transplant Proc 2010;42(6):2011–5. https://doi.org/10.1016/j.transproceed.2010.05.091.
[94] Suszynski TM, Rizzari MD, Scott WE, Eckman PM, Fonger JD, John R, Chronos N, Tempelman LA, Sutherland DE, Papas KK. Persufflation (gaseous oxygen perfusion) as a method of heart preservation. J Cardiothorac Surg 2013;8:105. https://doi.org/10.1186/1749-8090-8-105.
[95] Bunzl A, Burgen AS, Burns BD, Pedley N, Terroux KG. Methods for studying the reflex activity of the frog's spinal cord. Br J Pharmacol Chemother 1954;9(2):229–35. https://doi.org/10.1111/j.1476-5381.1954.tb00846.x.
[96] Burns BD, Robson JG, Smith GK. The survival of mammalian tissues perfused with intravascular gas mixtures of oxygen and carbon dioxide. Can J Biochem Physiol 1958;36(5):499–504.
[97] Flatmark A, Slaattelid O, Woxholt G. Gaseous persufflation during machine perfusion of human kidneys before transplantation. Eur Surg Res 1975;7(2):83–90. https://doi.org/10.1159/000127794.
[98] Hirche TO, Born T, Jungblut S, Sczepanski B, Kenn K, Köhnlein T, Hirche H, Wagner TO. Oxygen generation by combined electrolysis and fuel-cell technology: clinical use in COPD patients requiring long time oxygen therapy. Eur J Med Res 2008;13(10):451–8.
[99] Maleki T, Cao N, Song SH, Kao C, Ko S-CA, Ziaie B. An ultrasonically powered implantable micro-oxygen generator (IMOG). IEEE Trans Biomed Eng 2011;58(11):3104–11. https://doi.org/10.1109/TBME.2011.2163634.
[100] Oxygen Generator Cell Design for Future Submarines on JSTOR. (1996) Retrieved November 13, 2021, from https://www.jstor.org/stable/44725543.

Chapter 8

Encapsulation devices to enhance graft survival: The latest in the development of micro and macro encapsulation devices to improve clinical, xeno, and stem cell transplantation outcomes

Thomas Loudovaris

Immunology and Diabetes, St Vincent's Institute for Medical Research, Fitzroy, VIC, Australia; University of Arizona, Tucson, AZ, United States

Introduction

Many human diseases, such as diabetes, are caused by the failure of the body to produce an essential protein. Such diseases could be effectively treated if the afflicted individuals were to receive transplants of cells that supply the missing protein. To currently do this, such cells need to be protected from the host immune system by immunosuppressing the host with drugs that unfortunately also have a host of off-target effects. Utilizing a treatment protocol including immunosuppression, Pancreatic Islet cells which contain the insulin-producing Beta cells, have effectively treated patients suffering from Type 1 diabetes where they also have severe hypoglycaemic unawareness and many of these patients have become insulin-independent when given enough islets. With immunosuppression, most tissues and organs can be successfully transplanted between individuals with 1-year graft survival of greater than 90% for Kidney, Heart, and Liver transplants (Based on OPTN data as of March 15, 2021). The 10% failure rate increases over time and by 5 years the survival rate falls a further 15%. So as good as the immunosuppression has become it cannot protect all grafts and its protective effects diminish over time.

An alternative to the use of these toxic immunosuppressive drugs is to utilize an encapsulation or immunoisolation system that protects the transplanted cells from rejection without the need for immunosuppression. The concept of curing diabetes using pancreatic islets enclosed within protective membranes is not new and has often been described as a bioartificial pancreas. There are reports in the scientific literature reporting effective treatment of diabetes through encapsulated pancreatic islet cell transplantation [1,2]. Most of these "cures" have been short-lived and have suffered from membrane biocompatibility issues of the implantable device as well as lack of large-scale tissue supply. In addition, most approaches have focused on transplanting animal tissues into humans because of the severe shortage of human pancreatic islets. The barriers to transplantation of animal tissues (xenografts) into humans are significantly higher than transplantation of human tissues or cells, even with the use of encapsulation. In this chapter we will describe and discuss the use of encapsulation to provide a safe and effective transplant method to enhance graft survival without the use of systemic immunosuppression. This will include discussion of common encapsulation problems of biocompatibility, membrane permeability, immune protection, and long-term stability. It also focuses on the survival of therapeutic cells for treatment of deficiencies or diseases presenting experimental and clinical use of encapsulated cells to deliver proteins such as insulin.

Design of reactors and transplantation devices

Ideally, an encapsulated beta cell therapy should meet three requirements.

(1) Provide Immune protection against allograft rejection and autoimmunity, and hence will be a long-term therapy.
(2) It should have access to vascular supply to provide glucose regulation.
(3) Be minimally invasive and be retrievable ensuring patient safety.

Pancreas and Beta Cell Replacement. https://doi.org/10.1016/B978-0-12-824011-3.00007-2

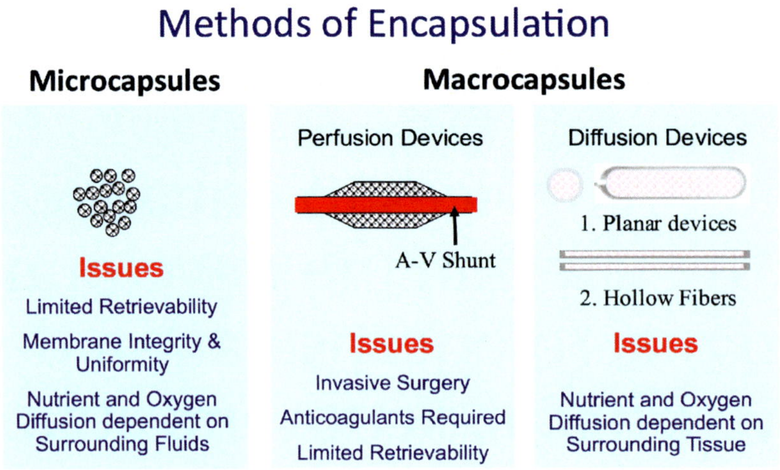

FIG. 1 Three common forms of cell encapsulation are depicted, microcapsules, microcapsule perfusion devices, and microcapsule diffusion devices.

Encapsulation within cell impermeable barriers has been developed as an alternative to the chronic immunosuppression required to protect transplanted cells or tissues. This transplant method is often referred to as immunoisolation because the transplanted tissue is isolated from the immune system by the membrane barrier. These cell impermeable barriers restrict the components of the immune system that cause rejection but allow the passage of therapeutic proteins and nutrition. There are two categories of the cell containing implantable devices; (A) microencapsulation, in which cells are encapsulated in gel spheres ranging from approximately 1 mm in diameter down to a thin coating around the cells, or (B) macroencapsulation in which cells are encapsulated in devices that are either attached to the vascular system, anastomosed to the avascular system as arteriovenous shunts [3], or implanted into soft tissues which then become slowly vascularised (see Fig. 1).

Microencapsulation

Microcapsules are usually made from alginate housing one or a small number of islets or islet cell clusters, although other materials have been used. The use of microencapsulated islets was first published in 1980 [4], before then microcapsules encapsulated molecules and enzymes [5] [6]. Microcapsules can vary in size from the conformal coating of islets, 150-300 μm up to 1.5 mm in diameter, varying due to the polymers used [7]. Microcapsules have the best surface area/volume of the device configurations if the spheroids as they can be kept as small as possible. Increasing sphere sizes increases the amount of biomaterial per islet/cell increasing the overall amount of material to be implanted. Current fabrication techniques for microencapsulation, such as air jet-driven and electrostatic methods have kept size small [8]. Conformal coating minimizes capsule size even further. A very thin immunoprotective biomaterial covers the cell cluster, conforming to their topography [9]. The most common biomaterials for conformal coatings are photo-crosslinked polyethylene glycol (PEG) [10] and hydroxyethyl methacrylate-methyl methacrylate (HEMA-MMA) [11] which can provide coatings in the range of 1–15 μm.

Macroencapsulation

Unlike microcapsules, macrocapsules house more than one islet and have the advantage that they can be fully retrieved. As mentioned engineered to be part of the vasculature [12] or create their own vasculature [13,14]. Intravascular macroencapsulation was championed in the late 80s to early 90s by Biohybrid Technologies Incorporated, but unlike extravascular macroencapsulation devices can suffer from thrombosis.

Macroencapsulation devices were called diffusion chambers when they were first used [15]. These planar devices are very similar to those currently in use, consisting of a sandwich of cell impermeable polytetrafluoroethylene (ptfe) membranes. Macrocapsules have been constructed from many polymers and hydrogels [16] and combinations of both [17]. Diffusion through hydrogels is different from the traffic of proteins and gases through the pores of solid polymers. With hydrogels, electrostatic and hydrophobic can dominate over size exclusion properties of the gel [18] compared to the solid polymers where size exclusion dictates diffusion properties. Hollow fibers were adopted many years ago as a form of macrocapsule [19,20] and had success in correcting diabetes in rats and this form of device has continued to the present [21] (Fig. 1).

All encapsulation devices face hurdles and investigators have different technologies. The major hurdles to their use are biocompatibility, membrane permeability, vascular integration, and immune protection.

Biocompatibility

When any material is implanted into a host, the hosts biological defenses treat the implant as a foreign invader and try to isolate it and immunologically destroy it. Since most implanted materials are made from non-living substances, they are impervious to most if not all the cytotoxic factors the defenses can throw at it. However, some of these factors and immunological cells can direct other cells to isolate the invader by forming an avascular fibrotic capsule around the device. This response is often referred to as the foreign body reaction and is an indicator of biocompatibility. The cellular reaction around the implant is mediated through a fluid biolayer which consists of water, solutes, proteins, and other molecules from the physiological fluids of the host. Therefore, many approaches to increase biocompatibility are to use or develop membranes that do not bind to these proteins or molecules. Another method of avoiding these issues is by engineering devices such as microcapsules which have the advantage in that they float freely in the peritoneal cavity, but they do eventually embed into the soft tissues of the peritoneal cavity. Not surprisingly, the chemical composition and the purity of the gelation materials of these devices has improved their biocompatibility. Another type of device is the perfusion macrocapsule, where blood flows through the device, these have been designed with non-thrombogenic materials to prevent blood clots that could potentially dislodge and cause embolism. Diffusion macrocapsules, either cylindrical or flat sheet devices are embedded directly into body tissues and these undergo more typical foreign body responses. The foreign body response to biomaterials has been well documented [22] and begins with non-specific protein adsorption, which does not occur in the normal physiological process of wound healing and may be the instigator in the foreign body reaction. This is followed by the adherence of monocytes, leukocytes, and platelets and in turn, leads to the upregulation of cytokines and proinflammatory processes. As a result of the chronic inflammation that follows, the adherent macrophages cannot phagocytose the considerably larger foreign body and fuse together to form multinucleated foreign body giant cells that may persist for the lifetime of the implant [22]. The final result is a device that becomes walled off by an avascular fibrous layer of connective tissue that can be greater than 200 μm thick rendering the device ineffective.

There are a number of means to address these events, targeting monocyte adhesion has been one such approach to prevent the foreign body reaction from occurring [23]. Monocyte adhesion to biomaterials in vitro has been able to be altered by modification of the surface chemistry [24]. However, in vivo, the foreign body reaction seems independent of just simple surface chemistry with different materials all eliciting a foreign body response [25]. From all the work that has already been undertaken that it is apparent that the initial step of non-specific protein adsorption is not prevented in vivo and that an absolutely inert biomaterial may not exist [23,26].

These studies indicate that altering the surface chemical properties of the implant material that eliminate the non-specific protein adsorption may positively influence biocompatibility [27]. Polyethylene glycol (PEG), has been found to be resistant to both protein and cell adhesion and has been used as a hydrogel or a component in microcapsules and in macrocapsules [28]. A number of groups have found that the topographical properties of the implant surface strongly influence the biocompatibility properties of the implant [29,30]. In screening studies in which various membranes were implanted subcutaneously in rats [31], a group of membranes were identified, that became neovascularized at the membrane-host tissue interface with a markedly reduced foreign body response (Fig. 2) [31,32].

These vascularizing membranes were found to have similar properties when implanted into humans and have been incorporated into devices for transplantation of living cells at high packing densities [33–35]. These membranes appear not to avoid protein or cell adhesion but appear to disrupt the formation of the continuous layer of multinucleated foreign body giant cells and the fibrous avascular layer of the foreign body response. It also appears that the macrophages contribute to the neovascularization as in vivo treatment with anti-macrophage antibodies diminishes the observed neovascularization. However, the vascularizing properties were found to be associated with the ability of cells to infiltrate these membranes, rendering them useless as a cellularly impermeable barrier. The solution was to develop a dual or bilayer membrane, with the outer layer providing vascularization and the inner layer providing the cell impermeable properties. This is the basis for the TheraCyte device or implant system (Fig. 3) which has been used and the results of its use published by many researchers. Another key component of an encapsulation device is chemical and physical stability especially in the long-term if the device is to be implanted for extended periods. Both layers of the TheraCyte device are made of expanded polytetrafluoroethylene (ePTFE). Expanded PTFE has a long history as an implantable material in humans and because it is chemically inert and stable it has been used as the basic material for vascular grafts with millions of ePTFE vascular grafts having been successfully implanted and functioned long-term worldwide. Studies with vascular ePTFE grafts found the ePTFE implant structure, like that of the vascularizing outer layer of the TheraCyte device, facilitates ingrowth of host tissue which

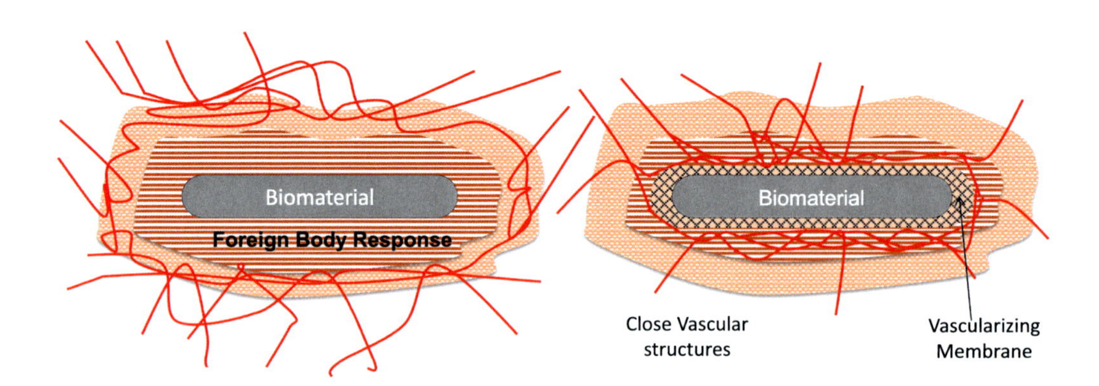

FIG. 2 A vascularizing membrane encourages close vascular structures to develop through the foreign body response and sit close to the encapsulating biomaterial/membrane.

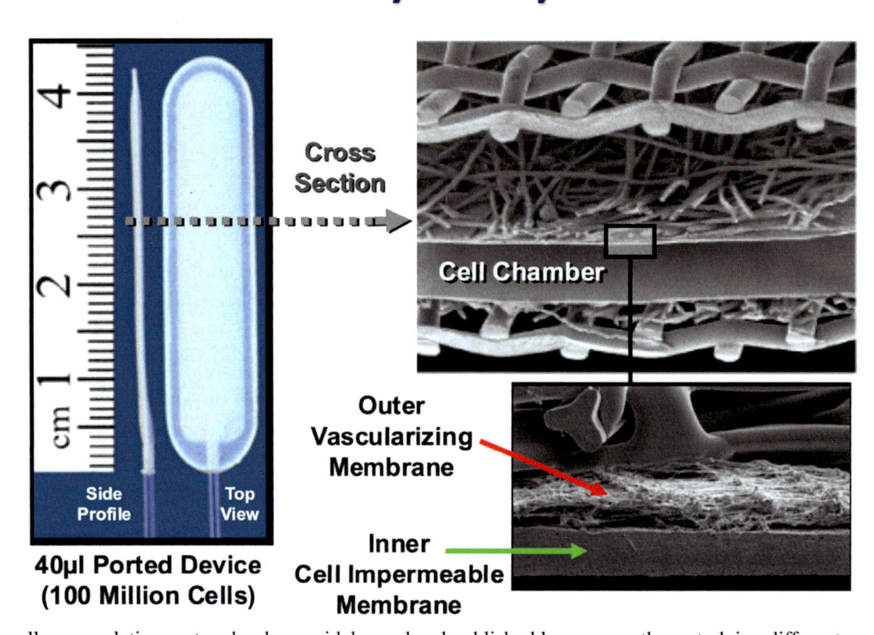

FIG. 3 The TheraCyte cell encapsulation system has been widely used and published by many authors studying different areas of research. The image on the left is a top and side view of a 40 μL device. The cross-sections on the left show the physical makeup of TheraCyte devices, with the important two inner membranes, one providing immune protection through cell impermeability and the other enhancing vascularization around the device. Also shown are two outer membranes that provide physical support to the inner membranes and the device itself.

increases the attachment to surrounding soft tissues, increasing fixation strength and decreasing any possibility of movement or even extrusion but still allowed easy removal [36]. A few companies, Viacyte, Beta-O2, and Procyon have also adopted the use of ePTFE membranes as the basis for the construction of their encapsulation devices. In contrast, the lack of physical and chemical stability of the hydrogels that are used for microencapsulation, has led to the design of biodegradable microcapsules [37]. The thinking behind biodegradability was that the hydrogel coating would degrade about the time the islets would begin failing in the patient and be cleared in the kidney. This treatment will require periodic replacement of the degraded capsules and the tissues they contain.

Site of implantation

The site of implantation is one of the most important aspects to the outcome of the implantable device in regard to how it will and how long it will be functional. We can utilize several ways to best implant devices to maximize outcomes. Some places in the body are immune privileged. At these sites, an inflammatory response would be detrimental and therefore, the body's immune system either cannot access these places or some of the functions of the immune system are suppressed. Privileged sites include the brain, the eye, the testis, and the thymus [38,39]. With microencapsulation, spheres are commonly placed in the peritoneal cavity where less than optimal nutrition for the encapsulated cells is provided by the peritoneal fluids. Intraperitoneal transplantation of microencapsulated islets has been found to restore normoglycemia in many diabetic animals [2,29,30,40–43]. The transport of insulin from the peritoneal cavity to the bloodstream is mainly by passive diffusion [44], not surprisingly, animals with intraperitoneal capsules lack a normal glucose tolerance test nor normal insulin responses to a physiological stimulus [45]. The abnormal function of the encapsulated tissues could be accounted for by the lack of close contact with the bloodstream. Microcapsules are however difficult to quantitatively retrieve after implantation in the peritoneal cavity and their fragility can lead to irritation and possible lymphatic and other duct blockages within the peritoneal cavity [46]. This has therefore led to research of biodegradable microcapsules, obviating the need for retrieval [47].

Unlike microcapsules, macroencapsulation devices are retrievable if complications arise and are less fragile. Efforts at implantation of tissues within macroencapsulation devices have been carried out for over 60 years at both intravascular and extravascular sites. Intravascular devices have been shown to function for at least 267 days [48]. However, placement of these devices requires vascular surgery and exposes the patient to potentially fatal complications due to thrombosis or the effects of the anticoagulant therapy used to reduce the risk of a thrombotic event. Although intravascular devices may not be able to be placed in immune-privileged sites a variation has been studied that experimentally permits xenotransplantation [49].

Extravascular devices have so far consisted of islets placed in hollow fibers and have functioned for extended periods [50]. However, because they usually are implanted into soft tissues, for example, subcutaneous tissues, they have suffered the most from lack of biocompatibility and the foreign body response [51] [52]. It was observed in numerous laboratories that when devices containing pancreatic islets were implanted in animals, diabetes was corrected transiently. Immune privileged sites have also been used for extravascular macroencapsulation devices for the delivery of neurotrophic factors to the brain [53–57] and the eye [58]. Immune privileged sites not only provide a reprieve from the foreign body response but also provides added protection for encapsulated xenografts [59].

Immune protection of encapsulated transplants

Immune privileged sites are not always practical and therefore many encapsulated cells have been placed at either subcutaneous or intraperitoneal locations. Therefore, the immunology of encapsulated cells at these sites. The permeability of the encapsulating membrane provides a balance between immune protection and nutrition of the encapsulated tissues and can be crucial to the function of the encapsulation system.

To minimize both the humoral and cellular immune response of the host, the encapsulating membrane should not only restrict direct contact with cells of the immune system and depending on if the implant tissue is from another species (xenograft) or from the same species (allograft), may also have to exclude the penetration of cytotoxic molecules into the capsule. Some of the molecules considered dangerous include antibodies, components of the complement system, cytokines, and other cytotoxic substances. Therefore, investigators have focused on controlling access to these molecules. Initially, most of the focus was to restrict antibodies and components of the complement system with a molecular mass restriction of usually 50,000 Da or higher [60]. However, for xenografts, this was found not to be immunoprotective and investigators then focused on the cytokines which meant a molecular mass restriction of even smaller molecules down to as small as 10,000 to 14,000 Da [61–63]. However, there is evidence that membrane barriers including these restrictive membranes do not prevent the exiting of insulin and other smaller molecules which can still elicit an immune response by the host [64].

Allograft protection

Even from very early studies with encapsulation devices, it was found that allografts readily survived as long as there was inhibition of cellular contact with the host [15,65]. This protection was maintained even if the animals were previously immunized with the transplanted cells [66]. Encapsulated islets were found not to stimulate allogeneic splenocytes

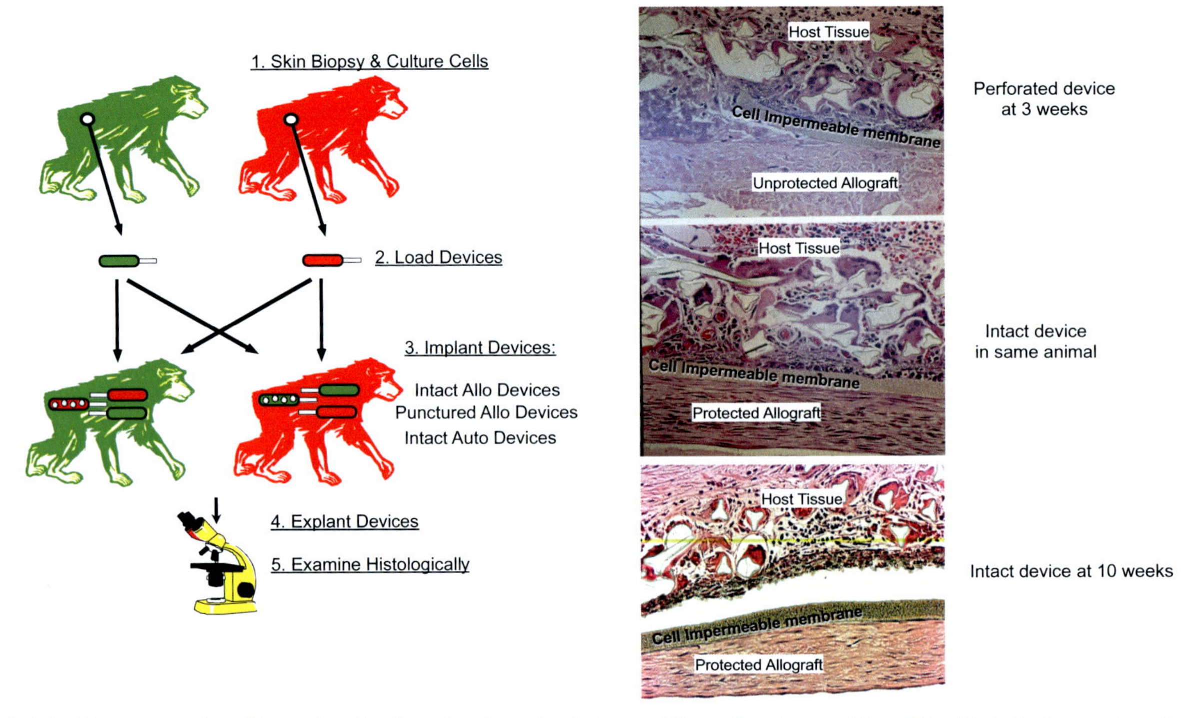

FIG. 4 Fibroblasts were cultured from skin biopsies taken from three baboons. When adequate quantities of fibroblasts had grown in culture, they were loaded into 40 μL TheraCyte devices and implanted subcutaneously into each of the baboons. Each baboon received seven devices, one containing autologous fibroblasts and six containing allogeneic fibroblasts, three each from unrelated (allogeneic) biopsied baboons, two of the devices were intact and one was perforated with a 30G needle. Devices were removed and examined histologically at 3 weeks and 10 weeks post-implantation, shown in the images on the right. Immune cells can be seen around the three devices shown, with the encapsulated tissue dead in the perforated devices but surviving/ protected in intact devices at 3 and 10 weeks post-implantation.

in vitro under conditions that unencapsulated islets would [67], further highlighting the importance of inhibiting cellular contact with the immune system for allograft survival. Studies with allografts in TheraCyte devices in humans also found long-term protection of allografts for greater than a year [68]. Studies in primates also demonstrated allograft protection in intact devices where the allograft was destroyed in intentionally perforated devices (Fig. 4). The molecular permeability of encapsulation devices may not be crucial. Encapsulation devices that have a high molecular permeability also will have the greatest capacity for nutritional diffusion as well as the diffusion of therapeutic proteins.

A presentation by Dr. David Scharp (Novocell, Inc. in Irvine California) at the American Diabetes Association's 65th Annual Scientific Sessions in San Diego, described a study conducted in diabetic baboons in which allogeneic islet cells, microencapsulated in PEG, were injected subcutaneously with low doses of cyclosporine immunosuppression for a period of 30 days. Histological examination of the subcutaneous implant sites revealed little immune reactivity and increased capillaries. Increased vascularization around implant devices has been observed before with the use of immunomodulatory drugs such as cyclosporine [69,70]. Dr. Scharp's presentation demonstrated allograft protection in a group of 15 of 16 baboons which demonstrated significant reductions in hemoglobin A1c, average and fasting blood glucose, glycaemic excursions, and insulin requirements. Also shown was that the baboons can be re-treated safely so there should be the possibility to do multiple implants safely without the permanent use of immunosuppressive drugs.

Microcapsules appear to be able to evade the immune system by their ability to remain unattached from host tissues, a situation also where they cannot be close to the bloodstream and may not be able to provide normal glucose regulation. Severe fibrotic responses have also been observed whenever microcapsules, especially containing xenogeneic islets, become overgrown or attach to host tissues, resulting in decreased function and death of the encapsulated cells as described earlier [71–74].

Xenograft protection

With allografts, the inhibition of cellular contact with the host's immune cells is sufficient to confer protection. With encapsulated xenografts inhibiting cell to cell contact with the host does not appear to be adequate [75–77]. The mechanism

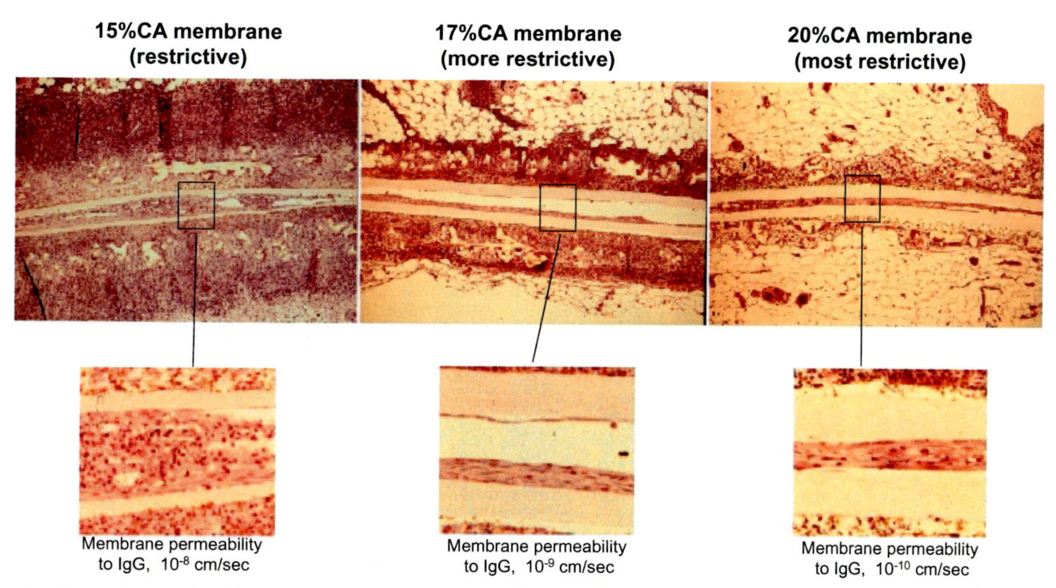

FIG. 5 Xenografts (3-week mouse fetal lung implants in rats) encapsulated in Cellulose Acetate membranes of varying permeabilities. Note the decrease in tissue density as permeability decreases.

of encapsulated xenograft destruction appears to be dependent on the presence of CD4 cells that cause a huge inflammatory response destroying the vital vasculature that the encapsulated cells contain and even elicits such severe reaction that co-encapsulated isografts are destroyed [29,30,78,79]. Compared to allografts, many more proteins that permeate from the encapsulated xenografted cells are recognized as foreign causing this response.

For human treatment, allogeneic tissue is in limited supply hence xenografts are an appealing source, hence the continued pursuit of encapsulated xenografts. Membrane permeability can be adjusted to increase xenograft survival (Fig. 5). Highly restrictive membranes reduce the degree of inflammation around the device, but the compromise is that there is less tissue that can be supported by these membranes. These severely restrict the permeability of molecules the size of IgG (10^{-10} to 10^{-8} cm/*sec*), this is compared to the permeability of IgG through TheraCyte device membranes (10^{-5} cm/sec). Even molecules the size of 17,000 Da are restricted by these membranes.

Decreasing permeability of the encapsulating membrane appears to provide greater protection by restricting the shed foreign xenograft antigens from migrating out and eliciting an immune response (Fig. 5). Tight membranes allow only small molecules which are considered less immunogenic and hence less inflammatory [80,81]. However, with less permeable membranes there is the compromise that less encapsulated tissue can be supported (Fig. 5). At the other end of the permeability spectrum if cells can passage through there is no immune protection (Fig. 4). Allograft protection with high cell density sits in between and is where the permeability of most macroencapsulation devices can be found (Fig. 6). Microencapsulation for xenografts has focused on restricting molecular mass even down to as small as 10,000 to 14,000 Da [61–63]. This is the point where not only do less immunogenic molecules permeate out but also restricts potentially harmful cytokines from entering. The density of the total encapsulated cell mass is low for microcapsules as they have been usually scattered throughout the peritoneum. However, when they do collect in the peritoneal cavity they do fibrosis and fail [82]. Wang et al. found rat islets encapsulated in their cylindrical device survived in C57BL/6 mice, which developed significant anti-xenograft IgG antibodies compared to encapsulated allografts implanted C57BL/6 mice [83]. It must be noted that this is an "immunocompetent" animal that we have found experimentally to be permissive for encapsulated xenografts, yet C57/BL/6 mouse is the most published mouse strain with encapsulation. C57BL/6 mice have a defect in inflammation due to a missense mutation in Nlrp12, whose expression is required for effective recruitment of neutrophils to inflammatory sites [84]. Neutrophils have been found to be important in the xenograft response [85–87].

The immune system plays a role in the inflammatory response to encapsulated xenograft since immune suppressive agents, such as Cyclosporin used for allograft organ transplantation, can overcome xenograft destruction as demonstrated by Fig. 7. However, immunological molecules such as antibodies and complements are not required for this destruction. In vivo injections of hyperimmune anti-xenograft serum failed to cause loss of encapsulated xenografts in immunodeficient rats [88]. Similarly, complement depletion with cobra venom factor failed to protect encapsulated xenografts [88]. The encapsulated xenograft response does involve immune cells, particularly CD4 + T cells and encapsulated xenografts have benefited from the use of a directed immune modulation [89]. This can be especially beneficial if it can be delivered

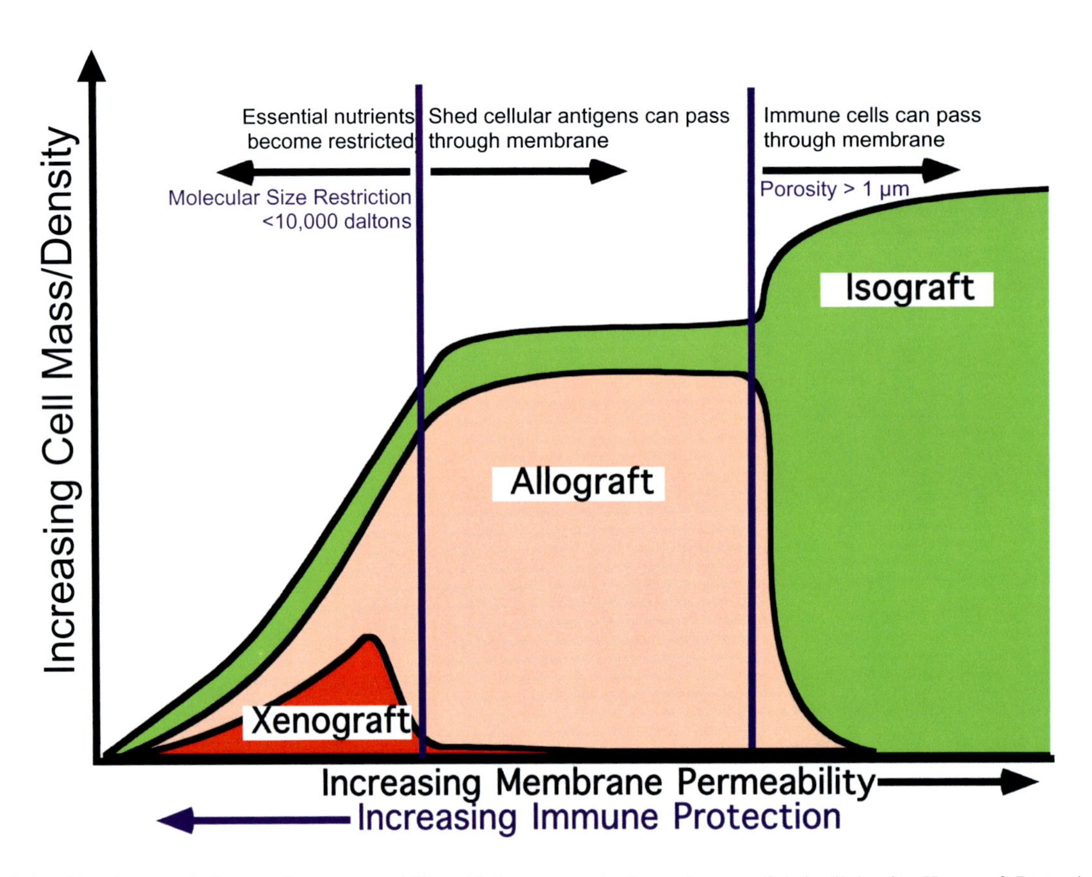

FIG. 6 Relationship of encapsulating membrane permeability with immune protection and encapsulated cell density. Xenograft Protection is achieved at the cost of restricting essential nutrients and cell density.

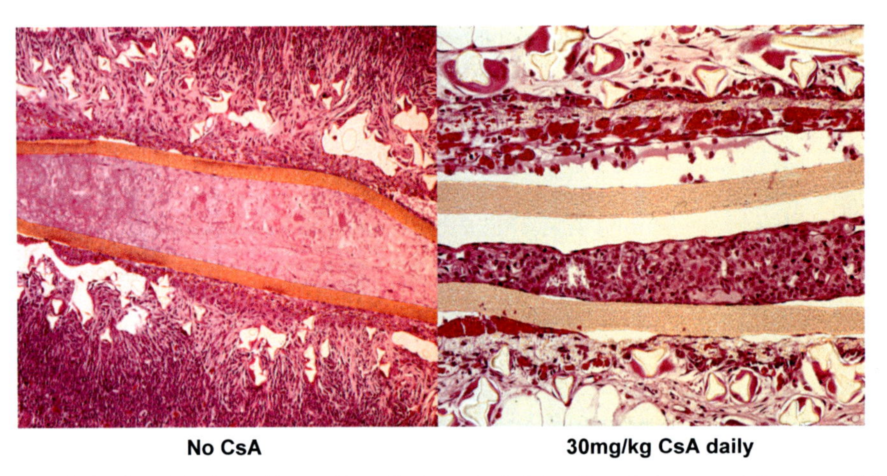

FIG. 7 Encapsulated xenografts implanted in Rats with or without Cyclosporin A (TheraCyte devices containing mouse fetal lung tissue at 3 weeks post-implantation).

locally and not systemically [90,91]. Fig. 8 shows Rat hybridoma cells GK1.5 which secrete anti-mouse CD4 that can protect themselves as well as co-encapsulated human cells in an immunocompetent mouse. Rat hybridomas secreting antibodies to various surface molecules on immune cells impact local environments that surround the encapsulation device (Table 1). Strong immune modulators appear to have the greatest amount of vascularisation. Macrophage depletion not only reduced immune inflammation it also decreased the amount of vasculature, showing the multiple functions of macrophages.

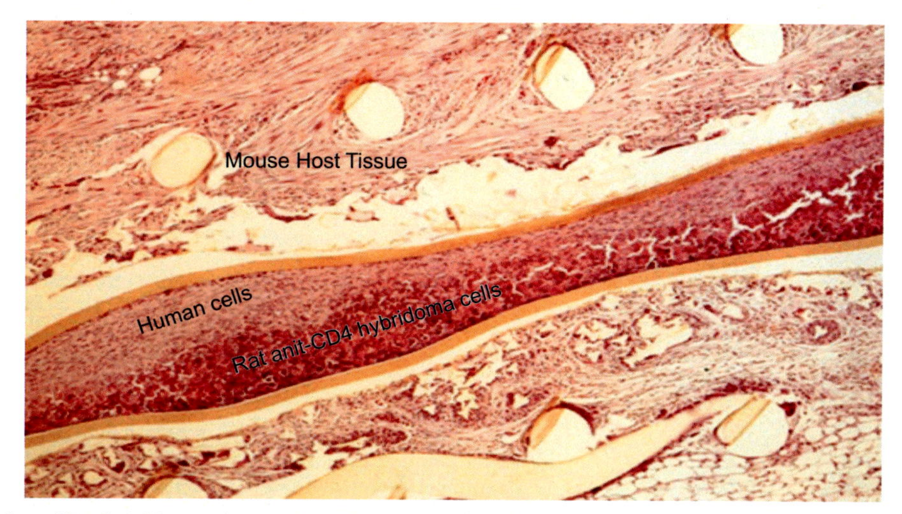

FIG. 8 Co-encapsulation of Rat hybridoma cells (GK1.5, anti-mouse CD4) and human fibroblast cells (MSU1.2), TheraCyte device, 3 weeks post-implantation in a NOD mouse.

TABLE 1 Antibody secreting hybridoma cells encapsulated in TheraCyte devices, 3 weeks post-implantation in Balb/C mice (*n* = 4).

Anti-mouse antibody secreted by rat hybridoma	Average vascular structures per linear cm (< 15 μm from membrane)	Immune response (Graded 1–6)
Anti-CD4	106.8	1.0
Anti-Lymphocyte function-associated antigen 1 (LFA-1)	95.2	2.2
Anti-Macrophage (MAC 1)	25.8	2.0
Anti-Intercellular adhesion molecule-1 (ICAM-1)	9.4	4.0
Rat lung cells (Control)	0.1	5.9

Immune Response Grading: 1, Low-level macrophage reaction, 2, Intermediate macrophage reaction or ordered fibroblasts, 3, High macrophage reaction or standard foreign body, scattered lymphocytes, 4, High macrophage reaction, pooled lymphocytes, 5, Heavy reaction, macrophages and lymphocytes, plasma cells, 6, Extremely high reaction, macrophages and lymphocytes, plasma cells.

Autoimmune protection

Allograft protection by encapsulation does not appear to be diminished in the presence of autoimmunity [67,92–95]. However, when encapsulated xenografts are tested in autoimmune animal models of diabetes, their functionality is further decreased, probably due to pre-existing autoimmune responses to beta-cells or preexisting xenoantibodies. Immune destruction and loss in function of microencapsulated islet tissues have been found to happen faster in autoimmune animal models [96,97].

Device vascularisation

The vasculature is key to the function of any encapsulation device and people have tried to develop intravascular devices [48] but extravascular is the more common approach. The extravascular approach can vary from engineering of the physical structure of the biomaterial to the delivery of vascular enhancing molecules such as VEG-F or a combination of both [13,14,23,31,32,69,98–113]. TheraCyte was one of the early adopters of vascularisation with a biomaterial approach [31,32,69,99,108,114]. With TheraCyte devices, the foreign body response is disorganized allowing for vascular structures to penetrate past outer physical supportive structures and reach the vascularizing membrane coating the cell impermeable membrane Fig. 9. This vascularisation is device structure driven [31] and the vascular structures can last for

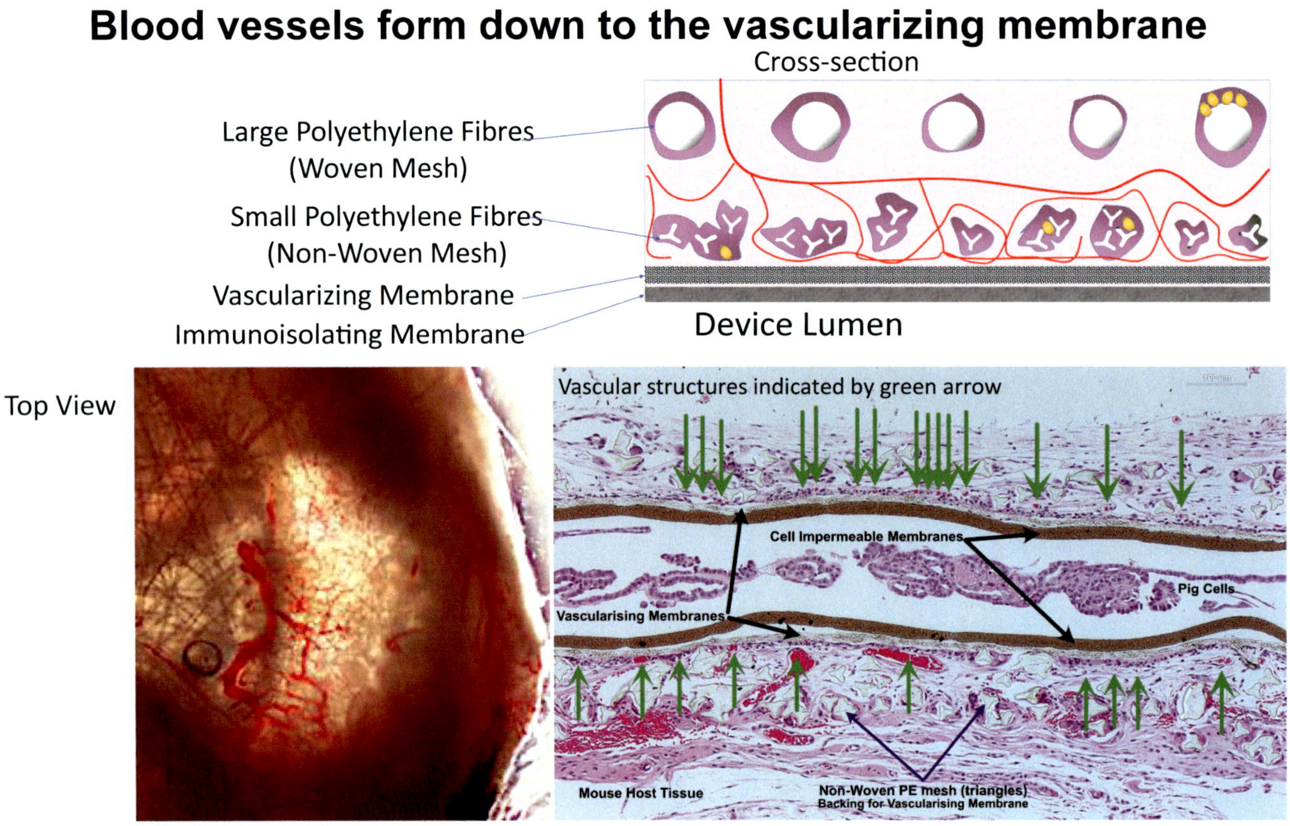

FIG. 9 Close Vascular Structures (CVS) form outside a TheraCyte device. PK-1 cells (pig kidney) were implanted for 4 weeks in NOD SCID mice.

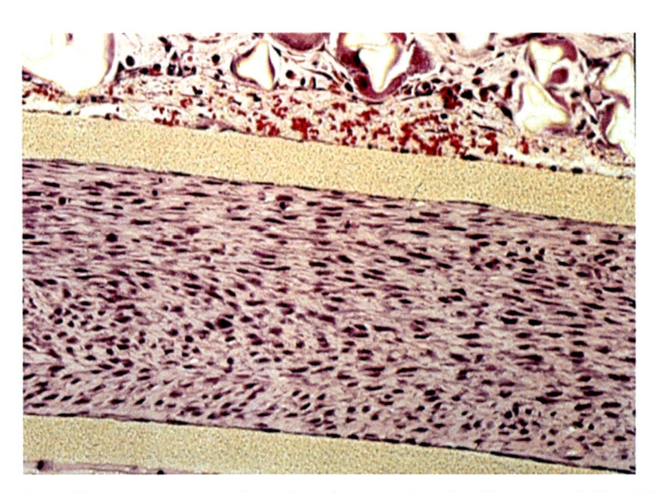

FIG. 10 Close Vascular Structures of the TheraCyte device allow for high cell density. Human cell line MSU1.2 was implanted for 4 weeks in athymic rats.

months around empty devices [114]. The close vascular structures allow for high cell densities of the encapsulated cells (Fig. 10). Others have also found that varying the pore size of the device material can modulate the response and the healing process [23,32] even shifting the foreign body response to more pro-healing M2 macrophage over the proinflammatory M1 macrophages [115,116]. In Table 1 we saw that local release of anti-macrophage suppressed both inflammation and vascularisation. Incorporating immune modulators into the biomaterial may allow for tuning the immune response to the implant [117,118].

Even with devices that vascularize, the extra vasculature takes weeks to develop, and the encapsulated cells may not survive or suffer huge losses during that period. Some devices lend themselves to be implanted, allowed to vascularize,

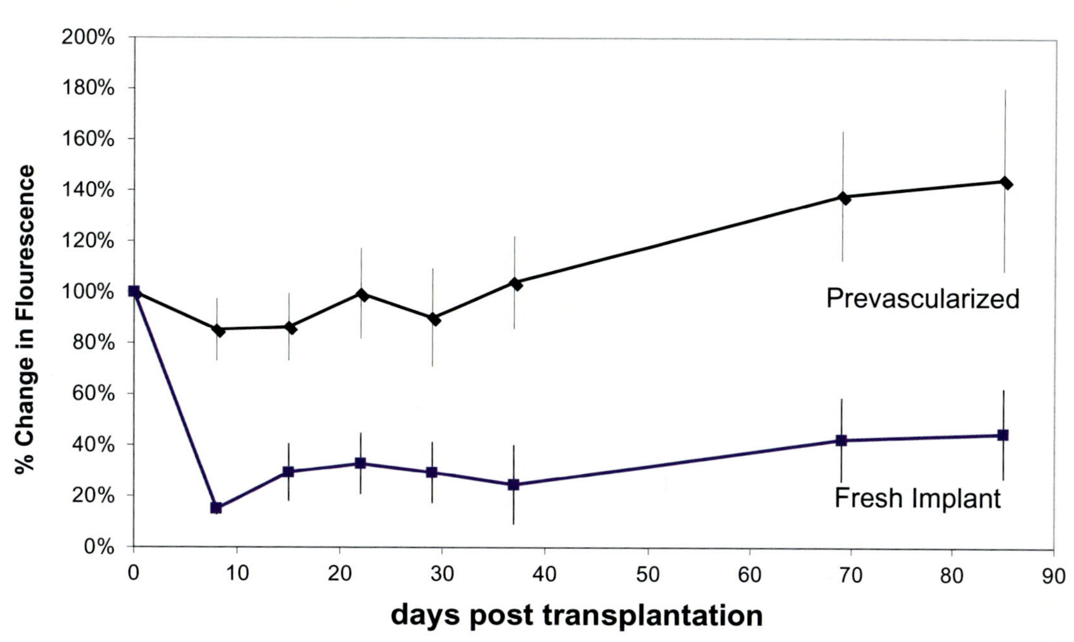

FIG. 11 Significant cell survival in pre-implanted devices compared to freshly-loaded devices. DsRed labeled human U2OS cells loaded into TheraCyte devices and then implanted into NOD-SCID mice ($n=4$, average and standard deviation shown) or loaded into TheraCyte devices pre-implanted for 3 weeks ($n=4$, average and standard deviation shown).

and then be loaded. Fig. 9 clearly shows a 90% loss in cells during the vascularization period, however, if the device is prevascularized losses are only 20%. Because of the growth potential of the cell line used in these experiments, after 10 weeks there were 40% more cells in the devices than initially loaded (Fig. 11).

High cell density and increased encapsulated cell mass are supported by an increased vascularization around the devices that is caused by the encapsulated cells themselves. Even though there are massive cell losses, devices with cells have a higher degree of vascularization than empty devices (Fig. 12). Additional vascularization occurs by the presence of cells in the devices even after there has been device-driven vascularization. The amount of cell-driven vascularization is also dependent on the implanted cells, with an adrenal cell line showing more vascularization than a kidney cell line (Fig. 12). The adrenal cell line secretes immunomodulatory factors, such glucocorticoids [119] suggesting that they too can impact the vasculature by possibly modifying the macrophage response [120]. Encapsulated Insulin-producing cells and mesenchymal

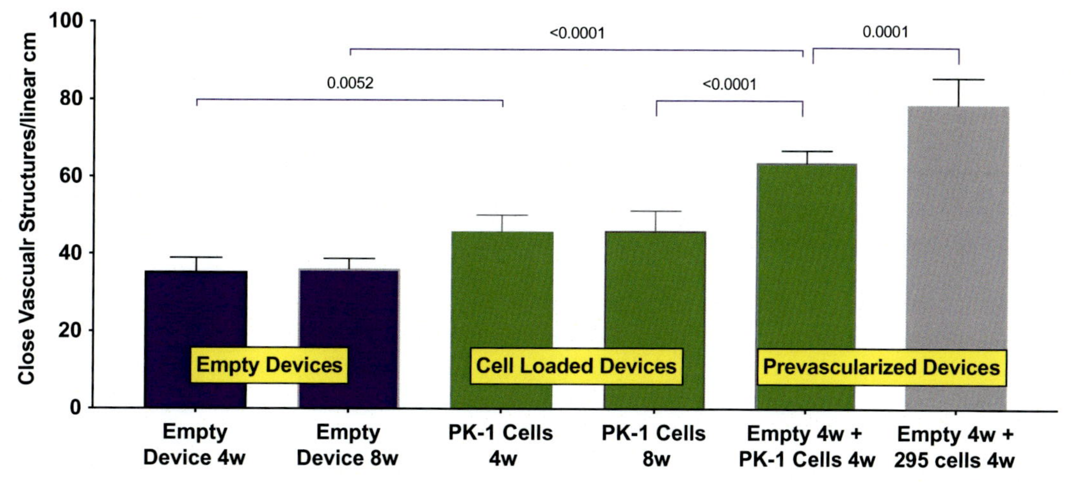

FIG. 12 Impact of cells and prevascularization on close vascular structures (CVS). PK-1 cells (pig kidney) and 295 cells (human adrenal) implanted in NOD SCID mice ($n=5$), (CVS within 15 μm of the TheraCyte vascularizing membrane surface were counted with the average number of vascular structures/cm of membrane length and standard deviation shown). *P* values from multi-comparison one-way ANOVA shown between column pairs.

Impact of added oxygen on encapsulated cells *in vitro*

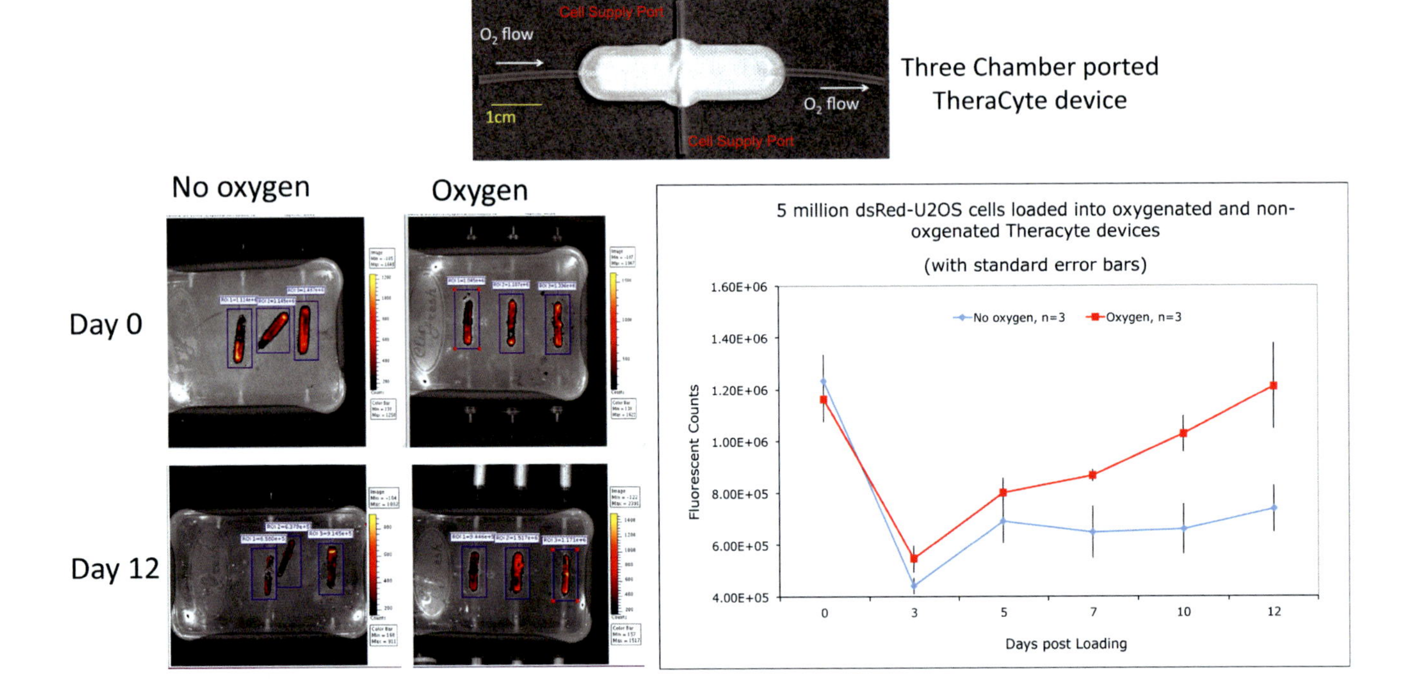

Devices were imaged using the IVIS Spectrum by Caliper

FIG. 13 Impact of oxygen supplementation of encapsulated bone osteosarcoma cell line U2-OS (fluorescenated with dsRed). 5×10^6 cells were loaded into the top chamber of a three-chamber device. Half the devices were supplemented with 40% oxygen/air mix through the middle chamber. Fluorescence was measured on an IVIS spectrum in vivo imaging system by PerkinElmir.

cells have also been found to improve wound healing in the local area of the implant [121] with vascular endothelial growth factor (VEGF) being one of the main molecules involved [122]. Attempts at recreating the cell-driven vascularization without cells have been made with the use of angiogenic factors such as VEGF [109–111].

Some devices do not lend themselves to be prevascularised and have looked at alternatives to get the encapsulated cells over the initial wound healing period. Because that initial healing period can be hypoxic people have looked at oxygen supplementation [123–131]. The Beta-Air device manufactured by Beta-O2 has transitioned an oxygen supplemented device from rodent studies [125] through to large animal studies [132] and into human clinical trials [133]. The Beta-O2 device is a circular flat sheet device with two oxygen refilling ports linked to the main device by tubing [132]. The main device is 6.8 cm in diameter and 1.8 cm thick. The thickness is due to two outer islet modules 600 μm thick each and a central oxygen module. In humans, the devices have been transplanted subcutaneously [133]. Similar to other flat sheet devices the outer membrane is made of 0.4 μm pore PTFE but without any published enhancements to improve vascularisation. In rodent studies, these devices have supported relatively high densities of islets of up to 4800 islets per square centimeter [125].

TheraCyte and Procyon have also made a multichambered device, where the central chamber has been used to deliver oxygen to the encapsulated cells located in a chamber above and below the oxygen chamber. Fig. 13 is an example of the benefits of supplemented oxygen for cells invitro. Here the growth rate of a flourescenated human bone osteosarcoma epithelial cell is observed in vitro, with half the devices supplemented with 40% oxygen. After an initial loss, the supplemented devices recovered to 100% of the loaded amount within 12 days compared to non-supplemented devices which remained closer to 50% of the loaded amount. Procyon is pursuing oxygen supplementation to enhance encapsulated cell implants.

Efficacy

When it comes to correcting diabetes with encapsulated cells there have been many successful studies performed in animals and it has been recently that some have progressed to clinical studies in humans. When it comes to human efficacy studies, encapsulated cells become important. Most of the successful animal studies have used islets, these islet sources are either insufficient or unsuitable for human implantation. This has been the major hurdle why many encapsulated methods have not proceeded to clinical trials.

Studies in non-obese diabetic (NOD) mice have found that encapsulate islets can correct diabetes using microcapsules [93,94,134–136], or hollow fiber macrocapsules [137]. Similarly, insulinoma cells encapsulated in TheraCyte devices have also corrected diabetic NOD mice [95]. The encapsulated cells were able to provide life-long glucose homeostasis, including providing normal glucose tolerance tests. Unlike NOD mice where autoimmune diabetes occurs spontaneously, other animal models require diabetes to be induced by chemical and/or surgical means. Both microcapsules and macrocapsules have been efficacious in correcting the diabetic state in rats [19,20,125,138–145], dogs [3,50,146–155], pigs [156–158] and primates [132,159–161]. Encapsulation has also been used to deliver proteins in non-diabetic models such as human growth hormone [35], erythropoietin [162], parathyroid hormones [68,163–167], neurotrophic factors [53,55,168–178].

There have been 28 clinical trials (Table 2) using encapsulated cells listed on clinicaltrials.gov, which covers many but not all encapsulated clinical trials. The first reported case of encapsulated islets in humans was in 1994, where type 1 diabetic patient had received microencapsulated human islets into the peritoneal cavity [2] or in hollow fiber macrocapsules that were transplanted subcutaneously [179]. Even though a short period of insulin independence was reported for the microcapsules, this was not a true test of immune protective effects of the encapsulation as the patient was also on immunosuppressive drugs. While no immune suppressants were used for the hollow fiber macrocapsules, the implant period was only 2 weeks. One year survival of allogeneic human parathyroid cells, encapsulated in TheraCyte devices was reported in non-immunosuppressed patients in 1996 [68,180]. The Beta-O2 prototype device has also been used in a clinical study and has been reported to maintain human islet survival for 10 months [181], also in type 1 diabetic patient but without immunosuppression. Insulin independence was not achieved because the encapsulated islet mass of 2100 islet equivalents per kilogram of body weight was suboptimal. However, it was reported that there was a reduction in daily insulin requirements and hemoglobin A1c levels. Table 2 lists trials with multiple encapsulation methods, microcapsules, flat sheet macrocapsules, and hollow fiber macrocapsules all were found to be safe and showed the ability to protect against immune rejection without immune suppression. Table 2 also demonstrates these devices can be manufactured in a quality acceptable by regulatory authorities suitable for human use.

Conclusion

Encapsulated cell therapy is a unique science combining both engineering and biology that must work harmoniously to be effective. The best device, no matter how biocompatible will not work if the encapsulated cells don't also function effectively. Encapsulation devices of many types with different symmetries and chemistries have been engineered for many years. People are still finding more symmetries and chemistries to improve encapsulation and provide a favorable environment for the encapsulated cells. The cells to be used in any of the various types of encapsulation devices, however, have been more elusive. Most tissues that have been used have been isolated from primary tissues and are of limited availability. Advances in various forms of stem cell differentiation have the potential to also become a promising source of tissue for encapsulation. One company, Viacyte, has taken advantage of the treatment of diabetes and is currently in clinical studies (Table 2). Stem cell biologists are trying to recreate the normal cells to replace the deficient or defective cells in the patient. However, an encapsulation device is an artificial situation, and these "normal cells" may not be the most efficient cells to encapsulate. Islets are a good example, they are highly vascularised mini-organelles with glucose-sensing capacity and insulin release is within the internal vascular structures. When encapsulated now glucose sensing and insulin release is with vascular structures that are external to the islet and the encapsulating material. It may be better to generate or modify stem cell creations to adapt and function more efficiently as encapsulated cells to correct the diseased state. While the clinical studies with current technologies provide hope and promise for patients, there is still scope for engineers and biologists to work together to improve encapsulated cell therapies into a product that can treat many patients.

TABLE 2 Encapsulated cell studies listed on ClinicalTrials.Gov.

Rank	NCT number	Title	Conditions	Interventions	Sponsor/Collaborators
1	NCT00063765	Evaluation of Safety of Ciliary Neurotrophic Factor Implants in the Eye	Retinitis Pigmentosa	Drug: Ciliary Neurotrophic Factor Implant NT-501	National Eye Institute (NEI)|National Institutes of Health Clinical Center (CC)
2	NCT00260234	Safety and Efficacy of PEG-Encapsulated Islet Allografts Implanted in Type I Diabetic Recipients	Diabetes Mellitus, Type 1	Biological: Allogeneic Cultured Islet Cells (human); Encapsulated	Novocell | Diabetes & Glandular Disease Research Associates, P.A., San Antonio, TX | CHRISTUS Health
3	NCT00447993	A Study of Encapsulated Cell Technology (ECT) Implant for Patients With Late Stage Retinitis Pigmentosa	Retinitis Pigmentosa	Drug: NT-501	Neurotech Pharmaceuticals
4	NCT00447980	A Study of Encapsulated Cell Technology (ECT) Implant for Participants With Early Stage Retinitis Pigmentosa	Retinitis Pigmentosa	Drug: NT-501	Neurotech Pharmaceuticals
5	NCT00447954	A Study of an Encapsulated Cell Technology (ECT) Implant for Patients With Atrophic Macular Degeneration	Macular Degeneration	Drug: NT-501 implant | Other: Sham Procedure	Neurotech Pharmaceuticals
6	NCT01163825	Encapsulated Cell Biodelivery of Nerve Growth Factor to Alzheimer¬¥s Disease Patients	Alzheimer's Disease	Drug: Nerve Growth Factor	NsGene A/S | Karolinska Institutet | Region Stockholm
7	NCT00790257	Safety and Efficacy Study of Encapsulated Human Islets Allotransplantation to Treat Type 1 Diabetes	Type 1 Diabetes Mellitus	Device: Encapsulated human islets in a "Monolayer Cellular Device"	Cliniques universitaires Saint-Luc- Universit √© Catholique de Louvain

Phases	Funded bys	Start date	Completion date	Locations
Phase 1	NIH	Jun-2003	Mar-2006	National Eye Institute (NEI), Bethesda, Maryland, United States
Phase 1/Phase 2	Industry | Other	Nov-2005	Dec-2007	CHRISTUS Santa Rosa Transplant Institute, San Antonio, Texas, United States | Diabetes & Glandular Disease Research Associates, P.A., San Antonio, Texas, United States
Phase 2	Industry	Jan-2007	Oct-2009	Retina-Vitreous Associates Medical Group, Beverly Hills, California, United States | University of California, Davis, Sacramento, California, United States | University of California, San Francisco, San Francisco, California, United States | Retina Group of Florida, Hollywood, Florida, United States | University of Florida, Jacksonville, Florida, United States | Ophthalmic Consultants of Boston, Boston, Massachusetts, United States | Kellogg Eye Center, Ann Arbor, Michigan, United States | University of Minnesota, Minneapolis, Minnesota, United States | NY University Medical Center, New York, New York, United States | Casey Eye Institue, Portland, Oregon, United States | The Hamilton Eye Institute, Memphis, Tennessee, United States | Retina Foundation of Southwest, Dallas, Texas, United States | University of Utah, Salt Lake City, Utah, United States
Phase 2	Industry	Jan-2007	Jul-2010	Retina-Vitreous Associates Medical Group, Beverly Hills, California, United States | University of California, Davis, Sacramento, California, United States | University of California, San Francisco, San Francisco, California, United States | Bascom Palmer Eye Insitute, Miami, Florida, United States | Ophthalmic Consultants of Boston, Boston, Massachusetts, United States | Kellogg Eye Center, Ann Arbor, Michigan, United States | University of Minnesota, Minneapolis, Minnesota, United States | NY University Medical Center, New York, New York, United States | Casey Eye Institue, Portland, Oregon, United States | The Hamilton Eye Institute, Memphis, Tennessee, United States | Retina Foundation of Southwest, Dallas, Texas, United States | University of Utah, Salt Lake City, Utah, United States
Phase 2	Industry	Jan-2007	Oct-2009	Retina-Vitreous Associates Medical Group, Beverly Hills, California, United States | Retina Group of Florida, Hollywood, Florida, United States | Bascom Palmer Eye Institute, Miami, Florida, United States | Ophthalmic Consultants of Boston, Boston, Massachusetts, United States | Beaumont Eye Institute, Royal Oaks, Michigan, United States | Retina Foundation of Southwest, Dallas, Texas, United States | Vitreoretinal Consultants, Houston, Texas, United States | University of Utah, Salt Lake City, Utah, United States
Phase 1	Industry | Other	Jan-2008		Karolinska University Hospital, Stockholm, Sweden
Phase 1	Other	Nov-2008	Sep-2015	University clinical Hospital Saint-Luc, Brussels, Belgium

TABLE 2 Encapsulated cell studies listed on ClinicalTrials.Gov—cont'd

Rank	NCT number	Title	Conditions	Interventions	Sponsor/Collaborators
8	NCT00940173	Open-label Investigation of the Safety and Effectiveness of DIABECELL(R) in Patients With Type I Diabetes Mellitus	Type 1 Diabetes	Device: DIABECELL(R)	Diatranz Otsuka Limited
9	NCT01379729	Bet Cell Therapy in Diabetes Type 1	Type 1 Diabetes	Other: Transplantation of encapsulated beta cells.	AZ-VUB | Universitair Ziekenhuis Brussel | Universitaire Ziekenhuizen Leuven
10	NCT01739829	Open-label Investigation of the Safety and Effectiveness of DIABECELL¬Æ in Patients With Type 1 Diabetes Mellitus	Type 1 Diabetes	Device: DIABECELL (R)	Diatranz Otsuka Limited
11	NCT01530659	Retinal Imaging of Subjects Implanted With Ciliary Neurotrophic Factor (CNTF)-Releasing Encapsulated Cell Implant for Early-stage Retinitis Pigmentosa	Retinitis Pigmentosa | Usher Syndrome Type 2 | Usher Syndrome Type 3	Drug: NT-501 | Procedure: Sham	Neurotech Pharmaceuticals | University of California, San Francisco | FDA Office of Orphan Products Development
12	NCT01550523	Pilot Immunotherapy Trial for Recurrent Malignant Gliomas	Malignant Glioma of Brain	Drug: IGF-1R/AS ODN | Device: biodiffusion chamber	Sidney Kimmel Cancer Center at Thomas Jefferson University | Thomas Jefferson University
13	NCT01736228	Open-label Investigation of the Safety and Efficacy of DIABECELL in Patients With Type 1 Diabetes Mellitus	Type 1 Diabetes	Device: DIABECELL	Living Cell Technologies
14	NCT01734733	Open-label Investigation of the Safety and Clinical Effects of NTCELL in Patients With Parkinson's Disease	Parkinson's Disease	Other: NTCELL	Living Cell Technologies
15	NCT02064309	An Open Label, Pilot Investigation, to Assess the Safety and Efficacy of Transplantation of Macro-encapsulated Human Islets Within the Bioartificial Pancreas Beta-Air in Patients With Type 1 Diabetes Mellitus	Long-standing Type 1 Diabetes Mellitus	Device: Beta-Air device for encapsulation of transplanted human islets	Uppsala University Hospital | Beta-O2 Technologies Ltd.
16	NCT01949324	A Phase 2 Multicenter Randomized Clinical Trial of CNTF for MacTel	Macular Telangiectasia Type 2	Biological: Ciliary neurotrophic factor (CNTF)|Procedure: Sham procedure | Device: NT-501 Implant | Procedure: NT-501 Implant procedure	Neurotech Pharmaceuticals | The Lowy Medical Research Institute Limited | The Emmes Company, LLC
17	NCT02239354	A Safety, Tolerability, and Efficacy Study of VC-01,Ñ¢ Combination Product in Subjects With Type I Diabetes Mellitus	Type 1 Diabetes Mellitus	Device: VC-01,Ñ¢ Combination Product	ViaCyte | California Institute for Regenerative Medicine (CIRM)

Phases	Funded bys	Start date	Completion date	Locations
Phase 1/Phase 2	Industry	Jul-2009	Oct-2013	Centre for Clinical Research and Effective Practice, Auckland, New Zealand
Phase 2	Other	May-2011	May-2018	UZ Brussel, Brussels, Belgium | UZ Leuven, Leuven, Belgium
Phase 1/Phase 2	Industry	Aug-2011	Jun-2014	Hospital Interzonal General de Agudos Eva Per √≥n, Buenos Aires, Argentina
Phase 2	Industry | Other | U.S. Fed	Jan-2012	Aug-2019	University of California, San Francisco, San Francisco, California, United States
Phase 1	Other	Feb-2012	May-2013	Thomas Jefferson University Hospital; Jefferson Hospital for Neurosciences, Philadelphia, Pennsylvania, United States
Phase 2	Industry	Nov-2012	Dec-2014	Hospital Interzonal General de Agudos Eva Per √≥n, San Martin, Buenos Aires, Argentina
Phase 1/Phase 2	Industry	Jul-2013	Jun-2020	Auckland City Hospital, Auckland, New Zealand
Phase 1/Phase 2	Other | Industry	Feb-2014	May-2022	Uppsala University Hospital, Uppsala, Sweden
Phase 2	Industry | Other	Apr-2014	May-2017	Jules Stein Eye Institute, Los Angeles, California, United States | Bascom Palmer, Miami, Florida, United States | Emory University, Atlanta, Georgia, United States | National Eye Institute, Bethesda, Maryland, United States | Massachusetts Eye and Ear Infirmary, Retina Service, Boston, Massachusetts, United States | University of Michigan, Kellogg Eye Center, Ann Arbor, Michigan, United States | Retina Associates of Cleveland, Inc., Beachwood, Ohio, United States | University of Wisconsin, Madison, Wisconsin, United States | Save Sight Institute, Sydney, New South Wales, Australia | Centre for Eye Research Australia, East Melbourne, Australia | Lions Eye Institute, Nedlands, Australia
Phase 1/Phase 2	Industry | Other	Sep-2014	Jun-2021	University of California at San Diego, San Diego, California, United States | University of Alberta Hospital, Edmonton, Alberta, Canada

TABLE 2 Encapsulated cell studies listed on ClinicalTrials.Gov—cont'd

Rank	NCT number	Title	Conditions	Interventions	Sponsor/Collaborators
18	NCT02228304	Study of the Intravitreal Implantation of NT-503-3 Encapsulated Cell Technology (ECT) for the Treatment of Recurrent Choroidal Neovascularization (CNV) Secondary to Age-related Macular Degeneration (AMD)	Macular Degeneration	Drug: NT-503-3 ECT implantation | Drug: Eylea¬Æ injected intravitreally administered every 8 weeks	Neurotech Pharmaceuticals
19	NCT02507583	Antisense102: Pilot Immunotherapy for Newly Diagnosed Malignant Glioma	Malignant Glioma | Neoplasms	Drug: IGF-1R/AS ODN; Surgery with tissue harvest and implantation 20 diffusion chambers in the rectus sheath with IGF-1R/AS ODN within 24 h of craniotomy, implanted for 48 h.	david andrews | Thomas Jefferson University
20	NCT02683629	Investigation of the Safety and Efficacy of NTCELL¬Æ [Immunoprotected (Alginate-Encapsulated) Porcine Choroid Plexus Cells for Xenotransplantation] in Patients With Parkinson's Disease	Parkinson's Disease	Biological: NTCELL Implantation | Other: Sham Surgery	Living Cell Technologies | Statistecol Consultants Limited

Phases	Funded bys	Start date	Completion date	Locations
Phase 1/Phase 2	Industry	Sep-2014	Apr-2016	Associated Retina Consultants, Ltd., Phoenix, Arizona, United States | Retina-Vitreous Associates Medical Group, Beverly Hills, California, United States | University of California, Irvine, The Gavin Herbert Eye Institute, Irvine, California, United States | Jacobs Retina Center at UCSD, La Jolla, California, United States | Colorado Retina Associates, Golden, Colorado, United States | Retina Health Center, Ft. Myers, Florida, United States | National Ophthalmic Research Institute, Ft. Myers, Florida, United States | Center for Retina and Macular Disease, Winter Haven, Florida, United States | Southeast Retina Center, PC, Augusta, Georgia, United States | Georgina Retina, P.C., Marietta, Georgia, United States | Illinois Retina Associates, S.C., Joliet, Illinois, United States | Illinois Retina Associates, Oak Park, Illinois, United States | Retina Associates of Kentucky, Lexington, Kentucky, United States | The Retina Group of Washington, Chevy Chase, Maryland, United States | Cumberland Valley Retina Consultants, PC, Hagerstown, Maryland, United States | Ophthalmic Consultants of Boston, Boston, Massachusetts, United States | University of Michigan, Kellogg Eye Center, Ann Arbor, Michigan, United States | William Beaumont Hospital-Ophthalmology Research, Royal Oak, Michigan, United States | Lifelong Vision Foundation, Chesterfield, Missouri, United States | Sierra Eye Associates, Reno, Nevada, United States | Retina Center of New Jersey, LLC, Bloomfield, New Jersey, United States | NJ Retina, New Brunswick, New Jersey, United States | Western Carolina Retinal Associates, Asheville, North Carolina, United States | Wake Forest Baptist Health Eye Center, Winston-Salem, North Carolina, United States | Cleveland Clinic, Cleveland, Ohio, United States | Retina Northwest, PC, Portland, Oregon, United States | Mid Atlantic Retina, Philadelphia, Pennsylvania, United States | Palmetto Retina Center, LLC, Florence, South Carolina, United States | Black Hills Regional Eye Institute, Rapid City, South Dakota, United States | Valley Retina Institute, PA, McAllen, Texas, United States | Medical Center Ophthalmology Associates, San Antonio, Texas, United States | Medical College of Wisconsin, Milwaukee, Wisconsin, United States | Soroka Medical Center, Beer Sheva, Israel | Hadassah-Hebrew University Medical Center, Jerusalem, Israel | Meir Medical Center, Kfara Saba, Israel | Rabin Medical Center, Petach-tikva, Israel | Kaplan Medical Center, Rehovot, Israel | Sourasky Medical Center, Tel-Aviv, Israel
Phase 1	Other	Sep-2015	Aug-2020	Thomas Jefferson University Hospital, Philadelphia, Pennsylvania, United States
Phase 2	Industry | Other	Feb-2016	May-2019	Auckland City Hospital, Auckland, New Zealand

TABLE 2 Encapsulated cell studies listed on ClinicalTrials.Gov—cont'd

Rank	NCT number	Title	Conditions	Interventions	Sponsor/Collaborators
21	NCT02862938	Study of NT-501 Encapsulated Cell Therapy for Glaucoma Neuroprotection and Vision Restoration	Glaucoma	Drug: NT-501 ECT implant | Other: Sham	Stanford University
22	NCT02939118	One-Year Follow-up Safety Study in Subjects Previously Implanted With VC-01,Ñ¢	Type 1 Diabetes Mellitus		ViaCyte
23	NCT03071965	Extension Study of NT-501 Ciliary Neurotrophic Factor (CNTF) Implant for Macular Telangiectasia (MacTel)	Macular Telangiectasia	Biological: Ciliary neurotrophic factor (CNTF)|Procedure: Surgery	Neurotech Pharmaceuticals
24	NCT03162926	A Safety and Tolerability Study of VC-02,Ñ¢ Combination Product in Subjects With Type 1 Diabetes Mellitus	Type 1 Diabetes Mellitus	Combination Product: VC-02 Combination Product (aka PEC-Direct)	ViaCyte
25	NCT03163511	A Safety, Tolerability, and Efficacy Study of VC-02,Ñ¢ Combination Product in Subjects With Type 1 Diabetes Mellitus and Hypoglycemia Unawareness	Type 1 Diabetes Mellitus With Hypoglycemia	Combination Product: VC-02 Combination Product	ViaCyte | California Institute for Regenerative Medicine (CIRM)|Horizon 2020 - European Commission
26	NCT04678557	A Study to Evaluate Safety, Engraftment, and Efficacy of VC-01 in Subjects With T1 Diabetes Mellitus	Type 1 Diabetes	Combination Product: VC-01 Combination Product	ViaCyte
27	NCT04541628	Safety & Efficacy of Encapsulated Allogeneic FVIII Cell Therapy in Hemophilia A	Hemophilia A	Combination Product: SIG-001	Sigilon Therapeutics, Inc.
28	NCT04577300	Dual Intravitreal Implantation of NT-501 Encapsulated Cell Therapy for Glaucoma	Glaucoma	Drug: NT-501 | Other: Sham comparator	Stanford University

Phases	Funded bys	Start date	Completion date	Locations
Phase 2	Other	Jun-2016	Dec-2023	Byers Eye Institute at Stanford University, Palo Alto, California, United States
	Industry	Nov-2016	Nov-2023	
Phase 2	Industry	May-2017	Aug-2021	Stein Eye Institute/David Geffen School of Medicine, Los Angeles, California, United States | University of Miami-Miller School of Medicine, Bascom Palmer Eye Institute, Miami, Florida, United States | Emory University School of Medicine, Dept of Ophthalmology, Emory University Eye Center, Atlanta, Georgia, United States | NIH Clinical Center, Rockville, Maryland, United States | Massachusetts Eye and Ear Infirmary, Boston, Massachusetts, United States | University of Michigan, Kellogg Eye Center, Ann Arbor, Michigan, United States | Retina Associates of Cleveland, Inc., Cleveland, Ohio, United States | University of Wisconsin-Madison, Department of Ophthalmology and Visual Sciences, Madison, Wisconsin, United States | Sydney Eye Hospital, Sydney, New South Wales, Australia | Centre for Eye Research Australia, East Melbourne, Victoria, Australia | Lions Eye Institute, Nedlands, Western Australia, Australia
Phase 1	Industry	Jul-2017	Feb-2018	University of Alberta, Edmonton, Alberta, Canada
Phase 1/Phase 2	Industry | Other	Jul-2017	Mar-2023	City of Hope National Medical Center, Duarte, California, United States | UCLA-UCI Alpha Stem Cell Clinic, Irvine, California, United States | UC Davis - Alpha Stem Cell Clinic, Sacramento, California, United States | University of California San Diego, San Diego, California, United States | Johns Hopkins University, Baltimore, Maryland, United States | University of Minnesota, Minneapolis, Minnesota, United States | Ohio State University, Columbus, Ohio, United States | University Hospital Brussels, Brussel, Belgium | University of Alberta, Edmonton, Alberta, Canada | University of British Columbia, Vancouver, British Columbia, Canada
Phase 1/Phase 2	Industry	Jun-2019	Apr-2023	AMCR Institute, Escondido, California, United States | Diablo Clinical Research, Walnut Creek, California, United States | Atlanta Diabetes Associates, Atlanta, Georgia, United States | Texas Diabetes & Endocrinology, Austin, Texas, United States
Phase 1/Phase 2	Industry	Sep-2020	Sep-2026	Clinical Study Site, Indianapolis, Indiana, United States | Clinical Study Site, Boston, Massachusetts, United States | Clinical Study Site, Seattle, Washington, United States | Clinical Study Site, London, United Kingdom | Clinical Study Site, Manchester, United Kingdom | Clinical Study Site, Southampton, United Kingdom
Phase 2	Other	May-2021	Dec-2023	Byers Eye Institute at Stanford University, Palo Alto, California, United States

References

[1] Calafiore R. Transplantation of microencapsulated pancreatic human islets for therapy of diabetes mellitus. A preliminary report. ASAIO J 1992;38(1):34–7.
[2] Soon-Shiong P, Heintz RE, Merideth N, Yao QX, Yao Z, Zheng T, Murphy M, Moloney MK, Schmehl M, Harris M, et al. Insulin independence in a type 1 diabetic patient after encapsulated islet transplantation. Lancet 1994;343(8903):950–1.
[3] Maki T, Ubhi CS, Sanchez-Farpon H, Sullivan SJ, Borland K, Muller TE, Solomon BA, Chick WL, Monaco AP. Successful treatment of diabetes with the biohybrid artificial pancreas in dogs. Transplantation 1991;51(1):43–51.
[4] Lim F, Sun AM. Microencapsulated islets as bioartificial endocrine pancreas. Science 1980;210(4472):908–10.
[5] Chang TM. Semipermeable microcapsules. Science 1964;146(3643):524–5.
[6] Chang TM. Immobilization of enzymes, adsorbents, or both within semipermeable microcapsules (artificial cells) for clinical and experimental treatment of metabolite-related disorders. Birth Defects Orig Artic Ser 1973;9(2):66–76.
[7] de Vos P, Lazarjani HA, Poncelet D, Faas MM. Polymers in cell encapsulation from an enveloped cell perspective. Adv Drug Deliv Rev 2014;67–68:15–34.
[8] Liu H, Wang Y, Wang H, Zhao M, Tao T, Zhang X, Qin J. A droplet microfluidic system to fabricate hybrid capsules enabling stem cell organoid engineering. Adv Sci (Weinheim, Ger) 2020;7(11):1903739.
[9] Sefton MV, May MH, Lahooti S, Babensee JE. Making microencapsulation work: conformal coating, immobilization gels and in vivo performance. J Control Release 2000;65(1–2):173–86.
[10] Sawhney AS, Pathak CP, Hubbell JA. Modification of islet of langerhans surfaces with immunoprotective poly(ethylene glycol) coatings via interfacial photopolymerization. Biotechnol Bioeng 1994;44(3):383–6.
[11] May MH, Sefton MV. Conformal coating of small particles and cell aggregates at a liquid-liquid interface. Ann N Y Acad Sci 1999;875:126–34.
[12] Sullivan SJ, Maki T, Carretta M, Ozato H, Borland K, Mahoney MD, Muller TE, Solomon BA, Monaco AP, Chick WL. Evaluation of the hybrid artificial pancreas in diabetic dogs. ASAIO J 1992;38(1):29–33.
[13] Mridha AR, Dargaville TR, Dalton PD, Carroll L, Morris MB, Vaithilingam V, Tuch BE. Prevascularized retrievable hybrid implant to enhance function of subcutaneous encapsulated islets. Tissue Eng Part A 2020.
[14] Skrzypek K, Nibbelink MG, Karbaat LP, Karperien M, van Apeldoorn A, Stamatialis D. An important step towards a prevascularized islet macroencapsulation device-effect of micropatterned membranes on development of endothelial cell network. J Mater Sci Mater Med 2018;29(7):91.
[15] Prehn RT, Weaver JM, Algire GH. The diffusion-chamber technique applied to a study of the nature of homograft resistance. J Natl Cancer Inst 1954;15(3):509–17.
[16] Espona-Noguera A, Ciriza J, Cañibano-Hernández A, Orive G, Hernández RMM, Saenz Del Burgo L, Pedraz JL. Review of advanced hydrogel-based cell encapsulation systems for insulin delivery in type 1 diabetes mellitus. Pharmaceutics 2019;11(11).
[17] Espona-Noguera A, Ciriza J, Cañibano-Hernández A, Villa R, Saenz Del Burgo L, Alvarez M, Pedraz JL. 3D printed polyamide macroencapsulation devices combined with alginate hydrogels for insulin-producing cell-based therapies. Int J Pharm 2019;566:604–14.
[18] Gehrke SH, Fisher JP, Palasis M, Lund ME. Factors determining hydrogel permeability. Ann N Y Acad Sci 1997;831:179–207.
[19] Whittemore AD, Chick WL, Galletti PM, Like AA, Colton CK, Lysaght MJ, Richardson PD. Effects of the hybrid artificial pancreas in diabetic rats. Trans Am Soc Artif Intern Organs 1977;23:336–41.
[20] Whittemore AD, Chick WL, Galletti PM, Mannick JA. Function of hybrid artificial pancreas in diabetic rats. Surg Forum 1977;28:93–7.
[21] Skrzypek K, Groot Nibbelink M, Liefers-Visser J, Smink AM, Stoimenou E, Engelse MA, de Koning EJP, Karperien M, de Vos P, van Apeldoorn A, Stamatialis D. A high cell-bearing capacity multibore hollow fiber device for macroencapsulation of islets of langerhans. Macromol Biosci 2020;20(8), e2000021.
[22] Anderson JM, Rodriguez A, Chang DT. Foreign body reaction to biomaterials. Semin Immunol 2008;20(2):86–100.
[23] Sussman EM, Halpin MC, Muster J, Moon RT, Ratner BD. Porous implants modulate healing and induce shifts in local macrophage polarization in the foreign body reaction. Ann Biomed Eng 2014;42(7):1508–16.
[24] Collier TO, Anderson JM. Protein and surface effects on monocyte and macrophage adhesion, maturation, and survival. J Biomed Mater Res 2002;60(3):487–96.
[25] Johansson CB, Hansson HA, Albrektsson T. Qualitative interfacial study between bone and tantalum, niobium or commercially pure titanium. Biomaterials 1990;11(4):277–80.
[26] Wilson CJ, Clegg RE, Leavesley DI, Pearcy MJ. Mediation of biomaterial-cell interactions by adsorbed proteins: a review. Tissue Eng 2005;11(1–2):1–18.
[27] Tan RP, Hallahan N, Kosobrodova E, Michael PL, Wei F, Santos M, Lam YT, Chan AHP, Xiao Y, Bilek MMM, Thorn P, Wise SG. Bioactivation of encapsulation membranes reduces fibrosis and enhances cell survival. ACS Appl Mater Interfaces 2020;12(51):56908–23.
[28] Chandy T, Mooradian DL, Rao GH. Evaluation of modified alginate-chitosan-polyethylene glycol microcapsules for cell encapsulation. Artif Organs 1999;23(10):894–903.
[29] Weber CJ, Zabinsi S, Koschitzky T, Rajotte R, Wicker L, Peterson L, D'Agati V, Reemtsma K. Microencapsulated dog and rat islet xenografts into streptozotocin-diabetic and NOD mice. Horm Metab Res Suppl 1990;25:219–26.
[30] Weber CJ, Zabinski S, Koschitzky T, Wicker L, Rajotte R, D'Agati V, Peterson L, Norton J, Reemtsma K. The role of CD4+ helper T cells in the destruction of microencapsulated islet xenografts in nod mice. Transplantation 1990;49(2):396–404.
[31] Brauker JH, Carr-Brendel VE, Martinson LA, Crudele J, Johnston WD, Johnson RC. Neovascularization of synthetic membranes directed by membrane microarchitecture. J Biomed Mater Res 1995;29(12):1517–24.

[32] Padera RF, Colton CK. Time course of membrane microarchitecture-driven neovascularization. Biomaterials 1996;17(3):277–84.

[33] Brauker J, Frost GH, Dwarki V, Nijjar T, Chin R, Carr-Brendel V, Jasunas C, Hodgett D, Stone W, Cohen LK, Johnson RC. Sustained expression of high levels of human factor IX from human cells implanted within an immunoisolation device into athymic rodents. Hum Gene Ther 1998;9(6):879–88.

[34] Carr-Brendel VE, Geller RL, Thomas TJ, Boggs DR, Young SK, Crudele J, Martinson LA, Maryanov DA, Johnson RC, Brauker JH. Transplantation of cells in an immunoisolation device for gene therapy. Methods Mol Biol 1997;63:373–87.

[35] Josephs SF, Loudovaris T, Dixit A, Young SK, Johnson RC. In vivo delivery of recombinant human growth hormone from genetically engineered human fibroblasts implanted within Baxter immunoisolation devices. J Mol Med (Berl) 1999;77(1):211–4.

[36] Greene D, Pruitt L, Maas CS. Biomechanical effects of e-PTFE implant structure on soft tissue implantation stability: a study in the porcine model. Laryngoscope 1997;107(7):957–62.

[37] Lanza RP, Jackson R, Sullivan A, Ringeling J, McGrath C, Kühtreiber W, Chick WL. Xenotransplantation of cells using biodegradable microcapsules. Transplantation 1999;67(8):1105–11.

[38] Deacon T, Schumacher J, Dinsmore J, Thomas C, Palmer P, Kott S, Edge A, Penney D, Kassissieh S, Dempsey P, Isacson O. Histological evidence of fetal pig neural cell survival after transplantation into a patient with Parkinson's disease. Nat Med 1997;3(3):350–3.

[39] Selawry H, Whittington K, Fajaco R. Effect of cyclosporine on islet xenograft survival in the BB/W rat. Transplantation 1986;42(5):568–70.

[40] Fan MY, Lum ZP, Fu XW, Levesque L, Tai IT, Sun AM. Reversal of diabetes in BB rats by transplantation of encapsulated pancreatic islets. Diabetes 1990;39(4):519–22.

[41] Fritschy WM, Gerrits PO, Wolters GH, Pasma A, Van Schilfgaarde R. Glycol methacrylate embedding of alginate-polylysine microencapsulated pancreatic islets. Biotech Histochem 1995;70(4):188–93.

[42] O'Shea GM, Sun AM. Encapsulation of rat islets of Langerhans prolongs xenograft survival in diabetic mice. Diabetes 1986;35(8):943–6.

[43] Soon-Shiong P. Treatment of type I diabetes using encapsulated islets. Adv Drug Deliv Rev 1999;35(2–3):259–70.

[44] De Vos P, De Haan B, Wolters GH, Van Schilfgaarde R. Factors influencing the adequacy of microencapsulation of rat pancreatic islets. Transplantation 1996;62(7):888–93.

[45] de Vos P, Vegter D, Strubbe JH, de Haan BJ, van Schilfgaarde R. Impaired glucose tolerance in recipients of an intraperitoneally implanted microencapsulated islet allograft is caused by the slow diffusion of insulin through the peritoneal membrane. Transplant Proc 1997;29(1–2):756–7.

[46] Opara EC, Mirmalek-Sani SH, Khanna O, Moya ML, Brey EM. Design of a bioartificial pancreas(+). J Investig Med 2010;58(7):831–7.

[47] Bowers DT, Olingy CE, Chhabra P, Langman L, Merrill PH, Linhart RS, Tanes ML, Lin D, Brayman KL, Botchwey EA. An engineered macroencapsulation membrane releasing FTY720 to precondition pancreatic islet transplantation. J Biomed Mater Res B Appl Biomater 2018;106(2):555–68.

[48] Sullivan SJ, Maki T, Borland KM, Mahoney MD, Solomon BA, Muller TE, Monaco AP, Chick WL. Biohybrid artificial pancreas: long-term implantation studies in diabetic, pancreatectomized dogs. Science 1991;252(5006):718–21.

[49] Atwater I, Yañez A, Cea R, Navia A, Jeffs S, Arraya V, Szpak-Glasman M, Leighton X, Goping G, Bevilacqua JA, Moreno R, Brito J, Arriaza C, Ommaya A. Cerebral spinal fluid shunt is an immunologically privileged site for transplantation of xenogeneic islets. Transplant Proc 1997;29(4):2111–5.

[50] Lanza RP, Borland KM, Lodge P, Carretta M, Sullivan SJ, Muller TE, Solomon BA, Maki T, Monaco AP, Chick WL. Treatment of severely diabetic pancreatectomized dogs using a diffusion-based hybrid pancreas. Diabetes 1992;41(7):886–9.

[51] Scharp DW, Mason NS, Sparks RE. Islet immuno-isolation: the use of hybrid artificial organs to prevent islet tissue rejection. World J Surg 1984;8(2):221–9.

[52] Colton CK, Avgoustiniatos ES. Bioengineering in development of the hybrid artificial pancreas. J Biomech Eng 1991;113(2):152–70.

[53] Katsuragi S, Ikeda T, Date I, Shingo T, Yasuhara T, Mishima K, Aoo N, Harada K, Egashira N, Iwasaki K, Fujiwara M, Ikenoue T. Implantation of encapsulated glial cell line-derived neurotrophic factor-secreting cells prevents long-lasting learning impairment following neonatal hypoxic-ischemic brain insult in rats. Am J Obstet Gynecol 2005;192(4):1028–37.

[54] Kim BJ, Choi JY, Choi H, Han S, Seo J, Kim J, Joo S, Kim HM, Oh C, Hong S, Kim P, Choi IS. Astrocyte-encapsulated hydrogel microfibers enhance neuronal circuit generation. Adv Healthc Mater 2020;9(5), e1901072.

[55] Moriarty N, Cabré S, Alamilla V, Pandit A, Dowd E. Encapsulation of young donor age dopaminergic grafts in a GDNF-loaded collagen hydrogel further increases their survival, reinnervation, and functional efficacy after intrastriatal transplantation in hemi-Parkinsonian rats. Eur J Neurosci 2019;49(4):487–96.

[56] Widmer HR. Combination of cell transplantation and glial cell line-derived neurotrophic factor-secreting encapsulated cells in Parkinson's disease. Brain Circ 2018;4(3):114–7.

[57] Yasuhara T, Borlongan CV, Date I. Ex vivo gene therapy: transplantation of neurotrophic factor-secreting cells for cerebral ischemia. Front Biosci 2006;11:760–75.

[58] Arranz-Romera A, Esteban-Pérez S, Molina-Martínez IT, Bravo-Osuna I, Herrero-Vanrell R. Co-delivery of glial cell-derived neurotrophic factor (GDNF) and tauroursodeoxycholic acid (TUDCA) from PLGA microspheres: potential combination therapy for retinal diseases. Drug Deliv Transl Res 2021;11(2):566–80.

[59] Ahn YH, Bensadoun JC, Aebischer P, Zurn AD, Seiger A, Björklund A, Lindvall O, Wahlberg L, Brundin P, Kaminski Schierle GS. Increased fiber outgrowth from xeno-transplanted human embryonic dopaminergic neurons with co-implants of polymer-encapsulated genetically modified cells releasing glial cell line-derived neurotrophic factor. Brain Res Bull 2005;66(2):135–42.

[60] Brissová M, Lacík I, Powers AC, Anilkumar AV, Wang T. Control and measurement of permeability for design of microcapsule cell delivery system. J Biomed Mater Res 1998;39(1):61–70.

[61] Broadhead KW, Biran R, Tresco PA. Hollow fiber membrane diffusive permeability regulates encapsulated cell line biomass, proliferation, and small molecule release. Biomaterials 2002;23(24):4689–99.

[62] Risbud M. Tissue engineering: implications in the treatment of organ and tissue defects. Biogerontology 2001;2(2):117–25.
[63] Risbud MV, Bhargava S, Bhonde RR. In vivo biocompatibility evaluation of cellulose macrocapsules for islet immunoisolation: Implications of low molecular weight cut-off. J Biomed Mater Res A 2003;66(1):86–92.
[64] Jones KS, Sefton MV, Gorczynski RM. In vivo recognition by the host adaptive immune system of microencapsulated xenogeneic cells. Transplantation 2004;78(10):1454–62.
[65] Weaver JM, Algire GH, Prehn RT. The growth of cells in vivo in diffusion chambers. II. The role of cells in the destruction of homografts in mice. J Natl Cancer Inst 1955;15(6):1737–67.
[66] Algire GH, Weaver JM, Prehn RT. Growth of cells in vivo in diffusion chambers. I. Survival of homografts in immunized mice. J Natl Cancer Inst 1954;15(3):493–507.
[67] Duvivier-Kali VF, Omer A, Parent RJ, O'Neil JJ, Weir GC. Complete protection of islets against allorejection and autoimmunity by a simple barium-alginate membrane. Diabetes 2001;50(8):1698–705.
[68] Tibell A, Rafael E, Wennberg L, Nordenström J, Bergström M, Geller RL, Loudovaris T, Johnson RC, Brauker JH, Neuenfeldt S, Wernerson A. Survival of macroencapsulated allogeneic parathyroid tissue one year after transplantation in nonimmunosuppressed humans. Cell Transplant 2001;10(7):591–9.
[69] Martinson LA, Brauker JH, Johnson RC, Loudovaris T. Methods for enhancing vascularization of implant devices; 1996. US Patent.
[70] Weber CJ, Hagler MK, Chryssochoos JT, Larsen CP, Pearson TC, Jensen P, Kapp JA, Linsley PS. CTLA4-Ig prolongs survival of microencapsulated rabbit islet xenografts in spontaneously diabetic Nod mice. Transplant Proc 1996;28(2):821–3.
[71] Clayton HA, James RF, London NJ. Islet microencapsulation: a review. Acta Diabetol 1993;30(4):181–9.
[72] Cole DR, Waterfall M, McIntyre M, Baird JD. Microencapsulated islet grafts in the BB/E rat: a possible role for cytokines in graft failure. Diabetologia 1992;35(3):231–7.
[73] Gotfredsen CF, Stewart MG, O'Shea GM, Vose JR, Horn T, Moody AJ. The fate of transplanted encapsulated islets in spontaneously diabetic BB/Wor rats. Diabetes Res 1990;15(4):157–63.
[74] Wijsman J, Atkison P, Mazaheri R, Garcia B, Paul T, Vose J, O'Shea G, Stiller C. Histological and immunopathological analysis of recovered encapsulated allogeneic islets from transplanted diabetic BB/W rats. Transplantation 1992;54(4):588–92.
[75] Brauker J, Martinson LA, Young SK, Johnson RC. Local inflammatory response around diffusion chambers containing xenografts. Nonspecific destruction of tissues and decreased local vascularization. Transplantation 1996;61(12):1671–7.
[76] Safley SA, Kenyon NS, Berman DM, Barber GF, Willman M, Duncanson S, Iwakoshi N, Holdcraft R, Gazda L, Thompson P, Badell IR, Sambanis A, Ricordi C, Weber CJ. Microencapsulated adult porcine islets transplanted intraperitoneally in streptozotocin-diabetic non-human primates. Xenotransplantation 2018;25(6), e12450.
[77] Weber CJ, Safley S, Hagler M, Kapp J. Evaluation of graft-host response for various tissue sources and animal models. Ann N Y Acad Sci 1999;875:233–54.
[78] Kobayashi T, Harb G, Rajotte RV, Korbutt GS, Mallett AG, Arefanian H, Mok D, Rayat GR. Immune mechanisms associated with the rejection of encapsulated neonatal porcine islet xenografts. Xenotransplantation 2006;13(6):547–59.
[79] Loudovaris T, Mandel TE, Charlton B. CD4+ T cell mediated destruction of xenografts within cell-impermeable membranes in the absence of CD8+ T cells and B cells. Transplantation 1996;61(12):1678–84.
[80] Clementi ME, Marini S, Condò SG, Giardina B. Antibodies against small molecules. Ann Ist Super Sanita 1991;27(1):139–43.
[81] Greenfield EA, DeCaprio J, Brahmandam M. Making weak antigens strong: cross-linking peptides to KLH with maleimide. Cold Spring Harb Protoc 2018;2018(10).
[82] De Vos P, Van Straaten JF, Nieuwenhuizen AG, de Groot M, Ploeg RJ, De Haan BJ, Van Schilfgaarde R. Why do microencapsulated islet grafts fail in the absence of fibrotic overgrowth? Diabetes 1999;48(7):1381–8.
[83] Wang X, Maxwell KG, Wang K, Bowers DT, Flanders JA, Liu W, Wang LH, Liu Q, Liu C, Naji A, Wang Y, Wang B, Chen J, Ernst AU, Melero-Martin JM, Millman JR, Ma M. A nanofibrous encapsulation device for safe delivery of insulin-producing cells to treat type 1 diabetes. Sci Transl Med 2021;13(596).
[84] Ulland TK, Jain N, Hornick EE, Elliott EI, Clay GM, Sadler JJ, Mills KA, Janowski AM, Volk AP, Wang K, Legge KL, Gakhar L, Bourdi M, Ferguson PJ, Wilson ME, Cassel SL, Sutterwala FS. Nlrp12 mutation causes C57BL/6J strain-specific defect in neutrophil recruitment. Nat Commun 2016;7:13180.
[85] Al-Mohanna F, Saleh S, Parhar RS, Khabar K, Collison K. Human neutrophil gene expression profiling following xenogeneic encounter with porcine aortic endothelial cells: the occult role of neutrophils in xenograft rejection revealed. J Leukoc Biol 2005;78(1):51–61.
[86] Devos T, Waer M, Billiau AD. Cross-species chemoattraction and xenograft failure: do neutrophils play a role? Transplantation 2004;78(12):1717–8.
[87] Mok D, Black M, Gupta N, Arefanian H, Tredget E, Rayat GR. Early immune mechanisms of neonatal porcine islet xenograft rejection. Xenotransplantation 2019;26(6), e12546.
[88] Loudovaris T, Charlton B, Hodgson RJ, Mandel TE. Destruction of xenografts but not allografts within cell impermeable membranes. Transplant Proc 1992;24(5):2291–2.
[89] McKenzie AW, Georgiou HM, Zhan Y, Brady JL, Lew AM. Protection of xenografts by a combination of immunoisolation and a single dose of anti-CD4 antibody. Cell Transplant 2001;10(2):183–93.
[90] Granicka LH, Kawiak JW, Glowacka E, Werynski A. Encapsulation of OKT3 cells in hollow fibers. ASAIO J 1996;42(5):M863–6.
[91] Lathuilière A, Mach N, Schneider BL. Encapsulated cellular implants for recombinant protein delivery and therapeutic modulation of the immune system. Int J Mol Sci 2015;16(5):10578–600.

[92] Faleo G, Lee K, Nguyen V, Tang Q. Assessment of immune isolation of allogeneic mouse pancreatic progenitor cells by a macroencapsulation device. Transplantation 2016;100(6):1211–8.

[93] Kobayashi T, Aomatsu Y, Iwata H, Kin T, Kanehiro H, Hisanaga M, Ko S, Nagao M, Nakajima Y. Indefinite islet protection from autoimmune destruction in nonobese diabetic mice by agarose microencapsulation without immunosuppression. Transplantation 2003;75(5):619–25.

[94] Kobayashi T, Aomatsu Y, Kanehiro H, Hisanaga M, Nakajima Y. Protection of NOD islet isograft from autoimmune destruction by agarose microencapsulation. Transplant Proc 2003;35(1):484–5.

[95] Loudovaris T, Jacobs S, Young S, Maryanov D, Brauker J, Johnson RC. Correction of diabetic nod mice with insulinomas implanted within Baxter immunoisolation devices. J Mol Med (Berl) 1999;77(1):219–22.

[96] Weber C, Ayres-Price J, Costanzo M, Becker A, Stall A. NOD mouse peritoneal cellular response to poly-L-lysine-alginate microencapsulated rat islets. Transplant Proc 1994;26(3):1116–9.

[97] Xu BY, Yang H, Serreze DV, MacIntosh R, Yu W, Wright Jr JR. Rapid destruction of encapsulated islet xenografts by NOD mice is CD4-dependent and facilitated by B-cells: innate immunity and autoimmunity do not play significant roles. Transplantation 2005;80(3):402–9.

[98] Bowers DT, Song W, Wang LH, Ma M. Engineering the vasculature for islet transplantation. Acta Biomater 2019;95:131–51.

[99] Brauker J, Martinson LA, Hill RS, Young SK, Carr-Brendel VE, Johnson RC. Neovascularization of immunoisolation membranes: the effect of membrane architecture and encapsulated tissue. Transplant Proc 1992;24(6):2924.

[100] Brown DL, Meagher PJ, Knight KR, Keramidaris E, Romeo-Meeuw R, Penington AJ, Morrison WA. Survival and function of transplanted islet cells on an in vivo, vascularized tissue engineering platform in the rat: a pilot study. Cell Transplant 2006;15(4):319–24.

[101] Groot Nibbelink M, Skrzypek K, Karbaat L, Both S, Plass J, Klomphaar B, van Lente J, Henke S, Karperien M, Stamatialis D, van Apeldoorn A. An important step towards a prevascularized islet microencapsulation device: in vivo prevascularization by combination of mesenchymal stem cells on micropatterned membranes. J Mater Sci Mater Med 2018;29(11):174.

[102] He Y, Xu Z, Fu H, Chen B, Wang S, Chen B, Zhou M, Cai Y. Combined microencapsulated islet transplantation and revascularization of aortorenal bypass in a diabetic nephropathy rat model. J Diabetes Res 2016;2016:9706321.

[103] Kaushiva A, Turzhitsky VM, Darmoc M, Backman V, Ameer GA. A biodegradable vascularizing membrane: a feasibility study. Acta Biomater 2007;3(5):631–42.

[104] Lembert N, Wesche J, Petersen P, Doser M, Zschocke P, Becker HD, Ammon HPT. Encapsulation of islets in rough surface, hydroxymethylated polysulfone capillaries stimulates VEGF release and promotes vascularization after transplantation. Cell Transplant 2005;14(2–3):97–108.

[105] Maciel J, Oliveira MI, Colton E, McNally AK, Oliveira C, Anderson JM, Barbosa MA. Adsorbed fibrinogen enhances production of bone- and angiogenic-related factors by monocytes/macrophages. Tissue Eng Part A 2014;20(1–2):250–63.

[106] Magisson J, Sassi A, Xhema D, Kobalyan A, Gianello P, Mourer B, Tran N, Burcez CT, Bou Aoun R, Sigrist S. Safety and function of a new prevascularized bioartificial pancreas in an allogeneic rat model. J Tissue Eng 2020;11, 2041731420924818.

[107] Marchioli G, Zellner L, Oliveira C, Engelse M, Koning E, Mano J, Karperien, Apeldoorn AV, Moroni L. Layered PEGDA hydrogel for islet of Langerhans encapsulation and improvement of vascularization. J Mater Sci Mater Med 2017;28(12):195.

[108] Martinson LA, Brauker JH, Johnson RC, Loudovaris T. Methods for enhancing vascularization of implant devices. US: B. Healthcare; 1995.

[109] Scheiner KC, Coulter F, Maas-Bakker RF, Ghersi G, Nguyen TT, Steendam R, Duffy GP, Hennink WE, O'Cearbhaill ED, Kok RJ. Vascular endothelial growth factor-releasing microspheres based on poly(ε-caprolactone-PEG-ε-caprolactone)-b-poly(L-lactide) multiblock copolymers incorporated in a three-dimensional printed poly(dimethylsiloxane) cell macroencapsulation device. J Pharm Sci 2020;109(1):863–70.

[110] Stiegler P, Matzi V, Pierer E, Hauser O, Schaffellner S, Renner H, Greilberger J, Aigner R, Maier A, Lackner C, Iberer F, Smolle-Jüttner FM, Tscheliessnigg K, Stadlbauer V. Creation of a prevascularized site for cell transplantation in rats. Xenotransplantation 2010;17(5):379–90.

[111] Trivedi N, Steil GM, Colton CK, Bonner-Weir S, Weir GC. Improved vascularization of planar membrane diffusion devices following continuous infusion of vascular endothelial growth factor. Cell Transplant 2000;9(1):115–24.

[112] Wang K, Yu LY, Jiang LY, Wang HB, Wang CY, Luo Y. The paracrine effects of adipose-derived stem cells on neovascularization and biocompatibility of a macroencapsulation device. Acta Biomater 2015;15:65–76.

[113] Weaver JD, Headen DM, Hunckler MD, Coronel MM, Stabler CL, García AJ. Design of a vascularized synthetic poly(ethylene glycol) macroencapsulation device for islet transplantation. Biomaterials 2018;172:54–65.

[114] Sörenby A, Rafael E, Tibell A, Wernerson A. Improved histological evaluation of vascularity around an immunoisolation device by correlating number of vascular profiles to glucose exchange. Cell Transplant 2004;13(6):713–9.

[115] Hady TF, Hwang B, Pusic AD, Waworuntu RL, Mulligan M, Ratner B, Bryers JD. Uniform 40-μm-pore diameter precision templated scaffolds promote a pro-healing host response by extracellular vesicle immune communication. J Tissue Eng Regen Med 2021;15(1):24–36.

[116] Kenneth Ward W. A review of the foreign-body response to subcutaneously-implanted devices: the role of macrophages and cytokines in biofouling and fibrosis. J Diabetes Sci Technol 2008;2(5):768–77.

[117] Gurevich DB, French KE, Collin JD, Cross SJ, Martin P. Live imaging the foreign body response in zebrafish reveals how dampening inflammation reduces fibrosis. J Cell Sci 2019;133(5).

[118] Mariani E, Lisignoli G, Borzi RM, Pulsatelli L. Biomaterials: foreign bodies or tuners for the immune response? Int J Mol Sci 2019;20(3).

[119] Rainey WE, Saner K, Schimmer BP. Adrenocortical cell lines. Mol Cell Endocrinol 2004;228(1–2):23–38.

[120] Diaz-Jimenez D, Kolb JP, Cidlowski JA. Glucocorticoids as regulators of macrophage-mediated tissue homeostasis. Front Immunol 2021;12, 669891.

[121] Aijaz A, Teryek M, Goedken M, Polunas M, Olabisi RM. Coencapsulation of ISCs and MSCs enhances viability and function of both cell types for improved wound healing. Cell Mol Bioeng 2019;12(5):481–93.

[122] Jośko J, Gwóźdź B, Jedrzejowska-Szypułka H, Hendryk S. Vascular endothelial growth factor (VEGF) and its effect on angiogenesis. Med Sci Monit 2000;6(5):1047–52.
[123] Brandhorst D, Brandhorst H, Johnson PRV. Strategies to improve the oxygen supply to microencapsulated islets. Transplantation 2020;104(2):237.
[124] Coronel MM, Liang JP, Li Y, Stabler CL. Oxygen generating biomaterial improves the function and efficacy of beta cells within a macroencapsulation device. Biomaterials 2019;210:1–11.
[125] Evron Y, Colton CK, Ludwig B, Weir GC, Zimermann B, Maimon S, Neufeld T, Shalev N, Goldman T, Leon A, Yavriyants K, Shabtay N, Rozenshtein T, Azarov D, DiIenno AR, Steffen A, de Vos P, Bornstein SR, Barkai U, Rotem A. Long-term viability and function of transplanted islets macroencapsulated at high density are achieved by enhanced oxygen supply. Sci Rep 2018;8(1):6508.
[126] Iwata H, Arima Y, Tsutsui Y. Design of bioartificial pancreases from the standpoint of oxygen supply. Artif Organs 2018;42(8):E168–e185.
[127] Lee EM, Jung JI, Alam Z, Yi HG, Kim H, Choi JW, Hurh S, Kim YJ, Jeong JC, Yang J, Oh KH, Kim HC, Lee BC, Choi I, Cho DW, Ahn C. Effect of an oxygen-generating scaffold on the viability and insulin secretion function of porcine neonatal pancreatic cell clusters. Xenotransplantation 2018;25(2), e12378.
[128] Liang JP, Accolla RP, Soundirarajan M, Emerson A, Coronel MM, Stabler C. Engineering a macroporous oxygen-generating scaffold for enhancing islet cell transplantation within an extrahepatic site. Acta Biomater 2021;130:268–80.
[129] McQuilling JP, Opara EC. Methods for incorporating oxygen-generating biomaterials into cell culture and microcapsule systems. Methods Mol Biol 2017;1479:135–41.
[130] McQuilling JP, Sittadjody S, Pendergraft S, Farney AC, Opara EC. Applications of particulate oxygen-generating substances (POGS) in the bioartificial pancreas. Biomater Sci 2017;5(12):2437–47.
[131] Papas KK, De Leon H, Suszynski TM, Johnson RC. Oxygenation strategies for encapsulated islet and beta cell transplants. Adv Drug Deliv Rev 2019;139:139–56.
[132] Ludwig B, Ludwig S, Steffen A, Knauf Y, Zimerman B, Heinke S, Lehmann S, Schubert U, Schmid J, Bleyer M, Schönmann U, Colton CK, Bonifacio E, Solimena M, Reichel A, Schally AV, Rotem A, Barkai U, Grinberg-Rashi H, Kaup FJ, Avni Y, Jones P, Bornstein SR. Favorable outcome of experimental islet xenotransplantation without immunosuppression in a nonhuman primate model of diabetes. Proc Natl Acad Sci U S A 2017;114(44):11745–50.
[133] Carlsson PO, Espes D, Sedigh A, Rotem A, Zimerman B, Grinberg H, Goldman T, Barkai U, Avni Y, Westermark GT, Carlbom L, Ahlström H, Eriksson O, Olerud J, Korsgren O. Transplantation of macroencapsulated human islets within the bioartificial pancreas βAir to patients with type 1 diabetes mellitus. Am J Transplant 2018;18(7):1735–44.
[134] Lum ZP, Tai IT, Krestow M, Norton J, Vacek I, Sun AM. Prolonged reversal of diabetic state in NOD mice by xenografts of microencapsulated rat islets. Diabetes 1991;40(11):1511–6.
[135] Safley SA, Kenyon NS, Berman DM, Barber GF, Cui H, Duncanson S, De Toni T, Willman M, De Vos P, Tomei AA, Sambanis A, Kenyon NM, Ricordi C, Weber CJ. Microencapsulated islet allografts in diabetic NOD mice and nonhuman primates. Eur Rev Med Pharmacol Sci 2020;24(16):8551–65.
[136] Weber C, Krekun S, Koschitzky T, Zabinski S, D'Agati V, Hardy M, Reemtsma K. Prolonged functional survival of rat-to-NOD mouse islet xenografts by ultraviolet-B (UV-B) irradiation plus microencapsulation of donor islets. Transplant Proc 1991;23(1 Pt 1):764–6.
[137] Altman JJ, Penfornis A, Boillot J, Maletti M. Bioartificial pancreas in autoimmune nonobese diabetic mice. ASAIO Trans 1988;34(3):247–9.
[138] Altman JJ, Houlbert D, Callard P, McMillan P, Solomon BA, Rosen J, Galletti PM. Long-term plasma glucose normalization in experimental diabetic rats with macroencapsulated implants of benign human insulinomas. Diabetes 1986;35(6):625–33.
[139] El-Halawani SM, Gabr MM, El-Far M, Zakaria MM, Khater SM, Refaie AF, Ghoneim MA. Subcutaneous transplantation of bone marrow derived stem cells in macroencapsulation device for treating diabetic rats; clinically transplantable site. Heliyon 2020;6(5), e03914.
[140] Jiang XF, Qian TL, Chen D, Lu HW, Xue P, Yang XW, Zhang LH, Hu YZ, Zhang DW. Correction of hyperglycemia in diabetic rats with the use of microencapsulated young market pig islets. Transplant Proc 2018;50(10):3895–9.
[141] Lim F, Sun AM. Microencapsulated islets in diabetic rats. Science 1981;213(4512):1146.
[142] Remuzzi A, Cornolti R, Bianchi R, Figliuzzi M, Porretta-Serapiglia C, Oggioni N, Carozzi V, Crippa L, Avezza F, Fiordaliso F, Salio M, Lauria G, Lombardi R, Cavaletti G. Regression of diabetic complications by islet transplantation in the rat. Diabetologia 2009;52(12):2653–61.
[143] Sun AM, O'Shea GM, Goosen MF. Injectable microencapsulated islet cells as a bioartificial pancreas. Appl Biochem Biotechnol 1984;10:87–99.
[144] Sun AM, Parisius W, Healy GM, Vacek I, Macmorine HG. The use, in diabetic rats and monkeys, of artificial capillary units containing cultured islets of Langerhans (artificial endocrine pancreas). Diabetes 1977;26(12):1136–9.
[145] Vinerean HV, Gazda LS, Hall RD, Rubin AL, Smith BH. Improved glucose regulation on a low carbohydrate diet in diabetic rats transplanted with macroencapsulated porcine islets. Cell Transplant 2008;17(5):567–75.
[146] Abalovich AG, Bacqué MC, Grana D, Milei J. Pig pancreatic islet transplantation into spontaneously diabetic dogs. Transplant Proc 2009;41(1):328–30.
[147] Araki Y, Solomon BA, Basile RM, Chick WL. Biohybrid artificial pancreas. Long-term insulin secretion by devices seeded with canine islets. Diabetes 1985;34(9):850–4.
[148] Calafiore R, Basta G, Falorni Jr A, Ciabattoni P, Brotzu G, Cortesini R, Brunetti P. Intravascular transplantation of microencapsulated islets in diabetic dogs. Transplant Proc 1992;24(3):935–6.
[149] Gabr MM, Zakaria MM, Refaie AF, Ismail AM, Khater SM, Ashamallah SA, Azzam MM, Ghoneim MA. Insulin-producing cells from adult human bone marrow mesenchymal stromal cells could control chemically induced diabetes in dogs: a preliminary study. Cell Transplant 2018;27(6):937–47.

[150] Gooch A, Zhang P, Hu Z, Loy Son N, Avila N, Fischer J, Roberts G, Sellon R, Westenfelder C. Interim report on the effective intraperitoneal therapy of insulin-dependent diabetes mellitus in pet dogs using "Neo-Islets," aggregates of adipose stem and pancreatic islet cells (INAD 012-776). PLoS One 2019;14(9), e0218688.

[151] Kin T, Iwata H, Aomatsu Y, Ohyama T, Kanehiro H, Hisanaga M, Nakajima Y. Xenotransplantation of pig islets in diabetic dogs with use of a microcapsule composed of agarose and polystyrene sulfonic acid mixed gel. Pancreas 2002;25(1):94–100.

[152] Lanza RP, Kühtreiber WM, Ecker DM, Marsh JP, Chick WL. Successful bovine islet xenografts in rodents and dogs using injectable microreactors. Transplant Proc 1995;27(6):3211.

[153] Lanza RP, Kühtreiber WM, Ecker DM, Marsh JP, Staruk JE, Chick WL. Xenotransplantation of bovine islets into dogs using biodegradable, injectable microreactors. Transplant Proc 1996;28(2):820.

[154] Soon-Shiong P, Feldman E, Nelson R, Komtebedde J, Smidsrod O, Skjak-Braek G, Espevik T, Heintz R, Lee M. Successful reversal of spontaneous diabetes in dogs by intraperitoneal microencapsulated islets. Transplantation 1992;54(5):769–74.

[155] Sun AM. Microencapsulation of pancreatic islet cells: a bioartificial endocrine pancreas. Methods Enzymol 1988;137:575–80.

[156] Neufeld T, Ludwig B, Barkai U, Weir GC, Colton CK, Evron Y, Balyura M, Yavriyants K, Zimermann B, Azarov D, Maimon S, Shabtay N, Rozenshtein T, Lorber D, Steffen A, Willenz U, Bloch K, Vardi P, Taube R, de Vos P, Lewis EC, Bornstein SR, Rotem A. The efficacy of an immunoisolating membrane system for islet xenotransplantation in minipigs. PLoS One 2013;8(8), e70150.

[157] Penfornis F, Icard P, Gotheil C, Boillot J, Cornec C, Barrat F, Altman JJ, Cochin JV. Bioartificial pancreas in pigs. Horm Metab Res Suppl 1990;25:200–2.

[158] Sweet IR, Yanay O, Waldron L, Gilbert M, Fuller JM, Tupling T, et al. Treatment of diabetic rats with encapsulated islets. J Cell Mol Med 2008;12(6b):2644–50.

[159] Dufrane D, Goebbels RM, Gianello P. Alginate macroencapsulation of pig islets allows correction of streptozotocin-induced diabetes in primates up to 6 months without immunosuppression. Transplantation 2010;90(10):1054–62.

[160] Elliott RB, Escobar L, Tan PL, Garkavenko O, Calafiore R, Basta P, Vasconcellos AV, Emerich DF, Thanos C, Bambra C. Intraperitoneal alginate-encapsulated neonatal porcine islets in a placebo-controlled study with 16 diabetic cynomolgus primates. Transplant Proc 2005;37(8):3505–8.

[161] Igarashi Y, D'Hoore W, Goebbels RM, Gianello P, Dufrane D. Beta-5 score to evaluate pig islet graft function in a primate pre-clinical model. Xenotransplantation 2010;17(6):449–59.

[162] Yanay O, Barry SC, Flint LY, Brzezinski M, Barton RW, Osborne WR. Long-term erythropoietin gene expression from transduced cells in bioisolator devices. Hum Gene Ther 2003;14(17):1587–93.

[163] Chen SH, Huang SC, Lui CC, Lin TP, Chou FF, Ko JY. Effect of theracyte-encapsulated parathyroid cells on lumbar fusion in a rat model. Eur Spine J 2012;21(9):1734–9.

[164] Chou FF, Huang SC, Chen SS, Wang PW, Huang PH, Lu KY. Treatment of osteoporosis with theracyte-encapsulated parathyroid cells: a study in a rat model. Osteoporos Int 2006;17(6):936–41.

[165] Hasse C, Schrezenmeir J, Stinner B, Schark C, Wagner PK, Neumann K, Rothmund M. Successful allotransplantation of microencapsulated parathyroids in rats. World J Surg 1994;18(4):630–4.

[166] Khryshchanovich V, Ghoussein Y. Allotransplantation of macroencapsulated parathyroid cells as a treatment of severe postsurgical hypoparathyroidism: case report. Ann Saudi Med 2016;36(2):143–7.

[167] Picariello L, Benvenuti S, Recenti R, Formigli L, Falchetti A, Morelli A, Masi L, Tonelli F, Cicchi P, Brandi ML. Microencapsulation of human parathyroid cells: an "in vitro" study. J Surg Res 2001;96(1):81–9.

[168] Emerich DF, Kordower JH, Chu Y, Thanos C, Bintz B, Paolone G, Wahlberg LU. Widespread striatal delivery of GDNF from encapsulated cells prevents the anatomical and functional consequences of excitotoxicity. Neural Plast 2019;2019:6286197.

[169] Emerich DF, Thanos CG. NT-501: an ophthalmic implant of polymer-encapsulated ciliary neurotrophic factor-producing cells. Curr Opin Mol Ther 2008;10(5):506–15.

[170] Fadia NB, Bliley JM, DiBernardo GA, Crammond DJ, Schilling BK, Sivak WN, Spiess AM, Washington KM, Waldner M, Liao HT, James IB, Minteer DM, Tompkins-Rhoades C, Cottrill AR, Kim DY, Schweizer R, Bourne DA, Panagis GE, Asher Schusterman 2nd M, Egro FM, Campwala IK, Simpson T, Weber DJ, Gause 2nd T, Brooker JE, Josyula T, Guevara AA, Repko AJ, Mahoney CM, Marra KG. Long-gap peripheral nerve repair through sustained release of a neurotrophic factor in nonhuman primates. Sci Transl Med 2020;12(527).

[171] Falcicchia C, Paolone G, Emerich DF, Lovisari F, Bell WJ, Fradet T, Wahlberg LU, Simonato M. Seizure-suppressant and neuroprotective effects of encapsulated BDNF-producing cells in a rat model of temporal lobe epilepsy. Mol Ther Methods Clin Dev 2018;9:211–24.

[172] Fransson A, Tornøe J, Wahlberg LU, Ulfendahl M. The feasibility of an encapsulated cell approach in an animal deafness model. J Control Release 2018;270:275–81.

[173] Garbayo E, Ansorena E, Lana H, Carmona-Abellan MD, Marcilla I, Lanciego JL, Luquin MR, Blanco-Prieto MJ. Brain delivery of microencapsulated GDNF induces functional and structural recovery in parkinsonian monkeys. Biomaterials 2016;110:11–23.

[174] Konerding WS, Janssen H, Hubka P, Tornøe J, Mistrik P, Wahlberg L, Lenarz T, Kral A, Scheper V. Encapsulated cell device approach for combined electrical stimulation and neurotrophic treatment of the deaf cochlea. Hear Res 2017;350:110–21.

[175] Mittoux V, Joseph JM, Conde F, Palfi S, Dautry C, Poyot T, Bloch J, Deglon N, Ouary S, Nimchinsky EA, Brouillet E, Hof PR, Peschanski M, Aebischer P, Hantraye P. Restoration of cognitive and motor functions by ciliary neurotrophic factor in a primate model of Huntington's disease. Hum Gene Ther 2000;11(8):1177–87.

[176] Moriarty N, Pandit A, Dowd E. Encapsulation of primary dopaminergic neurons in a GDNF-loaded collagen hydrogel increases their survival, re-innervation and function after intra-striatal transplantation. Sci Rep 2017;7(1):16033.

[177] Paolone G, Falcicchia C, Lovisari F, Kokaia M, Bell WJ, Fradet T, Barbieri M, Wahlberg LU, Emerich DF, Simonato M. Long-term, targeted delivery of GDNF from encapsulated cells is neuroprotective and reduces seizures in the pilocarpine model of epilepsy. J Neurosci 2019;39(11):2144–56.
[178] Wong FSY, Tsang KK, Chu AMW, Chan BP, Yao KM, Lo ACY. Injectable cell-encapsulating composite alginate-collagen platform with inducible termination switch for safer ocular drug delivery. Biomaterials 2019;201:53–67.
[179] Scharp DW, Swanson CJ, Olack BJ, Latta PP, Hegre OD, Doherty EJ, Gentile FT, Flavin KS, Ansara MF, Lacy PE. Protection of encapsulated human islets implanted without immunosuppression in patients with type I or type II diabetes and in nondiabetic control subjects. Diabetes 1994;43(9):1167–70.
[180] Tibell A, Rafael E, Wennberg L, Wernerson A, Sundberg B, Brauker JH, Geller RL, Johnson RC, Nordenstrõm J. Transplantation of macroencapsulated parathyroid tissue in man. Cell Transplant 1996;5.
[181] Ludwig B, Reichel A, Steffen A, Zimerman B, Schally AV, Block NL, Colton CK, Ludwig S, Kersting S, Bonifacio E, Solimena M, Gendler Z, Rotem A, Barkai U, Bornstein SR. Transplantation of human islets without immunosuppression. Proc Natl Acad Sci U S A 2013;110(47):19054–8.

Chapter 9

Transgenic pigs for islet xenotransplantation

Peter J. Cowan
Immunology Research Centre, St Vincent's Hospital Melbourne, Fitzroy, VIC, Australia; Department of Medicine, University of Melbourne, Parkville, VIC, Australia

Introduction

Porcine donors hold the promise of an inexhaustible alternative source of organs, tissues, and cells for transplantation into humans. Porcine islets are at the forefront of xenotransplantation, in terms of both efficacy in nonhuman primate (NHP) preclinical models and the state of progression to clinical trials. Islets from neonatal and adult porcine donors have been extensively tested in NHPs. Each type has pros and cons; for example, adult pig islets (APIs) are fragile but function immediately after transplantation, whereas neonatal pig islets (NPIs; also known as neonatal islet cell clusters or NICCs) take time to mature in vivo but are much easier to isolate and have greater proliferative capacity. The relative merits of NPIs and APIs are reviewed in detail elsewhere [1–3].

The longest reported survival and function of porcine islet xenografts in NHPs (>900 days) was achieved using wild type (WT) rather than genetically modified donors [4]. This is in stark contrast to the outcomes in preclinical solid organ xenotransplantation [5]. One could therefore argue that genetic modification of donor pigs to protect islet xenografts may be unnecessary for clinical success, at least initially. Balancing this is the fact that long-term reversal of diabetes in NHPs by WT porcine islets requires not only islet doses that are substantially higher than the clinical allogeneic islet equivalent—usually at least 100,000 versus 10,000–15,000 IEQ (islet equivalents)/kg body weight—but also an immunosuppressive regimen that would be unacceptable in the clinical setting [4,6]. When clinically applicable immunosuppressants were used in this model, the duration of insulin-independent normoglycemia was substantially reduced (maximum 129 days, median 44 days) [7].

Genetic modification of the donor pig provides the means to address both the efficacy and safety limitations of WT porcine islet xenotransplantation. Genetic strategies that are at various stages of development include "masking" the islets and strengthening their endogenous defenses to protect them from innate and adaptive immunity, improving the physiological compatibility of the islets and supercharging their function, and reducing or even eliminating the risk of transmission of porcine endogenous retrovirus (PERV) from the islet xenograft to the recipient. This chapter will describe each of these approaches in turn.

Protecting porcine islet xenografts from the innate immune response

The current clinical procedure for islet allotransplantation, described in detail in Chapters 5 and 6 in this book, is to infuse isolated islet preparations into the liver via the portal vein. The resulting nonphysiological interaction of processed tissue with blood can trigger a response termed the instant blood-mediated inflammatory reaction (IBMIR), which can rapidly destroy a significant proportion of the transplanted islet mass [8]. IBMIR is characterized by complement activation, platelet adherence, intravascular coagulation/thrombosis, release of proinflammatory cytokines and chemokines, and innate immune cell infiltration (reviewed in [9]). The intensity of IBMIR may be exacerbated for porcine islet xenografts by the presence of pre-formed antipig antibodies and the documented cross-species (pig/primate) molecular incompatibilities in the regulation of coagulation and innate immune cell activation [10]. Because IBMIR is the first major hurdle for islet xenografts, and innate immunity is more difficult to control with currently available pharmacological agents than adaptive immunity, it is unsurprising that most of the genetic modifications investigated to date are designed to attenuate the innate immune response.

Pancreas and Beta Cell Replacement. https://doi.org/10.1016/B978-0-12-824011-3.00006-0

Removing the carbohydrate targets of pre-formed xenoantibodies

Three carbohydrate moieties expressed on pig cell glycans—Gal, Neu5Gc, and Sda—have been identified as targets of preexisting antibodies in humans (Table 1). The genes for all three have been knocked out in pigs, either alone or in combination [11–14]. While the critical role of Gal as the major xenoantigen has been well established in pig-to-NHP models, the relative importance of Neu5Gc and Sda (alternatively referred to as SDa or Sd[a]) remains less clear, and the presence and level of anti-Neu5Gc and anti-Sda antibodies in humans appears to be more variable than that of anti-Gal. Nevertheless, in vitro incubation of human serum with peripheral blood mononuclear cells from knockout (KO) pigs has revealed that the major reduction in human antipig antibody binding caused by elimination of Gal is augmented by progressive deletion of Neu5Gc and Sda [15] (Fig. 1). Accordingly, the "triple-knockout" or TKO pig (GTKO/CMAH KO/B4GALNT2 KO) is increasingly viewed as the desired donor platform for clinical xenotransplantation, at least of solid organs [16,17]. It should be noted that one limitation in preclinical testing is that pig-to-Old World NHP models may not be ideal for assessing TKO organs and tissues. Unlike humans and New World NHPs, Old World NHPs have a functional *CMAH* gene [18] and do not make antibodies to Neu5Gc. There is also evidence to suggest that Old World NHPs possess pre-formed antibodies that recognize a putative new target (the so-called fourth xenoantigen) revealed by the deletion of Neu5Gc in CMAH KO pigs [14].

The contribution of Gal to the xenoantigenicity of pig islets has been studied for more than two decades. While an early study found that WT fetal pig islet xenografts were not susceptible to anti-Gal antibodies in a pig-to-mouse model [19], this does not appear to be the case for NPIs in the NHP model. Gal is strongly expressed on NPIs but only weakly on APIs [20]. GTKO and WT NPIs were evaluated in a series of NHP experiments by the Kirk group. In the first study [21], immunosuppressed diabetic rhesus monkeys received 50,000–60,000 IEQ/kg WT or GTKO NPIs intraportally ($n=5$ per group). WT NPIs induced intrahepatic inflammation and exhibited a high rate (80%) of primary nonfunction. GTKO NPIs generated significantly less inflammation than WT NPIs, reduced the rate of primary nonfunction to 20%, and sustained rejection-free survival for up to 249 days, with grafts eventually lost to cellular rejection (Fig. 2). Interestingly, the one WT NPI preparation that did enable insulin independence showed lower expression of Gal than the four that did not, further supporting a key role for this xenoantigen in the early failure of WT NPI xenografts.

The second study attempted to elucidate the mechanism of protection using a dual-islet transplant method in which 25,000–58,000 IEQ/kg WT or GTKO NPIs were transplanted into separate lobes of the liver in nonimmunosuppressed, nondiabetic rhesus monkeys [22]. This model was designed to directly compare different donor genotypes within a single recipient and thus control for differences between recipients, although it did not allow analysis of islet xenograft function. Xenografts were examined by immunohistochemistry at 1 and 24 h, revealing extensive deposition of platelets, antibody, and complement and infiltration by neutrophils and macrophages. Perhaps surprisingly, no significant differences in these readouts between GTKO and WT NPIs were observed at either time point. However, the possibility of systemic inflammatory signaling between graft sites could not be ruled out.

In a third study, WT and GTKO NPIs were compared to allogeneic (rhesus) islets in the dual-islet model [23]. WT NPIs showed increased apoptosis and IFNγ expression at 1 h and/or 24 h posttransplant compared to allogeneic islets, although other inflammatory markers were not significantly different. For the GTKO versus allogeneic comparison, the recipients were diabetic and immunosuppressed, and the analysis was extended to 7 days. Compared to the allogeneic islets, GTKO NPIs showed increased deposition of IgM and infiltration by macrophages and NK cells at day 7. Overall, the results of the three studies confirmed that IBMIR is more intense in the xenogeneic setting and that the GTKO modification improved the engraftment of NPIs, although the exact mechanism of protection remained unclear and there was considerable scope for further improvement. The effect of combining GTKO with additional protective genetic modifications is described in later sections.

TABLE 1 Carbohydrate xenoantigens expressed on pig cells.

Carbohydrate	Enzyme responsible for synthesis	Gene	Knockout
Galactose-α1,3-galactose (Gal)	α1,3-galactosyltransferase	*GGTA1*	GTKO
N-glycolylneuraminic acid (Neu5Gc)	Cytidine monophosphate-*N*-acetylneuraminic acid hydroxylase	*CMAH*	CMAH KO
Sda	β-1,4-*N*-acetyl-galactosaminyltransferase 2	*B4GALNT2*	B4GALNT2 KO

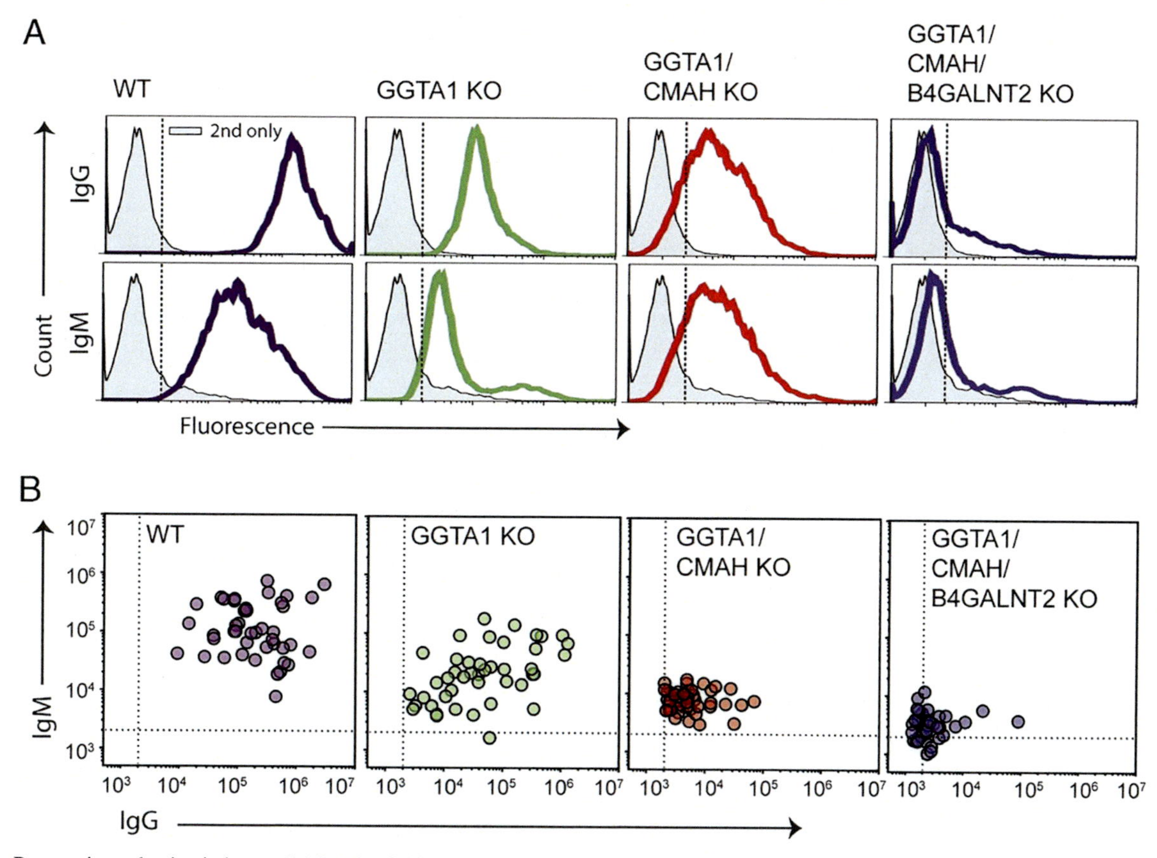

FIG. 1 Progressive reduction in human IgM and IgG binding to PBMCs from GTKO, GTKO/CMAH KO, and GGTA1/CMAH/B4GalNT2 KO (TKO) pigs. PBMC were incubated with serum from kidney transplant waitlisted patients and analyzed for IgM (y axes) and IgG (x axes) binding by flow cytometry. *Dotted lines* represent an MFI of 2000. (A) A single representative patient is screened on WT, GGTA1 KO, GGTA1/CMAH KO, GGTA1/CMAH/B4GalNT2 KO swine PBMC. (B) Sera from 44 randomly selected patients with unknown sensitization were incubated with PBMC from all 4 glycan backgrounds. *(Reproduced with permission from Martens GR, et al., Humoral reactivity of renal transplant-waitlisted patients to cells from GGTA1/CMAH/B4GalNT2, and SLA class I knockout pigs, Transplantation 101 (4) (2017):e86–e92 (https://journals.lww.com/transplantjournal/pages/articleviewer.aspx?year=2017&issue=04000&article=00019&type=Fulltext).)*

The possible importance of anti-Neu5Gc antibodies as a barrier to xenotransplantation has been reviewed recently [24]. Neu5Gc, also known as Hanganutziu–Deicher (H–D) antigen, is detectable on neonatal [25], juvenile [26], and adult [27] porcine islets. As noted earlier, the impact of eliminating Neu5Gc expression by knocking out the *CMAH* gene is difficult to test in the pig-to-Old World monkey model because these NHPs express Neu5Gc and do not make anti-Neu5Gc antibodies. Studies of CMAH KO porcine islets have therefore been limited to in vitro analyses. Lee et al. isolated islets from one-month-old pigs from two GTKO lines which differed only in CMAH [26]. Both lines were also transgenic for the human complement regulatory protein CD46. The GTKO/hCD46/CMAH KO and GTKO/hCD46/CMAH$^{+/+}$ islets were dissociated into single cells using trypsin and incubated with human serum to assess the binding of pre-formed human IgM and IgG antibodies. In addition, the islets were mixed in vitro with human blood to model IBMIR, with clotting time as the primary readout. The absence of Neu5Gc on islets had no significant effect on either human IgM/IgG binding or time to clotting. Although it is difficult to draw firm conclusions from these data, they suggest that pre-formed antibodies to Neu5Gc will not play a critical role in the immediate humoral response to GTKO porcine islets. Nevertheless, CMAH KO may have a part to play in attenuating the elicited antibody response in at least some individuals. Analysis of sera from humans transplanted with WT fetal pig islets in an early clinical trial [28] identified a dominant anti-αGal response but also revealed the development of anti-Neu5Gc antibodies in a subset of recipients [29].

Of the three recognized porcine carbohydrate xenoantigens, least is known about Sda (reviewed in [30]). The level of Sda expression on porcine islets has not been reported, and the effect of B4GALNT2 KO on their immunogenicity is unknown.

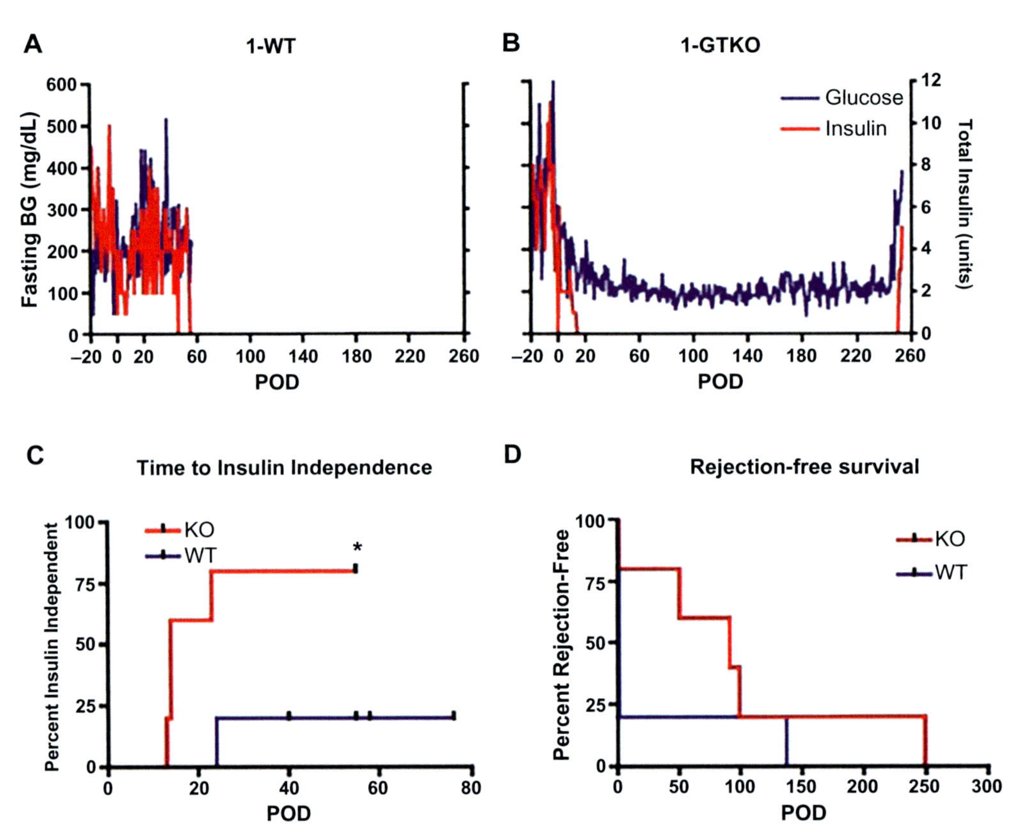

FIG. 2 Transplantation of Gal-deficient islets in rhesus monkeys results in superior graft function and more rapid return of euglycemia. (A) WT islet recipients most often experienced primary graft dysfunction, whereas transplantation of (B) GTKO NPIs resulted in prolonged insulin-independent normoglycemia, as shown in this representative pair. (C) Time to insulin independence was significantly shorter after transplantation of GTKO NPIs ($p=0.03$); however, duration of rejection-free survival (D) was similar for the two treatment groups ($p=0.3$). In (C), censored events represent recipient sacrifice. p Values calculated using log-rank test. *(Reproduced with permission from Thompson P. et al., Islet xenotransplantation using gal-deficient neonatal donors improves engraftment and function, Am J Transplant 11 (12) (2011):2593–2602 (John Wiley and Sons) © 2011 The American Society of Transplantation and the American Society of Transplant Surgeons.)*

Inhibiting the activation of complement by expressing human complement regulatory proteins (hCRPs)

Complement activation is a component of both IBMIR and the adaptive immune response and has thus been an important target for genetic modification in porcine islet xenotransplantation. Three membrane-bound human complement regulatory proteins (hCRPs) have been expressed in transgenic pigs: hCD46 and hCD55, which regulate complement mid-pathway at the level of the C3 and C5 convertases, and hCD59, which inhibits the formation of the terminal membrane attack complex. Van der Windt et al. transplanted immunosuppressed diabetic cynomolgus monkeys with 85,000–100,000 IEQ/kg hCD46-transgenic, WT, or GTKO APIs [31]. The degree of porcine C-peptide release, measured as a surrogate marker of early islet xenograft injury, suggested that IBMIR was not reduced by hCD46 expression. However, 4 of 5 recipients of hCD46 APIs achieved insulin independence for 87–396 days, in contrast to only 5–36 days for recipients of WT APIs. Thus transgenic complement regulation alone, while apparently insufficient to significantly attenuate IBMIR, was beneficial for longer-term survival. Interestingly, GTKO xenografts performed no better than WT xenografts in this study, possibly reflecting the already naturally low level of Gal expression on APIs.

The GTKO/hCRP combination

More recent pig-to-NHP studies of hCRP-transgenic islets have used donors that were also GTKO, with or without additional genetic modifications. Hawthorne et al. transplanted nondiabetic baboons with NPIs isolated from either WT pigs or GTKO transgenic pigs expressing hCD55, hCD59, and human α1,2-fucosyltransferase (H-transferase), which caps terminal oligosaccharides with the nonimmunogenic H-substance [32]. WT NPIs triggered profound IBMIR, resulting in their rapid destruction within hours of transplant. Treatment of two of the four recipients in this group with recombinant human antithrombin was ineffective at preventing intravascular thrombosis and xenograft destruction. In contrast, GTKO/hCD55/hCD59/hHT NPI xenografts ($n=5$) exhibited minimal evidence of IBMIR, with 1-h biopsies showing free NPIs and an absence of intravascular thrombosis. Nevertheless, a conventional immunosuppressive protocol (ATG, mycophenolate mofetil, and Tacrolimus) was unable to prevent cell-mediated rejection of the transgenic NPIs within 1 month [32]. Although the study design did not allow delineation of the relative contribution of each genetic modification, the results indicated that IBMIR can be addressed by a combinatorial genetic approach.

Supporting this, Bottino et al. observed a reduction in several indirect markers of IBMIR (porcine C-peptide release, perioperative dextrose requirements, serum IL-6) in 5 cynomolgus monkeys transplanted with APIs from GTKO/hCD46 pigs that also expressed other transgenes, compared to historical WT or hCD46-alone cohorts [33]. However, the apparent early protection did not translate to consistently prolonged graft survival. Although one xenograft functioned until the scheduled endpoint of 365 days, three others were lost within a week of transplant. The same group incubated NPIs from various genetically modified pigs with human whole blood in an in vitro model of IBMIR to examine the human response [34]. GTKO/hCD46 NPIs reduced activation of the alternative complement pathway compared to WT NPIs, but surprisingly accelerated coagulation. The reason for this unexpected finding remains unclear.

GTKO/hCD46 and GTKO NPIs were compared recently in the rhesus monkey dual-islet model [35,36]. Platelet, antibody, and complement (C3d, C4d) deposition was similar in GTKO/hCD46 and GTKO NPIs at 1 h, as was neutrophil infiltration. By 24 h, platelet deposition and neutrophil infiltration were significantly lower in GTKO/hCD46 islets, while antibody and complement staining remained similar [35]. The authors concluded that combining hCD46 expression with deletion of Gal provided a modest additional benefit by reducing early thromboinflammatory events.

Inhibiting coagulation and thrombosis

Tissue factor (TF) expression and intravascular coagulation are key features of IBMIR. The dual-islet model described before has provided useful in vivo mechanistic data on the early development of coagulation following transplantation of GTKO/hCD46 and GTKO NPIs [36]. Biologically inert synthetic microspheres of roughly the same size as the NPIs were used in one set of experiments to distinguish islet-mediated biological effects from the mechanical effects of portal vein embolization. Mediators and markers of coagulation (TF, factors XIIa and XIIIa, von Willebrand factor, thrombin, and fibrin) accumulated rapidly in the islets. Although the levels of most of these were not significantly different at 1 h or 24 h between GTKO/hCD46 and GTKO, fibrin and factor XIIIa were the exceptions, being modestly lower in GTKO/hCD46 NPIs at 24 h. These results indicate a small beneficial effect of hCRP expression on the coagulation component of IBMIR, possibly as a result of the many links between the complement and coagulation pathways [37]. However, IBMIR remained prominent, suggesting that additional transgenes to more directly regulate coagulation may be required.

Because pig islets are not normally directly exposed to the circulation, it is reasonable to expect—although this has not been directly examined—that they do not naturally express significant levels of anticoagulant molecules. Even if they did, it is known that not all porcine anticoagulant cell surface proteins interact efficiently with the human (and presumably NHP) coagulation system. Thrombomodulin (TBM) is the best-known example of this cross-species molecular incompatibility. TBM binds thrombin and converts thrombin's activity from procoagulant to anticoagulant by changing its substrate specificity from factors such as fibrinogen to protein C. Activated protein C interacts with various cofactors to shut down coagulation by irreversibly inactivating factors Va and VIIIa. Porcine TBM binds human thrombin but is a poor cofactor for human protein C activation [38]. On the other hand, porcine endothelial protein C receptor (EPCR), which facilitates activation of protein C by the TBM/thrombin complex, appears to function efficiently across the species divide [39]. The cross-species compatibility of tissue factor pathway inhibitor α (TFPIα) is less clearly defined. TFPIα binds factor Xa and factor VIIa/TF and is the key regulator of the initiation phase of coagulation. Two groups have reported that porcine TFPIα is functionally compatible with all components of the human TF pathway [40,41], whereas another study suggested that its interaction with human factor VIIa/TF is defective [42]. Whatever the case regarding compatibility, overexpressing one or more of human TBM, TFPI, and EPCR on the porcine islet surface would seem to be a logical approach to attenuating IBMIR. The thromboregulatory enzyme CD39, which degrades proinflammatory (ATP) and prothrombotic (ADP) extracellular nucleotides, could be added to this list, although the in vitro data to support CD39 overexpression to protect against IBMIR are mixed. When islets from hCD39-transgenic or WT mice were incubated with human blood, clotting time was significantly delayed by the presence of hCD39 [43]. In contrast, the time to clotting of human blood induced by GTKO/hCD46/hCD39 and GTKO/hCD46 NPIs was not significantly different [34].

Analysis of the in vivo impact of anticoagulant transgenes on porcine islet xenograft survival is limited to one NHP study to date [33]. APIs from various transgenic pigs on a common GTKO/hCD46 background were transplanted at doses of

50,000–100,000 IEQ/kg into immunosuppressed diabetic cynomolgus monkeys. The donors additionally expressed hCD39 ("3-GE," $n=1$ transplant), hTFPI/CTLA4-Ig ("4-GE," $n=2$), or hCD39/hTFPI/CTLA4-Ig ("5-GE," $n=2$). Although interpretation of the data was limited by factors such as the small group sizes, the variable islet dose, and the presence or absence of the CTLA4-Ig transgene, some tentative general conclusions could be drawn. First, as noted earlier, some markers of IBMIR were modestly but significantly reduced 2 h after transplant in recipients of the multimodified APIs compared to recipients of WT or hCD46-alone APIs. Second, there was no consistent long-term effect of the transgenes on islet xenograft survival. Although porcine C-peptide was detected at least intermittently in all recipients, graft survival was highly variable (5 days in the 3-GE recipient, 0 and 365 days in the 4-GE group, and three and 160 days in the 5-GE group).

Other strategies to control inflammation and innate immune cell activity

Reducing antibody binding, complement activation, and coagulation are well-advanced genetic approaches that are expected to have an impact on inflammation in porcine islet xenotransplantation. Other antiinflammatory strategies that are at an earlier stage of development include transgenic expression of the cytoplasmic ubiquitin-editing enzyme A20, which is a natural suppressor of islet inflammation. In vitro, adenovirus-mediated overexpression of human A20 in NPIs attenuated activation of NF-κB in response to TNFα, with no apparent detrimental effects on islet function [44]. In a pig-to-mouse model in which transduced WT NPIs were transplanted under the renal capsule of immunodeficient mice, this inhibition of inflammation translated to improved islet xenograft function, including a more rapid return to normoglycemia [44]. However, it remains to be determined whether A20 can be transgenically expressed at a sufficient level to protect porcine islet xenografts from IBMIR in the NHP model. Although human A20 transgenic pigs have been generated [45], the level of expression in islets (and any consequent functional protection) has not been reported.

Two other antiinflammatory molecules that have been investigated in pig-to-mouse models are soluble TNFα receptor (sTNFα-R) and heme oxygenase-1 (HO-1). sTNFα-R acts as a sink for TNFα, while HO-1 has antioxidant, antiapoptotic, and antiinflammatory properties. Yan et al. transplanted APIs from transgenic pigs expressing either sTNFα-R or HO-1 in a pig-to-nude mouse renal subcapsular model [46]. In this model, WT xenografts showed a dense neutrophil and macrophage infiltrate at day 28, and the majority failed before day 40, which the authors attributed to a prolonged inflammatory response to early islet injury and death. Both sTNFα-R and HO-1 xenografts exhibited reduced cellular infiltration and a higher rate of survival. In addition, early apoptosis was reduced in HO-1 islets. The same group used adenoviral delivery to express sTNFα-R and/or HO-1 in WT APIs, which were subsequently transplanted under the kidney capsule in a humanized mouse model [47]. Control xenografts all failed by day 10, whereas approximately half of the xenografts in the sTNFα-R, HO-1, and sTNFα-R/HO-1 groups survived to the predetermined endpoint of 45 days. Prolonged survival appeared to be associated with reduced early inflammation and apoptosis. Like A20, whether these results can be translated to the more challenging pig-to-NHP intraportal model remains to be determined.

Another goal in this area is to inhibit the direct phagocytic and cytotoxic activity of human innate immune cells toward porcine islet cells. Infiltration by monocytes/macrophages and NK cells is an early feature of pig-to-NHP islet xenotransplantation [23,36]. One of the mechanisms used by macrophages to distinguish self from nonself is the SIRPα/CD47 signaling pathway. Macrophages express the inhibitory receptor SIRPα, whereas other cells express its ligand CD47; engagement of SIRPα by CD47 sends an important "don't eat me" signal to the macrophage [48]. However, pig CD47 does not interact efficiently with human or baboon SIRPα, resulting in macrophage phagocytosis of cultured porcine cells in vitro [49] and of porcine hematopoietic cells in vivo [50,51]. This incompatibility has been corrected by transgenic expression of hCD47 in the pig, improving engraftment of hematopoietic cells, and prolonging survival of porcine skin xenografts in baboons [50,51]. Although the effect of hCD47 expression on porcine islet xenograft survival and function is yet to be determined, its potential benefits have been demonstrated in a rodent model in which islets were chemically engineered to express mouse CD47 (mCD47) on their surface [52]. Rat islets presenting mCD47 were protected from phagocytosis by mouse macrophages and showed significantly reduced injury and cellular infiltration in an in vitro model of IBMIR. In vivo, incorporation of mCD47 on mouse islets significantly improved their engraftment and function in a syngeneic intraportal transplant model [52].

Expression of an inhibitory signal on porcine cells has also been investigated as a means to attenuate the antixenograft activity of human NK cells. The nonclassical HLA class I molecule HLA-E inhibits human NK cells via the CD94/NKG2A receptor. Transgenic expression of hHLA-E has been shown to protect porcine endothelial cells from human NK cell-mediated cytotoxicity in vitro [53]. There is also evidence to suggest that hHLA-E provides modest protection in pig lung, heart, and limb ex vivo xenoperfusion models [54–56]. However, the expression of hHLA-E on porcine islets and consequent effects on islet xenograft survival and function await further analysis.

Protecting porcine islet xenografts from adaptive immunity

The innate immune response is a key determinant of the strength of the adaptive immune response [57]. Therefore, although the innate immune system is the primary target of the approaches outlined before, many of these approaches will also beneficially impact on protection from adaptive immunity, e.g., by reducing or eliminating anti-Gal and anti-non-Gal responses. Complementary transgenic strategies have been designed to more specifically avert or blunt the adaptive immune response and thus reduce the level of immunosuppression required to prevent islet xenograft rejection. These include local secretion of immunosuppressive molecules, downregulation or modification of swine leukocyte antigen (SLA) class I and/or II, and expression of T cell inhibitory signaling proteins.

Local immunosuppression by secretion of immunomodulatory molecules

LEA29Y (belatacept) is a high-affinity variant of the costimulatory blockade molecule CTLA4-Ig, which inhibits the activation and proliferation of effector T cells. Wolf and colleagues generated transgenic pigs expressing secreted LEA29Y from the islet-specific pig insulin promoter [58,59]. To test the "local immunosuppressive" properties of this molecule, LEA29Y-transgenic or WT NPIs were transplanted under the kidney capsule of diabetic, nonimmunosuppressed immunodeficient mice reconstituted with human immune cells. All WT NPIs were rejected before maturing sufficiently to reverse hyperglycemia; in contrast, 70% of recipients of LEA29Y NPIs became normoglycemic within 4 months of transplant, and the xenografts showed remarkably few infiltrating human leukocytes [59] (Fig. 3). Survival of the LEA29Y NPIs in the presence of a functional (albeit not completely normal) human immune system, and in the absence of systemic immunosuppression, was demonstrated for more than 6 months. This identifies the LEA29Y transgene as a promising candidate for translation to the pig-to-NHP model.

Nottle et al. also took the approach of targeting T cells, in this case by expressing a chimeric antihuman CD2 monoclonal antibody (mAb) in transgenic pigs [60]. CD2 was chosen because it is expressed on all T cells but particularly highly on $CD8^+$ and $CD4^+$ effector memory T cells [61]. This mAb had previously been shown to deplete human T cells in vivo and inhibit their costimulation in vitro [62]. The anti-CD2 mAb transgene was precisely knocked into the *GGTA1* gene using a high-fidelity CRISPR system [63], thus generating GTKO/anti-CD2 transgenic pigs in a single step. The mAb was detectable in the serum of the transgenic pigs but importantly had no apparent detrimental effects on their immunocompetence or general health, presumably because of its specificity for human/NHP (and not pig) CD2. The tissue expression pattern of the mAb in the transgenic pigs has not yet been reported.

Downregulation or modification of SLA class I and II

The role of SLA as a target of human antipig humoral and cellular responses has been reviewed recently [64]. It has become increasingly evident that at least some anti-HLA antibodies cross-react with SLA, and that human $CD8^+$ and $CD4^+$ T cells can recognize SLA class I and II, respectively. SLA I expression has been eliminated or reduced in pigs by gene editing to inactivate the genes for SLA I [65] or β2-microglobulin [66,67], which is required for SLA I expression

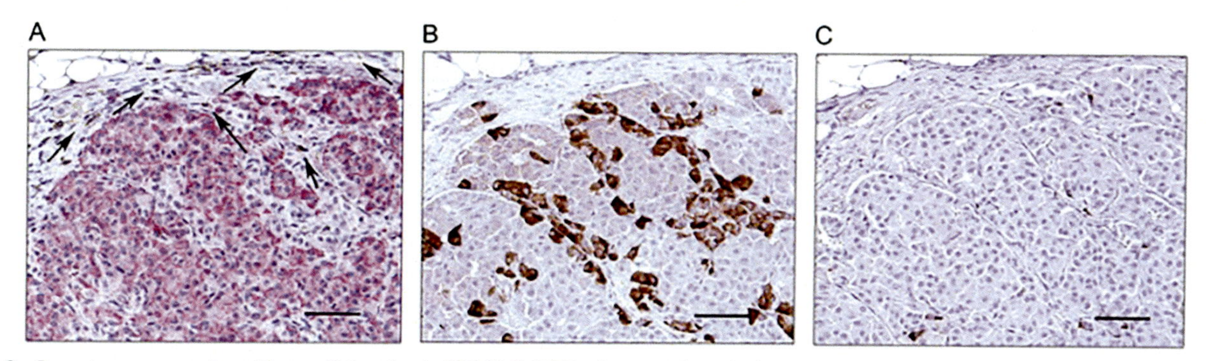

FIG. 3 Long-term preservation of beta cell function in HU-SRC-SCID mice transplanted with LEA29Y expressing NPICCs. Xenograft was harvested 29 weeks posttransplantation and stained for (A) human CD45 (*brown*) and insulin (*red*), (B) glucagon (*brown*), and (C) somatostatin (*brown*). Local LEA29Y expression preserved ICC structure and restricted human immune infiltration to the periphery of the transplantation site. Scale bar: 100 μm. *(Reproduced from Buerck LW, Schuster M, Oduncu FS, Baehr A, Mayr T, Guethoff S, et al. LEA29Y expression in transgenic neonatal porcine islet-like cluster promotes long-lasting xenograft survival in humanized mice without immunosuppressive therapy. Sci Rep 7(1) (2017):3572 under a Creative Commons License (https://creativecommons.org/licenses/by/4.0/).)*

on the cell surface. SLA II expression (which is technically more difficult to eliminate than SLA I expression) has been downregulated by inactivating the gene for the class II transactivator (CIITA) [68] or transgenically expressing a dominant-negative form (CIITA-DN) [69]. However, among the risks associated with these approaches are that the edited pigs will be immunocompromised, and xenografts derived from them may be more susceptible to human NK cells and viral infection. In the case of islet xenotransplantation, it may be possible to avoid the need to modify the donor pigs by instead modifying the isolated islets before transplantation. Two studies using allogeneic or xenogeneic cells provide support for this approach. In the first study, CRISPR/Cas9 was used to ablate HLA class Ia and class II expression and express a set of immunomodulatory factors in human pluripotent stem cells (hPSCs); this combination significantly increased the resistance of the hPSCs (and that of endothelial and vascular smooth muscle cells derived from them) to allogeneic T cells [70]. In the second study, porcine NPIs were dispersed into single cells, transduced with lentiviral vectors expressing short hairpin RNAs targeting β2M or CIITA, and re-formed into NPIs in bioreactors [71]. SLA I and SLA II expression in the transduced cells was downregulated by approximately 70% and 50%, respectively, providing significant protection against human humoral and T cell-mediated injury in vitro without increasing susceptibility to human NK cells. However, there would appear to be major practical hurdles to scaling up such a process sufficiently for clinical application.

A suggested alternative approach to ablating or substantially reducing expression of SLA class I and II is to use epitope mutagenesis to eliminate SLA antigenicity while retaining SLA function. Ladowski et al. [72] incubated human sera with transfected cells expressing various mutants of SLA-DQ (class II) to demonstrate that mutation of a single amino acid in SLA-DQ was sufficient to reduce human IgM and IgG binding for the majority of sera tested. The same group showed that mutation of a single residue in a particular SLA class I molecule reduced the binding of IgG present in allosensitized human sera [73]. These results suggest that it may be possible to engineer porcine donors to improve their histocompatibility, even for highly HLA-sensitized patients with a strong SLA crossmatch.

Expression of T cell inhibitory and apoptotic signals

The programmed cell death protein 1 (PD-1) receptor plays an important role in the maintenance of peripheral self-tolerance. Engagement of PD-1 expressed on T cells with its ligand, programmed death ligand-1 (PD-L1), inhibits IL-2 production and T cell proliferation. This "immune-evasive" property has been harnessed by expressing PD-L1 on transplanted cells to protect them from T cell-mediated rejection. Yoshihara et al. [74] demonstrated that PD-L1 overexpressed on human islet-like organoids protected them from rejection in diabetic immunocompetent mice for more than 50 days, in the absence of immunosuppression. Han et al. [70] reported that PD-L1 was a key component in a set of genetic modifications (deletion of HLA class I and CIITA, and overexpression of PD-L1, HLA-G, and CD47) that rendered hPSCs hyporesponsive to both adaptive and innate immune challenges. Interestingly, the T cell inhibitory effect of PD-L1 appeared to be limited to the $CD8^+$ T cell subset [70].

Human PD-L1 (hPD-L1) has recently been expressed in transgenic pigs [75]. In vitro coculture assays indicated that the presence of hPD-L1 on pig cells reduced both the proliferation response of human $CD4^+$ T cells and the sensitivity of the pig cells to lysis by human cytotoxic effector cells. Importantly, the transgenic pigs were healthy and showed no evidence of compromised immunity, supporting the earlier suggestion that hPD-L1 is functionally incompatible with porcine PD-1 [76]. Immunohistochemical analysis of the pancreas demonstrated expression of hPD-L1 on islets, although no in vivo (transplant) studies have yet been reported.

Tumor necrosis factor-related apoptosis-inducing ligand (TRAIL) can suppress T cell activity by inhibiting proliferation and inducing apoptosis. Kemter et al. generated transgenic pigs expressing human TRAIL (huTRAIL) [77]. Although huTRAIL-expressing pig cells did not induce apoptosis of primary human T cells in vitro, they did reduce T cell proliferation. However, the antiproliferative effect was only observed for huTRAIL-expressing dendritic cells and not for other porcine cells (fibroblasts), which may limit the efficacy of this approach in porcine islet xenotransplantation.

Improving the compatibility and function of porcine islets

Porcine insulin was used for decades to treat type 1 diabetes in humans before the development of recombinant human insulin. Porcine and human insulin differ by only one amino acid, but this is sufficient to induce antiporcine insulin antibodies in some individuals. To address this issue, Yang et al. [78] generated cloned pigs from gene-edited porcine cells in which the differing amino acid was replaced with its human equivalent. The resulting pigs were shown to express human but not porcine insulin, with no obvious detrimental effects on their health. In a further step, Cho et al. [79] generated transgenic pigs on an insulin gene KO background using a rat insulin promoter–human proinsulin construct. Human insulin and C-peptide were detected in these pigs, but the level of insulin was relatively low, suggesting that direct replacement of the complete pig proinsulin gene with its human counterpart may be a better option.

Porcine and human islets contain similar levels of insulin, but their secretion of insulin in response to glucose stimulation is quantitatively quite different: 2–3 fold and 10–12 fold, respectively [80]. This is likely to be a contributing factor to the requirement for comparatively large numbers of porcine islets to restore normoglycemia in pig-to-NHP models. To increase the glucose responsiveness of porcine islets, Mourad et al. [80] exploited two molecules that amplify insulin secretion by different pathways: glucagon-like peptide 1 (GLP-1), which binds to its receptor on β cells causing a rise in intracellular cAMP; and type 3 muscarinic receptor (M3R), which triggers an increase in intracellular calcium when activated by acetylcholine. Porcine islets (NPI and API) were transduced with adenoviral vectors encoding GLP-1 (modified to increase its half-life) and/or a constitutively active form of M3R. Expression of active M3R increased glucose-stimulated insulin secretion (GSIS) 2-fold, whereas expression of GLP-1 had no effect. Interestingly, coexpression of M3R and GLP-1 increased GSIS 4-fold, bringing the secretory function of porcine islets much closer to that of human islets. It remains to be determined whether these encouraging results can be translated to improved islet xenograft function in a transgenic pig-to-NHP model.

Reducing infectious risk

The risk of transfer of potentially pathogenic microorganisms from a xenograft to its human recipient is a key consideration in clinical porcine islet xenotransplantation. Most candidate bacteria, fungi, and viruses can be excluded or eliminated by careful husbandry, including vaccination, selective breeding, and maintenance of donor pigs in designated pathogen-free facilities. One notable exception is porcine endogenous retrovirus (PERV), which is carried in the genome of all pigs as two subtypes (PERV-A and PERV-B) that have been shown to infect certain human cell lines in vitro. A third subtype (PERV-C) is also of concern despite being capable of infecting only pig cells, because PERV-A/C recombinants show far greater infectivity toward human cells than the parental PERV-A [81]. Fortunately, PERV-C is not present in all pigs and can therefore be eliminated by selective breeding. The absence to date of preclinical or clinical evidence for in vivo transmission of PERV from porcine xenografts [82] also provides some reassurance, with the caveat that baboons and cynomolgus monkeys do not express a functional PERV receptor [83]. Furthermore, antiretroviral drugs used for the treatment of HIV have been shown to be effective against PERV replication in vitro at nanomolar concentrations [84]. Nevertheless, several genetic strategies have been developed to minimize or even eliminate the risk of PERV transmission. A number of groups have generated transgenic pigs expressing small interfering RNAs to downregulate the expression of PERV [85,86]. More recently, gene editing was employed to inactivate all 25 copies of PERV in primary porcine cells that were subsequently used to clone PERV KO pigs [87]. Whether this genome-wide elimination of PERV activity is essential for clinical islet xenotransplantation remains a matter of contention [88].

The next generation of highly modified donor pigs

CRISPR/Cas9 gene editing technology has dramatically simplified and accelerated the production of transgenic pigs with multiple genetic modifications [5,89]. It permits the simultaneous introduction of a raft of precise changes to the porcine genome, including the insertional inactivation of detrimental genes by integration of protective transgenes [60]. One pertinent example is a "9-gene" pig containing three knockouts (TKO, i.e., GTKO, CMAH KO, B4GALNT2 KO) and six human transgenes (hCD46, hCD55, hTBM, hEPCR, hCD47, hHO-1) [90]. The transgenes were inserted as two multigene blocks into the *GGTA1* and *CMAH* loci. The TKO phenotype and expression of the six transgenes was confirmed in heart, lung, kidney, and liver, although pancreatic expression was not reported. Even more striking are TKO pigs containing a single cassette of nine human transgenes (hCD46, hCD55, hCD59, hTBM, hTFPI, hCD39, hCD47, hHLA-E, hβ2-microglobulin), plus inactivating mutations in all PERV copies in the genome [91] (Fig. 4). These pigs expressed all of the transgenes except hTBM in various tissues and appeared to be healthy and fertile. In vitro studies demonstrated that endothelial cells from the engineered pigs were less thrombogenic than WT cells when mixed with human blood and were more resistant to human complement-dependent cytotoxicity, human NK cell-mediated killing, and phagocytosis by human macrophages [91]. The pancreatic phenotype of these pigs was not reported.

The challenge now is to establish what combination of modifications is ideal, which may differ for different organs and tissues. The multitransgenic pigs described before [90,91] appear to have been designed primarily for xenotransplantation of solid organs rather than islets, since in both lines the anticoagulant transgenes were expressed from an endothelial-specific promoter.

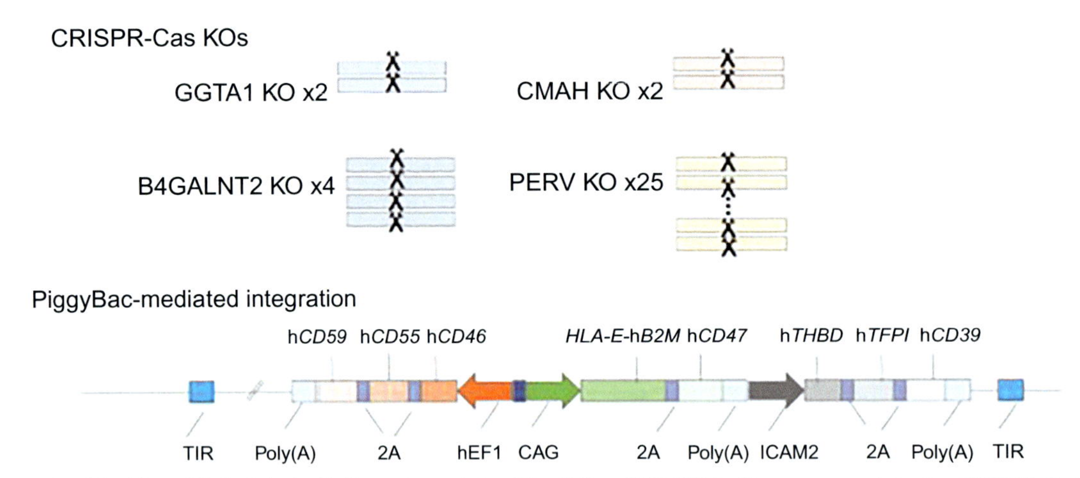

FIG. 4 Schematic of the 42 modified alleles in highly engineered pigs. The 3KO and PERVKO edits were generated using CRISPR–Cas9 with gRNAs targeting the 2 copies of *GGTA1*, 2 copies of *CMAH*, 4 copies of *B4GALNT2*, and 25 copies of the PERV element. The 9TG modification was generated using PiggyBac-mediated random integration of the nine human transgenes into the pig genome. The transgenes are expressed from three transcription cassettes, with each cassette expressing two to three genes linked by the porcine teschovirus 2A (2A) peptide. TIR, terminal inverted repeats of the PiggyBac transposon; hEF1α, CAG and ICAM2, promoters; Poly(A), polyadenylation signal. *(Reprinted from Yue Y. et al., Extensive germline genome engineering in pigs, Nat Biomed Eng © 2021, by permission from Springer Nature Customer Service Centre GmbH: Springer Nature, The Author(s), under exclusive licence to Springer Nature Limited 2020.)*

Which transgenes are relevant for alternative transplant sites and modalities?

The focus in this chapter so far has been on genetic modifications for "naked" porcine islets delivered into the liver, since this is current clinical practice for allotransplantation and has been used in most preclinical pig-to-NHP studies. However, the use of alternative transplant sites to avoid the relatively inhospitable environment of the liver is under active investigation. For example, brown adipose tissue [92] and dissected peritoneal pouches [93] have recently been validated as alternative sites in syngeneic mouse models, although translation to large animals is yet to be demonstrated. Subcutaneous delivery has also been revisited, with the demonstration that transplantation of islets in an optimized matrix can promote islet engraftment and survival in small and large animal models [94]. Most of the previously described protective genetic modifications for intraportal porcine islet xenografts would be expected to benefit "extrahepatic" grafts; the likely exceptions would be the human anticoagulant transgenes.

It is a somewhat different story for encapsulation, which has long been proposed as a means of protecting porcine islet xenografts from the immune response and thus reducing or even eliminating the requirement for immunosuppression (reviewed in [95]). Clinical trials of microencapsulated juvenile WT porcine islets have demonstrated safety [96] but only very limited efficacy [97]. Intuitively, some genetic modifications might be expected to improve the engraftment and function of encapsulated islets; these include transgenes designed to increase resistance to hypoxia [46] or to amplify insulin secretion in response to glucose [80]. The benefit of other modifications will depend on the permeability of the encapsulation membrane, i.e., whether it permits entry of antibodies, complement components, and inflammatory cytokines, and exit of secreted immunomodulatory molecules. Others, including transgenes for anticoagulants or molecules like hCD47 that depend on cell–cell contact, are unlikely to be of benefit to encapsulated islets.

Can genetic modification to protect islet xenografts have a detrimental impact on islet function?

Data from mouse models suggest that the highly tuned metabolism of the pancreatic β cell may render it susceptible to unforeseen side effects of islet-targeted genetic modifications. For example, transgenic expression of various molecules specifically in mouse islets can result in the spontaneous development of diabetes [98]. This may also apply to certain knockout mutations. Of potential relevance to islet xenotransplantation, CMAH KO mice fed on a high fat diet exhibited β cell dysfunction [99]. However, there have been no reports to date of xeno-related genetic modifications (or combinations thereof) that detrimentally affect glucose metabolism in pigs. Wijkstrom et al. [100] specifically addressed this question by examining GTKO/hCD46 pigs expressing a variety of transgenes (one or more of hCD39, hTFPI, and pig CTLA4-Ig) from the rat insulin promoter; none of the pigs showed evidence of defective β cell function in vitro. Although an initial study indicated higher basal and stimulated insulin and glucagon secretion in GTKO pigs compared to WT pigs [101], this clearly did not affect the ability of GTKO islet xenografts to maintain normoglycemia in NHP recipients [21]. Furthermore, Salama et al. [102] demonstrated normal glucose metabolism in GTKO/CMAH KO pigs and showed that NPIs isolated from the pigs were normal in yield and function. Whether islet xenografts from more highly modified transgenic lines such as those reported recently [90,91] will function normally remains to be determined.

Summary

Genetic modification of the donor pig offers a great opportunity to drive porcine islet xenotransplantation to the clinic. The road ahead is not straightforward, however. Determining the optimal transgene and knockout combination remains a challenging prospect. The preclinical pig-to-NHP model is expensive, time consuming, and not always informative (viz. testing of TKO islets in Old World monkey recipients). For some modifications (e.g., GTKO for NPIs), there is clear evidence of protection in the NHP model; for others (e.g., LEA29Y), the supporting evidence is restricted to data from small animal models; and for yet others (e.g., the anticoagulant transgenes), a logical argument can be made for their inclusion, but there is as yet no compelling in vivo data to support this. Nevertheless, steady progress continues to be made and there is considerable cause for cautious optimism.

References

[1] Nagaraju S, Bottino R, Wijkstrom M, Trucco M, Cooper DK. Islet xenotransplantation: what is the optimal age of the islet-source pig? Xenotransplantation 2015;22(1):7–19.

[2] Brandhorst H, Johnson PR, Brandhorst D. Pancreatic islets: methods for isolation and purification of juvenile and adult pig islets. Adv Exp Med Biol 2016;938:35–55.

[3] Kemter E, Denner J, Wolf E. Will genetic engineering carry xenotransplantation of pig islets to the clinic? Curr Diab Rep 2018;18(11):103.

[4] Shin JS, Min BH, Kim JM, Kim JS, Yoon IH, Kim HJ, et al. Failure of transplantation tolerance induction by autologous regulatory T cells in the pig-to-non-human primate islet xenotransplantation model. Xenotransplantation 2016;23(4):300–9.

[5] Cowan PJ, Tector AJ. The resurgence of xenotransplantation. Am J Transplant 2017;17(10):2531–6.

[6] Shin JS, Kim JM, Kim JS, Min BH, Kim YH, Jang JY, et al. Long-term control of diabetes in immunosuppressed nonhuman primates (NHP) by the transplantation of adult porcine islets. Am J Transplant 2015;15:2837–50.

[7] Kim JM, Hong SH, Chung H, Shin JS, Min BH, Kim HJ, et al. Long-term porcine islet graft survival in diabetic non-human primates treated with clinically available immunosuppressants. Xenotransplantation 2020;28(2), e12659.

[8] Eich T, Eriksson O, Lundgren T. Visualization of early engraftment in clinical islet transplantation by positron-emission tomography. N Engl J Med 2007;356(26):2754–5.

[9] Nilsson B, Ekdahl KN, Korsgren O. Control of instant blood-mediated inflammatory reaction to improve islets of Langerhans engraftment. Curr Opin Organ Transplant 2011;16(6):620–6.

[10] Cowan PJ, Robson SC. Progress towards overcoming coagulopathy and hemostatic dysfunction associated with xenotransplantation. Int J Surg 2015;23(Pt B):296–300.

[11] Phelps CJ, Koike C, Vaught TD, Boone J, Wells KD, Chen SH, et al. Production of alpha 1,3-galactosyltransferase-deficient pigs. Science 2003;299(5605):411–4.

[12] Lutz AJ, Li P, Estrada JL, Sidner RA, Chihara RK, Downey SM, et al. Double knockout pigs deficient in N-glycolylneuraminic acid and galactose alpha-1,3-galactose reduce the humoral barrier to xenotransplantation. Xenotransplantation 2013;20(1):27–35.

[13] Kwon DN, Lee K, Kang MJ, Choi YJ, Park C, Whyte JJ, et al. Production of biallelic CMP-Neu5Ac hydroxylase knock-out pigs. Sci Rep 2013;3:1981.

[14] Estrada JL, Martens G, Li P, Adams A, Newell KA, Ford ML, et al. Evaluation of human and non-human primate antibody binding to pig cells lacking GGTA1/CMAH/beta4GalNT2 genes. Xenotransplantation 2015;22(3):194–202.

[15] Martens GR, Reyes LM, Butler JR, Ladowski JM, Estrada JL, Sidner RA, et al. Humoral reactivity of renal transplant-waitlisted patients to cells from GGTA1/CMAH/B4GalNT2, and SLA class I knockout pigs. Transplantation 2017;101:e86–92.

[16] Ladowski J, Martens G, Estrada J, Tector M, Tector J. The desirable donor pig to eliminate all xenoreactive antigens. Xenotransplantation 2019;26(4), e12504.

[17] Cooper DKC, Ezzelarab M, Iwase H, Hara H. Perspectives on the optimal genetically engineered pig in 2018 for initial clinical trials of kidney or heart xenotransplantation. Transplantation 2018;102(12):1974–82.

[18] Springer SA, Diaz SL, Gagneux P. Parallel evolution of a self-signal: humans and new world monkeys independently lost the cell surface sugar Neu5Gc. Immunogenetics 2014;66(11):671–4.

[19] McKenzie IF, Koulmanda M, Mandel TE, Sandrin MS. Pig islet xenografts are susceptible to "anti-pig" but not Gal alpha(1,3)Gal antibody plus complement in Gal o/o mice. J Immunol 1998;161(10):5116–9.

[20] Rayat GR, Rajotte RV, Hering BJ, Binette TM, Korbutt GS. In vitro and in vivo expression of Galalpha-(1,3)Gal on porcine islet cells is age dependent. J Endocrinol 2003;177(1):127–35.

[21] Thompson P, Badell IR, Lowe M, Cano J, Song M, Leopardi F, et al. Islet xenotransplantation using gal-deficient neonatal donors improves engraftment and function. Am J Transplant 2011;11(12):2593–602.

[22] Martin BM, Samy KP, Lowe MC, Thompson PW, Cano J, Farris AB, et al. Dual islet transplantation modeling of the instant blood-mediated inflammatory reaction. Am J Transplant 2015;15(5):1241–52.
[23] Samy KP, Davis RP, Gao Q, Martin BM, Song M, Cano J, et al. Early barriers to neonatal porcine islet engraftment in a dual transplant model. Am J Transplant 2018;18(4):998–1006.
[24] Tector AJ, Mosser M, Tector M, Bach JM. The possible role of anti-Neu5Gc as an obstacle in xenotransplantation. Front Immunol 2020;11:622.
[25] Omori T, Nishida T, Komoda H, Fumimoto Y, Ito T, Sawa Y, et al. A study of the xenoantigenicity of neonatal porcine islet-like cell clusters (NPCC) and the efficiency of adenovirus-mediated DAF (CD55) expression. Xenotransplantation 2006;13(5):455–64.
[26] Lee W, Hara H, Ezzelarab MB, Iwase H, Bottino R, Long C, et al. Initial in vitro studies on tissues and cells from GTKO/CD46/NeuGcKO pigs. Xenotransplantation 2016;23(2):137–50.
[27] Komoda H, Miyagawa S, Kubo T, Kitano E, Kitamura H, Omori T, et al. A study of the xenoantigenicity of adult pig islets cells. Xenotransplantation 2004;11(3):237–46.
[28] Groth CG, Korsgren O, Tibell A, Tollemar J, Moller E, Bolinder J, et al. Transplantation of porcine fetal pancreas to diabetic patients. Lancet 1994;344(8934):1402–4.
[29] Blixt O, Kumagai Braesch M, Tibell A, Groth CG, Holgersson J. Anticarbohydrate antibody repertoires in patients transplanted with fetal pig islets revealed by glycan arrays. Am J Transplant 2009;9(1):83–90.
[30] Byrne G, Ahmad-Villiers S, Du Z, McGregor C. B4GALNT2 and xenotransplantation: a newly appreciated xenogeneic antigen. Xenotransplantation 2018;25(5), e12394.
[31] van der Windt DJ, Bottino R, Casu A, Campanile N, Smetanka C, He J, et al. Long-term controlled normoglycemia in diabetic non-human primates after transplantation with hCD46 transgenic porcine islets. Am J Transplant 2009;9(12):2716–26.
[32] Hawthorne WJ, Salvaris EJ, Phillips P, Hawkes J, Liuwantara D, Burns H, et al. Control of IBMIR in neonatal porcine islet xenotransplantation in baboons. Am J Transplant 2014;14(6):1300–9.
[33] Bottino R, Wijkstrom M, van der Windt DJ, Hara H, Ezzelarab M, Murase N, et al. Pig-to-monkey islet xenotransplantation using multi-transgenic pigs. Am J Transplant 2014;14(10):2275–87.
[34] Nagaraju S, Bertera S, Tanaka T, Hara H, Rayat GR, Wijkstrom M, et al. In vitro exposure of pig neonatal isletlike cell clusters to human blood. Xenotransplantation 2015;22(4):317–24.
[35] Samy KP, Gao Q, Davis RP, Song M, Fitch ZW, Mulvihill MS, et al. The role of human CD46 in early xenoislet engraftment in a dual transplant model. Xenotransplantation 2019;26(6), e12540.
[36] Song M, Fitch ZW, Samy KP, Martin BM, Gao Q, Patrick Davis R, et al. Coagulation, inflammation, and CD46 transgene expression in neonatal porcine islet xenotransplantation. Xenotransplantation 2021;28(3), e12680.
[37] Luo S, Hu D, Wang M, Zipfel PF, Hu Y. Complement in hemolysis- and thrombosis- related diseases. Front Immunol 2020;11:1212.
[38] Roussel JC, Moran CJ, Salvaris EJ, Nandurkar HH, d'Apice AJ, Cowan PJ. Pig thrombomodulin binds human thrombin but is a poor cofactor for activation of human protein C and TAFI. Am J Transplant 2008;8(6):1101–12.
[39] Salvaris EJ, Moran CJ, Roussel JC, Fisicaro N, Robson SC, Cowan PJ. Pig endothelial protein C receptor is functionally compatible with the human protein C pathway. Xenotransplantation 2020;27(2), e12557.
[40] Lee KF, Salvaris EJ, Roussel JC, Robson SC, d'Apice AJ, Cowan PJ. Recombinant pig TFPI efficiently regulates human tissue factor pathways. Xenotransplantation 2008;15(3):191–7.
[41] Choi CY, Kim YH, Bae J, Lee SJ, Kim HK, Park CG, et al. Pig tissue factor pathway inhibitor alpha fusion immunoglobulin inhibits pig tissue factor activity in human plasma moderately more efficiently than the human counterpart. Biotechnol Lett 2017;39(11):1631–8.
[42] Ji H, Li X, Yue S, Li J, Chen H, Zhang Z, et al. Pig BMSCs transfected with human TFPI combat species incompatibility and regulate the human TF pathway in vitro and in a rodent model. Cell Physiol Biochem 2015;36(1):233–49.
[43] Dwyer KM, Mysore TB, Crikis S, Robson SC, Nandurkar H, Cowan PJ, et al. The transgenic expression of human CD39 on murine islets inhibits clotting of human blood. Transplantation 2006;82(3):428–32.
[44] Zammit NW, Seeberger KL, Zamerli J, Walters SN, Lisowski L, Korbutt GS, et al. Selection of a novel AAV2/TNFAIP3 vector for local suppression of islet xenograft inflammation. Xenotransplantation 2020;28(3), e12669.
[45] Oropeza M, Petersen B, Carnwath JW, Lucas-Hahn A, Lemme E, Hassel P, et al. Transgenic expression of the human A20 gene in cloned pigs provides protection against apoptotic and inflammatory stimuli. Xenotransplantation 2009;16(6):522–34.
[46] Yan JJ, Yeom HJ, Jeong JC, Lee JG, Lee EW, Cho B, et al. Beneficial effects of the transgenic expression of human sTNF-alphaR-Fc and HO-1 on pig-to-mouse islet xenograft survival. Transpl Immunol 2016;34:25–32.
[47] Lee HS, Lee JG, Yeom HJ, Chung YS, Kang B, Hurh S, et al. The introduction of human heme oxygenase-1 and soluble tumor necrosis factor-alpha receptor type I with human IgG1 Fc in porcine islets prolongs islet xenograft survival in humanized mice. Am J Transplant 2016;16(1):44–57.
[48] Murata Y, Kotani T, Ohnishi H, Matozaki T. The CD47-SIRPalpha signalling system: its physiological roles and therapeutic application. J Biochem 2014;155(6):335–44.
[49] Nomura S, Ariyoshi Y, Watanabe H, Pomposelli T, Takeuchi K, Garcia G, et al. Transgenic expression of human CD47 reduces phagocytosis of porcine endothelial cells and podocytes by baboon and human macrophages. Xenotransplantation 2020;27(1), e12549.
[50] Tena A, Kurtz J, Leonard DA, Dobrinsky JR, Terlouw SL, Mtango N, et al. Transgenic expression of human CD47 markedly increases engraftment in a murine model of pig-to-human hematopoietic cell transplantation. Am J Transplant 2014;14(12):2713–22.
[51] Tena AA, Sachs DH, Mallard C, Yang YG, Tasaki M, Farkash E, et al. Prolonged survival of pig skin on baboons after administration of pig cells expressing human CD47. Transplantation 2017;101(2):316–21.

[52] Shrestha P, Batra L, Tariq Malik M, Tan M, Yolcu ES, Shirwan H. Immune checkpoint CD47 molecule engineered islets mitigate instant blood-mediated inflammatory reaction and show improved engraftment following intraportal transplantation. Am J Transplant 2020;20(10):2703–14.
[53] Weiss EH, Lilienfeld BG, Muller S, Muller E, Herbach N, Kessler B, et al. HLA-E/human beta2-microglobulin transgenic pigs: protection against xenogeneic human anti-pig natural killer cell cytotoxicity. Transplantation 2009;87(1):35–43.
[54] Laird CT, Burdorf L, French BM, Kubicki N, Cheng X, Braileanu G, et al. Transgenic expression of human leukocyte antigen-E attenuates GalKO. hCD46 porcine lung xenograft injury. Xenotransplantation 2017;24(2):e12294.
[55] Abicht JM, Sfriso R, Reichart B, Langin M, Gahle K, Puga Yung GL, et al. Multiple genetically modified GTKO/hCD46/HLA-E/hbeta2-mg porcine hearts are protected from complement activation and natural killer cell infiltration during ex vivo perfusion with human blood. Xenotransplantation 2018;25(5), e12390.
[56] Puga Yung G, Bongoni AK, Pradier A, Madelon N, Papaserafeim M, Sfriso R, et al. Release of pig leukocytes and reduced human NK cell recruitment during ex vivo perfusion of HLA-E/human CD46 double-transgenic pig limbs with human blood. Xenotransplantation 2018;25(1):e12357.
[57] Jain A, Pasare C. Innate control of adaptive immunity: beyond the three-signal paradigm. J Immunol 2017;198(10):3791–800.
[58] Klymiuk N, van Buerck L, Bahr A, Offers M, Kessler B, Wuensch A, et al. Xenografted islet cell clusters from INSLEA29Y transgenic pigs rescue diabetes and prevent immune rejection in humanized mice. Diabetes 2012;61(6):1527–32.
[59] Buerck LW, Schuster M, Oduncu FS, Baehr A, Mayr T, Guethoff S, et al. LEA29Y expression in transgenic neonatal porcine islet-like cluster promotes long-lasting xenograft survival in humanized mice without immunosuppressive therapy. Sci Rep 2017;7(1):3572.
[60] Nottle MB, Salvaris EJ, Fisicaro N, McIlfatrick S, Vassiliev I, Hawthorne WJ, et al. Targeted insertion of an anti-CD2 monoclonal antibody transgene into the GGTA1 locus in pigs using FokI-dCas9. Sci Rep 2017;7(1):8383.
[61] Podesta MA, Binder C, Sellberg F, DeWolf S, Shonts B, Ho SH, et al. Siplizumab selectively depletes effector memory T cells and promotes a relative expansion of alloreactive regulatory T cells in vitro. Am J Transplant 2020;20(1):88–100.
[62] Brady JL, Sutherland RM, Hancock M, Kitsoulis S, Lahoud MH, Phillips PM, et al. Anti-CD2 producing pig xenografts effect localized depletion of human T cells in a huSCID model. Xenotransplantation 2013;20(2):100–9.
[63] Fisicaro N, Salvaris EJ, Philip GK, Wakefield MJ, Nottle MB, Hawthorne WJ, et al. FokI-dCas9 mediates high-fidelity genome editing in pigs. Xenotransplantation 2020;27(1), e12551.
[64] Ladowski JM, Hara H, Cooper DKC. The role of SLAs in xenotransplantation. Transplantation 2021;105(2):300–7.
[65] Reyes LM, Estrada JL, Wang ZY, Blosser RJ, Smith RF, Sidner RA, et al. Creating class I MHC-null pigs using guide RNA and the Cas9 endonuclease. J Immunol 2014;193(11):5751–7.
[66] Sake HJ, Frenzel A, Lucas-Hahn A, Nowak-Imialek M, Hassel P, Hadeler KG, et al. Possible detrimental effects of beta-2-microglobulin knockout in pigs. Xenotransplantation 2019;26(6), e12525.
[67] Hein R, Sake HJ, Pokoyski C, Hundrieser J, Brinkmann A, Baars W, et al. Triple (GGTA1, CMAH, B2M) modified pigs expressing an SLA class I(low) phenotype-effects on immune status and susceptibility to human immune responses. Am J Transplant 2020;20(4):988–98.
[68] Fu R, Fang M, Xu K, Ren J, Zou J, Su L, et al. Generation of GGTA1−/−beta2M−/-CIITA−/− pigs using CRISPR/Cas9 technology to alleviate xenogeneic immune reactions. Transplantation 2020;104(8):1566–73.
[69] Hara H, Witt W, Crossley T, Long C, Isse K, Fan L, et al. Human dominant-negative class II transactivator transgenic pigs—effect on the human anti-pig T-cell immune response and immune status. Immunology 2013;140(1):39–46.
[70] Han X, Wang M, Duan S, Franco PJ, Kenty JH, Hedrick P, et al. Generation of hypoimmunogenic human pluripotent stem cells. Proc Natl Acad Sci U S A 2019;116(21):10441–6.
[71] Carvalho Oliveira M, Valdivia E, Verboom M, Yuzefovych Y, Sake HJ, Pogozhykh O, et al. Generating low immunogenic pig pancreatic islet cell clusters for xenotransplantation. J Cell Mol Med 2020;24(9):5070–81.
[72] Ladowski JM, Martens GR, Reyes LM, Hauptfeld-Dolejsek V, Tector M, Tector J. Examining epitope mutagenesis as a strategy to reduce and eliminate human antibody binding to class II swine leukocyte antigens. Immunogenetics 2019;71(7):479–87.
[73] Martens GR, Ladowski JM, Estrada J, Wang ZY, Reyes LM, Easlick J, et al. HLA class I-sensitized renal transplant patients have antibody binding to SLA class I epitopes. Transplantation 2019;103(8):1620–9.
[74] Yoshihara E, O'Connor C, Gasser E, Wei Z, Oh TG, Tseng TW, et al. Immune-evasive human islet-like organoids ameliorate diabetes. Nature 2020;586(7830):606–11.
[75] Buermann A, Petkov S, Petersen B, Hein R, Lucas-Hahn A, Baars W, et al. Pigs expressing the human inhibitory ligand PD-L1 (CD 274) provide a new source of xenogeneic cells and tissues with low immunogenic properties. Xenotransplantation 2018;25(5), e12387.
[76] Plege A, Borns K, Beer L, Baars W, Klempnauer J, Schwinzer R. Downregulation of cytolytic activity of human effector cells by transgenic expression of human PD-ligand-1 on porcine target cells. Transpl Int 2010;23(12):1293–300.
[77] Kemter E, Lieke T, Kessler B, Kurome M, Wuensch A, Summerfield A, et al. Human TNF-related apoptosis-inducing ligand-expressing dendritic cells from transgenic pigs attenuate human xenogeneic T cell responses. Xenotransplantation 2012;19(1):40–51.
[78] Yang Y, Wang K, Wu H, Jin Q, Ruan D, Ouyang Z, et al. Genetically humanized pigs exclusively expressing human insulin are generated through custom endonuclease-mediated seamless engineering. J Mol Cell Biol 2016;8(2):174–7.
[79] Cho B, Lee EJ, Ahn SM, Kim G, Lee SH, Ji DY, et al. Production of genetically modified pigs expressing human insulin and C-peptide as a source of islets for xenotransplantation. Transgenic Res 2019;28(5–6):549–59.
[80] Mourad NI, Perota A, Xhema D, Galli C, Gianello P. Transgenic expression of glucagon-like Peptide-1 (GLP-1) and activated muscarinic receptor (M3R) significantly improves pig islet secretory function. Cell Transplant 2017;26(5):901–11.
[81] Harrison I, Takeuchi Y, Bartosch B, Stoye JP. Determinants of high titer in recombinant porcine endogenous retroviruses. J Virol 2004;78(24):13871–9.

[82] Denner J. Why was PERV not transmitted during preclinical and clinical xenotransplantation trials and after inoculation of animals? Retrovirology 2018;15(1):28.
[83] Mattiuzzo G, Takeuchi Y. Suboptimal porcinc endogenous retrovirus infection in non-human primate cells: implication for preclinical xenotransplantation. PLoS One 2010;5(10), e13203.
[84] Argaw T, Colon-Moran W, Wilson C. Susceptibility of porcine endogenous retrovirus to anti-retroviral inhibitors. Xenotransplantation 2016;23(2):151–8.
[85] Dieckhoff B, Petersen B, Kues WA, Kurth R, Niemann H, Denner J. Knockdown of porcine endogenous retrovirus (PERV) expression by PERV-specific shRNA in transgenic pigs. Xenotransplantation 2008;15(1):36–45.
[86] Ramsoondar J, Vaught T, Ball S, Mendicino M, Monahan J, Jobst P, et al. Production of transgenic pigs that express porcine endogenous retrovirus small interfering RNAs. Xenotransplantation 2009;16(3):164–80.
[87] Niu D, Wei HJ, Lin L, George H, Wang T, Lee IH, et al. Inactivation of porcine endogenous retrovirus in pigs using CRISPR-Cas9. Science 2017;357(6357):1303–7.
[88] Denner J, Scobie L, Schuurman HJ. Is it currently possible to evaluate the risk posed by PERVs for clinical xenotransplantation? Xenotransplantation 2018;25(4), e12403.
[89] Cowan PJ, Hawthorne WJ, Nottle MB. Xenogeneic transplantation and tolerance in the era of CRISPR-Cas9. Curr Opin Organ Transplant 2019;24(1):5–11.
[90] Cooper DKC, Hara H, Iwase H, Yamamoto T, Li Q, Ezzelarab M, et al. Justification of specific genetic modifications in pigs for clinical organ xenotransplantation. Xenotransplantation 2019;26(4), e12516.
[91] Yue Y, Xu W, Kan Y, Zhao HY, Zhou Y, Song X, et al. Extensive germline genome engineering in pigs. Nat Biomed Eng 2021;5(2):134–43.
[92] Xu K, Xie R, Lin X, Jia J, Zeng N, Li W, et al. Brown adipose tissue: a potential site for islet transplantation. Transplantation 2020;104(10):2059–64.
[93] Kumano K, Vasu S, Liu Y, Lo ST, Mulgaonkar A, Pennington J, et al. Grafting islets to a dissected peritoneal pouch to improve transplant survival and function. Transplantation 2020;104(11):2307–16.
[94] Yu M, Agarwal D, Korutla L, May CL, Wang W, Griffith NN, et al. Islet transplantation in the subcutaneous space achieves long-term euglycaemia in preclinical models of type 1 diabetes. Nat Metab 2020;2(10):1013–20.
[95] Cooper DK, Matsumoto S, Abalovich A, Itoh T, Mourad NI, Gianello PR, et al. Progress in clinical encapsulated islet xenotransplantation. Transplantation 2016;100(11):2301–8.
[96] Morozov VA, Wynyard S, Matsumoto S, Abalovich A, Denner J, Elliott R. No PERV transmission during a clinical trial of pig islet cell transplantation. Virus Res 2017;227:34–40.
[97] Matsumoto S, Abalovich A, Wechsler C, Wynyard S, Elliott RB. Clinical benefit of islet xenotransplantation for the treatment of type 1 diabetes. EBioMedicine 2016;12:255–62.
[98] Sutherland RM, Mountford JN, Allison J, Harrison LC, Lew AM. The non-immune RIP-Kb mouse is a useful host for islet transplantation, as the diabetes is spontaneous, mild and predictable. Int J Exp Diabetes Res 2002;3(1):37–45.
[99] Kavaler S, Morinaga H, Jih A, Fan W, Hedlund M, Varki A, et al. Pancreatic beta-cell failure in obese mice with human-like CMP-Neu5Ac hydroxylase deficiency. FASEB J 2011;25(6):1887–93.
[100] Wijkstrom M, Bottino R, Iwase H, Hara H, Ekser B, van der Windt D, et al. Glucose metabolism in pigs expressing human genes under an insulin promoter. Xenotransplantation 2015;22(1):70–9.
[101] Casu A, Echeverri GJ, Bottino R, van der Windt DJ, He J, Ekser B, et al. Insulin secretion and glucose metabolism in alpha 1,3-galactosyltransferase knock-out pigs compared to wild-type pigs. Xenotransplantation 2010;17(2):131–9.
[102] Salama A, Mosser M, Leveque X, Perota A, Judor JP, Danna C, et al. Neu5Gc and alpha1-3 GAL xenoantigen knockout does not affect glycemia homeostasis and insulin secretion in pigs. Diabetes 2017;66(4):987–93.

Chapter 10

Xenogeneic pancreatic islet cell transplantation—Application of pig cells and techniques for clinical islet cell xenotransplantation

Jong-Min Kim[a,b,c,d,e,f], Rita Bottino[g,h], and Chung-Gyu Park[a,b,c,d,e,f]

[a]*Xenotransplantation Research Center, Seoul National University Hospital, Seoul, Korea,* [b]*Department of Microbiology and Immunology, Seoul National University Hospital, Seoul, Korea,* [c]*Institute of Endemic Diseases, Seoul National University Hospital, Seoul, Korea,* [d]*Cancer Research Institute, Seoul National University Hospital, Seoul, Korea,* [e]*Department of Biomedical Sciences, Seoul National University Graduate School, Seoul National University Hospital, Seoul, Korea,* [f]*Biomedical Research Institute, Seoul National University Hospital, Seoul, Korea,* [g]*Imagine Pharma, Devon, PA, United States,* [h]*Allegheny Health Network, Pittsburgh, PA, United States*

Introduction

Porcine islets as an alternative source for beta cell replacement

Patients with type 1 diabetes (T1D) often require multiple daily injections of recombinant human insulin to control their blood glucose level and to prevent diabetic ketoacidosis. This can be an unpleasant and burdensome experience. Potentially life-threatening side effects such as impaired awareness of hypoglycemia (IAH) have been found to occur in 19.5% of patients with T1D [1]. Exogenous insulin therapy does not fully mimic the action of the endogenous endocrine system and in some patients, severe hypoglycemia continues even with the best insulin treatment. Alternatively, transplantation of pancreatic islets containing insulin-producing beta cells from deceased human donors can restore the homeostasis of blood glucose levels in the recipient [2]. Therefore islet allotransplantation is now considered the ideal therapy for patients with T1D and IAH. However, islet allotransplantation is not widely applicable due to the shortage of human donors. This is especially true since it often requires several pancreas donors to provide islets for transplantation. Consequently, islet xenotransplantation has been proposed as a fundamental solution to overcome this challenge since pigs can supply an unlimited number of islets for transplantation [3]. Over the past few decades the challenges to successful islet xenotransplantation, mostly due to naturally occurring incompatibilities between species, have been evaluated and addressed.

The miniature pig is regarded as an ideal donor animal for xenotransplantation [4], because their organ size and physiology are similar to that of humans, they produce many offspring with a short gestation period, and there are less ethical issues compared to NHPs which are genetically closer to human. Past medical experiences have demonstrated the useful compatibility between pig and human insulin. Pig insulin, which has only one amino acid difference compared to human insulin, had been primarily used in treating type 1 diabetes patients for several decades before the development of recombinant human insulin [5].

Most xenotransplantation studies have, therefore, focused on the pig as a donor animal. However, a number of other potential alternatives have been proposed. Teleost (bony) fish, such as tilapia, have also been considered as a potential source of islets for xenotransplantation. Fish possess macroscopically visible distinct islet organs called Brockmann bodies (BB) [6] which can be inexpensively isolated, lack αGal (α-1,3-galactose) expression on the islets, and are resistant to hypoxia. Tilapia BB have demonstrated long-term normoglycemia in diabetic nude mice and in diabetic cynomolgus monkeys. This has been achieved despite tilapia insulin differing structurally from human insulin by 17 amino acids. To address this difference Wright et al. produced transgenic tilapia whose islets stably express physiological levels of humanized insulin [7], thus providing a feasible alternative source of cells for potential clinical therapy.

Significant progresses with long-term islet graft survival have been made in the field of preclinical islet xenotransplantation NHP studies over the past two decades. (Table 1) However, considerable immunological hurdles still exist in islet

Pancreas and Beta Cell Replacement. https://doi.org/10.1016/B978-0-12-824011-3.00005-9

TABLE 1 Long-term graft survival of porcine islets in NHPs.

Reference	Pig islet source (recipient)	Max. graft survival (days)	Immunosuppressive regimen
Hering et al. [8]	WT adult (C)	>187	Anti-IL-2R+anti-CD154+FTY720+rapamycin
Cardona et al. [9]	WT neonatal (R)	>260	Anti-IL-2R+anti-CD154+CTLA4-Ig+rapamycin
van der Windt et al. [10]	hCD46-Tg adult (C)	396	ATG+anti-CD154+MMF
Thompson et al. [11]	GTKO neonatal (R)	249	Anti-CD154+anti-LFA-1+CTLA-4-Ig+MMF
Bottino et al. [12]	Multi-Tg adult (C)	365	ATG+anti-CD154+MMF
Thompson et al. [13]	WT neonatal (R)	114	Anti-IL-2R+anti-LFA-1+CTLA-4-Ig+MMF+LFA-3-Ig
Thompson et al. [14]	WT neonatal (R)	>203	Anti-IL-2R+anti-CD40+CTLA-4-Ig+rapamycin
Shin et al. [15]	WT adult (R)	>603	ATG+CVF+anti-TNF+anti-CD154+rapamycin(+Treg)
Shin et al. [16]	WT adult (R)	1000	ATG+CVF+anti-TNF+anti-CD154 → anti-CD40+rapamycin(+Treg)
Shin et al. [17]	WT adult (R)	>320	ATG+CVF+anti-TNF+anti-CD40+rapamycin+tacrolimus
Kim et al. [18]	WT adult (R)	>222	ATG+ anti-TNF, IL-1, IL-6+IVIg+tacrolimus+Belimumab+rapamycin+tofacitinib
Sun et al. [19]	WT adult (C)	804	Alginate encapsulated, intraperitoneal
Dufrane et al. [20]	WT adult (C)	180	Alginate encapsulated subcutaneous with monolayer device

(C), cynomolgus; (R), Rhesus; *ATG*, antithymocyte globulin; *CVF*, cobra venom factor; *GTKO*, α1,3-galactosyl transferase-gene knockout; *MMF*, mycophenolate mofetil; *WT*, wild type.

xenotransplantation. Before the adoption of clinical islet xenotransplantation, its efficacy and safety must be rigorously validated in NHP preclinical studies [21]. The International Xenotransplantation Association (IXA) released a consensus statement on conditions for undertaking clinical trials of pig islet xenotransplantation in 2009 and updated it in 2019 [22]. According to the IXA statement, the clinical trial of pig islet xenotransplantation is acceptable when transplantation reproducibly reverses diabetes for more than 6 months in the pig-to-NHP preclinical study [22].

This chapter provides an overview of pig-to-NHP islet xenotransplantation studies with a focus on the *sources of porcine xenoislets, hurdles to successful transplantation, techniques to control the instant blood-mediated inflammatory reaction (IBMIR)* and *immunological rejection, clinical application* of xenoislets cell transplantation, and *future perspectives*.

Sources of porcine xenoislets

The optimum age of donor pig (embryonic, fetal, neonate, young, or adult) used to provide islets for preclinical/clinical xenotransplantation remains unsettled (Table 2) [23,24]. In general, adult pigs are considered ideal donors because they can provide a large number of islets which are also more structurally stable than other sources of porcine islets [25]. Adult islets are capable of secreting insulin within a few hours after transplantation. However, the difficulty of isolation, fragility of islets, high cost, and the long time required to raise pigs are major disadvantages [26]. Conversely, embryonic, fetal, and neonatal pig islets are easy to isolate and they also have the potential to further develop into endocrine tissue after transplantation [27]. In addition, the lower immunogenicity of these islets (except αGal expression) and revascularization of the host endothelium can be advantageous [28]. However, the major drawback of these islets is that exogenous insulin must be administered in order to control blood glucose levels after transplantation until in vivo islet function begins [29]. At present, only islet xenografts isolated from neonate (<5 days or 10–22 days) and adult pigs have shown the ability to reverse diabetes in NHP or humans [15,18,30,31].

TABLE 2 Characteristics of islets from donor groups (pigs) of different ages.

	Embryonic and fetal	Neonatal	Young adult and adult
Islet size(μm)	<50 (embryonic) 80 (fetal)	50–150	50–100 (young adult) 100–200 (adult)
Composition of b cells after in vitro culture	– (embryonic) 10% (fetal)	25%	>70%
Islet yield per pancreas (IEQs)	– (embryonic) ~8000 (fetal)	5000–10,000 IEQ/g of pancreas	500–1500 IEQ/g of pancreas (young adult) 1000–5000 IEQ/g of pancreas (adult)
Isolation and purification	None required	None required or easy	Very difficult (young adult) Difficult (adult)
Tissue culture	Resistant to hypoxia Resistance against ischemic injury Nonendocrine cells disappear	Resistant to hypoxia. Resistance against ischemic injury Nonendocrine cells disappear	Fragile Difficult to culture
In vivo functioning	Delayed >4–6 months (embryonic) Delayed >8 weeks (fetal)	Delayed >4 weeks More responsive to glucose vs fetal	Within hours
Cost	Low	Low	High
Gal antigen	Clear expression	Clear expression	Negligible
Sterile environment	Sterile procurement	Sterile procurement	Lack of sterile environment
Pig required per patient	60 anlages (embryonic) 70–100 (fetal) Ethical problem	7–10	2–4
Risk of pathogen transmission	Low	–	–
Tumorigenicity	Available	–	–
Immunogenicity	Relatively reduced immunogenicity		Human natural IgM, IgG, anti-nonGal antibodies, anti-*N*-glycolylneuraminic acid (NeuGc) antibodies

Modified from Park CG, Bottino R, Hawthorne WJ. Current status of islet xenotransplantation. Int J Surg 2015; 23(Pt B):261–266.

Embryonic and fetal tissues

Embryonic pig pancreatic tissue and fetal pig islet-like cell clusters (ICC) can be easily obtained from pregnant sows at precisely defined stages of their pregnancy. To prepare embryonic pig pancreatic tissue, the embryos are transferred in cold phosphate buffered saline with minimal warm ischemia time at gestational age of 42 days. Embryonic pancreatic tissues are then extracted under a light microscope and maintained in culture medium at 4°C pending transplantation [29]. To prepare ICC, the pancreas of fetus is removed at gestational age of 66–86 days and undergoes enzyme-mediated digestion and simple culture (4–9 days) until transplantation [27,32,33]. Insulin secretion is delayed after transplantation in vivo; however, the maturation time for ICC (at least 2 months) is significantly shorter than that of embryonic pancreatic tissue (typically >4 months).

Neonatal tissues

Generally, 1–5 day old newborn pigs are selected for neonatal islet cell preparation. After 4–9 days of in vitro culture, 10–50,000 neonatal pancreatic islet aggregates can generally be obtained from pancreatic tissue of a single newborn pig, consisting of fully differentiated pancreatic endocrine cells (~35%) and endocrine precursor cells (~57%) [27,32,34]. Although insulin secretion is also delayed after transplantation in vivo, the maturation time of neonatal pancreatic islet is about 4 weeks. The humoral xenoantigenicity of neonatal pancreatic islet is mainly associated with the expression of N-linked sugars including sialic acid and α-gal antigens in 11%–19% of the islet cells [35]; however, the transplanted neonatal pancreatic islets showed a lower T cell response than what was induced by adult porcine islets in patients with insulin-dependent diabetes mellitus [36]. The results of several long-term studies including diabetes management after transplantation of wild-type neonatal pancreatic islets into immunosuppressed primates have demonstrated the utility of neonatal pigs as an alternative pig source for future clinical applications [9,14,37,38].

Young adult and adult tissues

One research group reported that islet cells of young-adult (6- to 12-month-old) pigs dissociated into small aggregates and single cells, and exhibited inferior functional properties than adult (>2 year) islets both in vitro and in vivo [25]. In general, adult pigs over 2 years of age (especially retired breeders) are preferred as suitable donors due to a large number of islets with intact morphology, large size, and stability [25,39]. Although isolated adult pig islets showed relatively low insulin secretion and low islet recovery in vitro, transplanted adult pig islets have shown a strong ability to treat diabetes in diabetic recipients in vivo [40]. So, adult pig islets have a better potential for functional engraftment and prolonged survival in islet xenotransplantation.

Hurdles to successful transplantation

Instant blood-mediated inflammatory reaction

Clinical islet allotransplantation and porcine islet xenotransplantation are both carried out by infusing the cells via the portal vein to engraft transplanted islets in the liver. As such, the islets have direct contact with the recipient blood, which results in the instant blood-mediated inflammatory reaction (IBMIR) and induces massive early islet loss. The IBMIR is a complex and multifaceted phenomenon. It activates the complement system and coagulation pathway, platelets quickly bind to islets, and neutrophils and monocytes are infiltrated [41–43]. The IBMIR takes place shortly after infusion of islets into the portal vein [42]. In clinical islet allotransplantation, it is estimated that about 60%–70% of islets are lost due to IBMIR within several minutes after transplantation [44].

Direct exposure of islets to recipients' blood results in a rapid activation of the coagulation pathway since isolated islets express tissue factor which interacts with coagulation factor VII [45]. Large amounts of thrombin and fibrin are produced from prothrombin and fibrinogen, respectively, through the coagulation cascades. In addition, islets promote platelet activation by releasing von Willebrand factor, which acts as a bridge between the glycoprotein of platelet and the islet's exposed collagen [46]. As a result, the infused islets become lodged within the blood clot and lose their function [47]. The activation of coagulation is much stronger in islet xenotransplantation than in allotransplantation due to species-specific incompatibilities of several antithrombotic modulators, including tissue factor pathway inhibitor and thrombomodulin [48].

Alternative complement pathways are another major player in IBMIR [49]. Complement regulators such as decay accelerating factor (DAF or CD55) relieve spontaneous low level complement activation under physiological conditions. However, porcine complement regulators do not inhibit human complement effectors effectively due to species-specific incompatibility in islet xenotransplantation. The membrane attack complex (MAC) composed of activated complement components can directly lyse infused porcine islets [50,51]. Moreover, complement activation produces C3a and C5a which are very potent chemoattractants, and it recruits and activates inflammatory cells including neutrophils and monocytes rapidly [52].

Once the above-mentioned mechanisms are activated, stimulated islets with thrombin activate portal endothelial cells and produce platelet-activating factor (PAF) which is a potent neutrophil chemoattractant [53]. Activated platelets release a soluble form of CD40 ligand and activate CD40-expressing neutrophils [54]. Islets are also destroyed through the mechanism of isolation stress in which islets produce proinflammatory cytokines, including HMGB-1, IL-8, TNF-α, IL-1β, and IL-6 [55]. Consequently, the recruited neutrophils and monocytes are activated, and they then infiltrate the islets and release cytotoxic granules which induce lysis of the islet. These cytokines can activate innate cells. Ultimately, cytokines and innate cells can stimulate the subsequent adaptive immune responses of T and B cells [56]. In this context, reducing islet damage during isolation and culture period is of significant importance to prevent activation of the adaptive immune responses.

Acute cell-mediated rejection

T cell-mediated rejection is likely to present a significant barrier to porcine islet xenotransplantation. Friedman et al. showed that human $CD4^+$ T cell infiltration was observed to reject porcine islet cell clusters in a humanized mouse, although $CD8^+$ T cells were observed within the grafts at later time points [57]. These data demonstrate that human $CD4^+$ T cells play a critical role in porcine islet cell cluster xenograft rejection. Acute cellular rejection occurs during the first 24h to 20days after transplantation in diabetic primates resulting in a massive infiltration of macrophages and T cells ($CD4^+$ and $CD8^+$ T cells) in the grafts [8,56]. The human $CD4^+$ T cells were the major phenotype of activated T cell clones reactive against pig islet antigens in in vitro studies [58]. Additionally, the T cell-mediated response possibly induces numerous other cellular responses such as natural killer cell (NK cell), B cell, and innate responses. T cells play a major role in the cellular rejection against pig islets.

Acute antibody-mediated rejection

In adult pig islets, αGal is expressed in only 5% of islet cells, and α1,3-galactosyltransferase activity was also undetectable [59,60]. However, αGal expression on pig islets is age dependent, and both fetal and islet cell clusters and neonatal pig islets clearly express a relatively higher level of αGal antigens (up to 11%–19% of total islets) [59]. The expression of the Neu5Gc antigen is detectable in neonatal and adult pig islets. Pig islet Neu5Gc and sialic acid antigens were implicated in vitro in complement-dependent injury of islets by human antibodies and contribute clearly to the antigenicity of pig islets [35,60]. B4GALNT2 is another non-Gal antigen that leads to humoral rejection and is thought to be important in vascularized tissue xenotransplantation [61]. However, there is no data so far of its importance in porcine islets, thus more studies are needed to confirm the extent of B4GALNT2 expression in porcine islets.

Chronic rejection

Still, if the islet xenografts escape the acute damage due to IBMIR and acute humoral and cellular response, they will be subject to chronic rejection after the transplanted islet becomes vascularized. The humoral and cellular responses against xenografts are the major hurdles for long-term graft survival in pig-to-NHP islet transplantation. $CD4^+$ T and $CD8^+$ T cells dominantly infiltrate the grafts during acute and chronic rejection [8]. In addition, Kang et al. reported a positive correlation between the level of D-dimer and anti-Gal antibody. Fibrinogen deposition on porcine islet grafts that had survived long term was associated with subsequent graft loss in diabetic NHP [62]. To achieve long-term graft survival it is essential to target costimulatory molecules such as CD40L and B7, as demonstrated in many NHP models. Several studies have reported that some cases of graft rejection occurred coincidently with increased serum levels of aGal and non-Gal IgG or antiporcine islet-specific antibodies, suggesting the potential use of one of these parameters as a potential predictive biomarker of rejection, although the exact role of B cells and their antibodies reactive to porcine islets has not been completely elucidated in the NHP model [63]. In addition, non-Gal antibody generation also appear to correlate with T and B cells, marginal zone B cells, and natural killer cells residing in the spleen [64]. Nevertheless, the precise roles of individual cellular components in acute and chronic rejection remain to be clarified.

Techniques to control IBMIR and immunological rejection

Genetic modification of the source pig

As we have observed, there are multiple avenues responsible for the loss of pig islets after transplantation including IBMIR, acute humoral and cellular rejection, and chronic rejection. Many, but not all, are problematic in the context of islet allotransplantation as well. To overcome these problems, pigs have been developed with several genetic modifications (Table 3). The general strategies to introduce genetic modification in pigs are through (i) knocking out or knocking down genes for polysaccharide antigens (aGal, 5NeuGC, and B4GalNT2); (ii) knocking in human complement regulatory proteins (CD46, CD55, and CD59), human coagulation factors (TFPI, CD39, and thrombomodulin), cellular immune response regulatory proteins (CTLA4-Ig, HLA-E/human b2-microglobulin, and $LEA29Y^{ins}$), antiinflammation and antiapoptosis (human A20, human heme-oxygenase-1, and human signal regulatory protein α); (iii) knocking out PERV by siRNA due to safety concern for infectious disease, respectively; or (iv) combinations of the earlier mentioned genetic modifications. In diabetic NHP studies, van der Windt et al. (human CD46 transgenic pig) and Bottino et al. (human CD46/CD39/TFPI transgenic pig) achieved milestone long-term survival over 1 year of xenogeneic islets in conjunction with immune suppressants [10,12].

Encapsulation and new techniques for protecting islets

Islet encapsulation is based on the premise of protecting the islets from harm so that they do not require immunosuppression which may be toxic toward the islets themselves or potentially lead to complications in the patient. Encapsulation systems can be divided into two major categories: intravascular and extravascular. Intravascular devices require anastomosis in the

TABLE 3 Currently available genetically engineered pigs for islet xenotransplantation.

Target	Gene modification
Complement regulation	Human CD46 (membrane cofactor protein) [65] Human CD55 (decay accelerating factor) [66] Human CD59 (MAC-inhibitory protein) [67]
Anticoagulation	Human tissue factor pathway inhibitor [68] Human thrombomodulin [69] Human CD39 (ectonucleoside triphosphate diphosphohydrolase-1) [70]
Prevention of humoral response by polysaccharide antigen deletion or masking	GTKO [71] 5NeuGCKO [72] B4GalNT2KO [73] Human H-transferase gene expression [74] *N*-acetylglucosaminyltransferase III transgene [75]
Suppression of cellular immune response	Porcine CTLA4-Ig (CD152) [76] LEA29Y^{Ins} [77] CIITA-DN (MHC class II transactivator knockdown, resulting in swine leukocyte antigen class II knockdown) [78] HLA-E/human b2-microglobulin (inhibits human natural killer cell cytotoxicity) [79]
Antiinflammatory and antiapoptotic gene expression	Human heme-oxygenase1 [80] Human A20 (tumor necrosis factor-a-induced protein 3) [81] Human signal regulatory protein α [82]
PERV activation	PERV siRNA [83]

recipient's circulatory system. Extravascular devices do not require anastomosis and therefore have an advantage over intravascular devices in terms of preclinical and clinical applications.

In intravascular devices, islets are engrafted in a large semipermeable chamber containing a number of small diameter artificial capillaries made of a polyacrylonitrile-polyvinylchloride copolymer [84]. The device is connected to the vessel of the recipient by vascular anastomosis and requires antithrombotic therapy. An intravascular implantation of fetal rabbit islets contained in macrocapsules connected into the arterial-venous fistulas was carried out in T1D patients without immunosuppression in 2008 [85]. A significant reduction in exogenous insulin demands was seen in 14 of the 19 recipients at a 2-year follow-up. Due to this encouraging achievement, the clinical application of the intravascular device may be possible once new biomaterials become available.

Extravascular devices are categorized into two main types by their sizes: macroencapsules (large numbers of islet graft within a tubular diffusion chamber or planar chamber) [86] and microencapsules (individual or small groups of islets in a spherical hydrogel polymer with a stable mechanical structure) [87].

In recent preclinical study and clinical trials of macroencapsulated porcine islets, a monolayer of alginate with a human decellularized collagen matrix and adult porcine islets implanted subcutaneously demonstrated control of hyperglycemia for more than 6 months in diabetic NHPs without immunosuppression [20]. Additionally, subcutaneous implantation of adult porcine islets and mesenchymal stem cells (which were coencapsulated in the same monolayer device to improve the vascularization and oxygenation of the macroencapsulated islets) into diabetic NHPs showed even better control of hyperglycemia [88]. Valdes-Gonzalez et al. reported a longitudinal clinical study of 23 T1D patients transplanted with encapsulated neonatal porcine islet with homologous Sertoli cells and without immunosuppression. The macrodevices (6 cm × 0.8 cm) were implanted subcutaneously for 2 months to allow tissue ingrowth and vascularization before islet-Sertoli cell infusion. More than half of patients showed a reduction in their exogenous insulin requirement and detectable porcine C-peptide in their urine along with a limitation of tightly controlled treatment to validate their results [89].

The biocompatibility of pig islets microencapsulated in an alginate matrix has been confirmed for more than 6 months in subcapsular transplantation of nondiabetic NHPs [90] and their capacity to achieve hyperglycemic control and reduce insulin requirements in diabetic NHPs [19,91]. However, the clinical application of microencapsulated porcine islets is not yet supported by more solid preclinical achievements due to a lack of standardized formulations which contribute greatly to current reported lab-to-lab variation in biocompatibility and immune protection of microencapsulation systems [92]. A clinical trial of commercial microencapsulated pig islets (Diabecell) was conducted in New Zealand [30]. The study consisted of the transplantation of wild-type neonatal (7–22 days) porcine islets into the peritoneal cavity in 14 patients with unstable type 1 diabetes. The number of hypoglycemic unawareness episodes was reduced at 1 year after transplantation compared with pretransplantation. However, importantly the reduction in HbA1c and insulin doses was marginal. A follow-up clinical trial of encapsulated neonate porcine islet xenotransplantation was conducted in Argentina. After encapsulated neonatal porcine islet transplantation, the HbA1c was reduced to <7%. These improvements in the status of the patients have to date been maintained for more than 2 years [31]. While outcomes of many clinical islet allotransplantation studies focus on insulin independence, a reduction of severe hypoglycemic episodes and of IAH provides important therapeutic advantages to the patient even if exogenous insulin is required [93]. The same is likely true of islet xenotransplantation.

Although encouraging results were obtained in clinical trials of encapsulated porcine islets, additional studies are underway to further bolster preclinical data: (1) genetically engineered islet source (genetic engineering of the pig was discussed in "Genetic modification of the source pig" section); (2) a conformal coating method which may increase capsule stability, minimize capsule thickness and size and graft volume, and allow graft transplantation into the liver through the portal vein [94–97]; (3) implantation sites other than the peritoneal cavity in order to provide a simple and safe implanting/removal operation, immune protection, a physiological route for insulin delivery, a sufficient blood and oxygen supply, enough space for a large volume of encapsulated islets, and compatibility with immunoisolation systems [98]; (4) the development of more biocompatible capsule to minimize foreign body reactions such as alginate purity, polymer composition and morphology [99]; (5) methods for preventing xenograft rejection by localized immunosuppression using chemotherapeutic agents such as incorporation of rapamycin into the PEG layer of alginate microcapsule [100], macromolecular heparin conjugation [101] and by biological agents as immunomodulators such as curcumin [102], a synthetic compound with antiinflammatory properties, and incorporation of stromal-derived factor 1α (SDF-1α, also known as CXCL12) which recruit regulatory T cells [103].

Control of IBMIR

If we reduce the effects of IBMIR, we can prevent early islet loss and reduce islet numbers required for the reversal of diabetes. To reduce these effects, anticoagulants such as heparin-like molecules such as an unfractionated heparin, low molecular weight heparin, and low molecular weight dextran sulfate (LMW-DS) are used to prevent coagulation associated with IBMIR. Cooper and his colleagues used LMW-DS instead of heparin [10,12]. These heparin molecules act as cofactors that interact with antithrombin III to inactivate thrombin and coagulation factor Xa. The antithrombotic effect of dextran is mediated through its binding of erythrocytes, platelets, and vascular endothelium, increasing their electronegativity and thus reducing erythrocyte aggregation and platelet adhesiveness, additionally dextrans also reduce factor VIII-Ag Von Willebrand factor, thereby decreasing platelet function. Since there are several connections between the complement and the coagulation systems [104,105], LMW-DS had been shown to reduce complement activation as well [51]. Clopidogrel, a popular antiplatelet agent was used in one of the studies to reduce platelet activation and IBMIR [23].

Inhibitors of proinflammatory cytokines (TNF-α, IL-1β, and IL-6) were used to reduce early islet loss [55]. Many long-term survival islet xenotransplantation in NHP studies used etanercept [9,11,13,14,106] or adalimumab [15,17] as a TNF-α blocker. An IL-1β receptor antagonist (anakinra) and IL-6 blocker (tocilizumab) were used in preclinical islet xenotransplantation NHP studies [18].

For the inhibition of the complement system, eculizumab can be used in humans, but in NHPs this drug does not work. One group has validated the efficacy in NHP using cobra venom factor (CVF) to block the function of complement [15,17]. CVF is a potent homologue of C3b which forms a complex with the complement factor Bb, and the CVF-Bb complex acts as a C3 convertase [107]. Since this homologous C3 convertase is potent and almost depletes the available C3 within 17 h, pretreatment with CVF may prevent complement-dependent lysis of the infused islet cells [108]. However, CVF is inapplicable to humans due to its immunogenicity and is not clinically available. Factor H (Human soluble complement regulatory protein) which dissociates C3 convertase complex of the alternative pathway has been shown to reduce complement activation and protect the islets from damage in pig-to-NHP islet xenotransplantation [49]. IVIg has functions of immunomodulation and has been reported to inhibit complement-mediated inflammation and C5b-9 MAC [109]. It has been used to control IBMIR in preclinical NHP studies of porcine islet xenotransplantation [18].

Control of immunological rejection

Many groups have shown long-term graft survival in pig to NHP islet transplantation models using T cell depletion or anergy (ATG, Anti-IL-2R), costimulatory blockade (anti CD40L, anti CD40, anti-LFA-1, LFA-3 Ig, CTLA4-Ig), T cell homing agent (fingolimod), tolerance concept (regulatory T cell infusion), and conventional immunosuppressants (tacrolimus, sirolimus, MMF) (Table 1). According to the IXA's consensus, a preclinical trial would be considered a success leading to the clinic if four of six diabetic NHPs receiving porcine islets maintained normoglycemia for at least 6 months without any, or with a significantly reduced level of, exogenous insulin therapy with a clinically available immunosuppressive regimen [110]. Some groups have already shown that porcine islet-transplanted diabetic NHPs treated with an anti-CD154 mAb-based immunosuppressant regimen maintain normoglycemia for more than 6 months [9,10,15]. However, anti-CD154 mAb were restricted due to unanticipated thromboembolism in humans [111]. As an alternative to the anti-CD154 mAb therapy, an anti-CD40 mAb (Chi220)-based protocol showed prolonged porcine islet graft survival in NHPs, with a maximum of 203 days (mean 90.8 days) [14]. Another clone of anti-CD40 mAb (2C10R4) was also not as effective as the anti-CD154 mAb regimen at preventing graft loss [17]. Kim et al. reported long-term porcine islet graft survival in consecutive diabetic NHPs ($n=7$; >222, >200, 181, 89, 62, 55, and 34 days) without severe adverse events by utilizing a clinically available immunosuppressive regimen (ATG, adalimumab, anakinra, tocilizumab, IVIg, tacrolimus, belimumab, rapamycin, tofacitinib) [18]. Although this study could contribute greatly to the initiation of islet xenotransplantation clinical trials, since the immunosuppressive protocol is stronger than that of islet allotransplantation, an additional method such as a genetically engineered pig or newly developed anti-CD40L mAb without thromboembolism would likely be necessary to increase the survival of porcine islets graft to fulfill IXA guidelines.

Clinical application of xenoislet cell transplantation

DPF pigs

Designated pathogens to be screened for the clinical trials of porcine islet product were listed in IXA consensus statement in 2009 [112] and in 2016. Briefly summarized, donor animal pathogen screening strategy should be geographically appropriate, product specific, adaptive, and dynamic, and PERV-C negative pig could be considered preferable. PERV pig selection criteria should be primarily based on low PERV expression levels and lack of infectivity [113]. It has been reported that PERV could infect human cells in vitro in highly controlled circumstances but no strong evidence of infection in vivo has been reported. Unlike other pathogens that can be "bred out," PERV is integrated in the genome of pig, and the recombination of PERV A and PERV C is the most risky, so pig herds can be bred to be low in PERV C [114]. DPF list and screening strategy for the encapsulated porcine islet clinical trial in New Zealand in 2014 have been published [115]. In 2018 the DPF list for the Spring Point Project in the United States was also published [116]. Since infectious diseases for pig are influenced by geography, it is necessary to establish a list of DPF and screening methods with national authorities of each country.

Regulation

Clinical trials of xenotransplantation should be conducted only in countries where regulations and adequate resources empower a national health authority to effectively regulate these trials. In 2001 the World Health Organization (WHO) released a guidance document to facilitate consideration and implementation of a strategy for international cooperation and coordination on xenogeneic infection/disease surveillance and response [117]. The FDA is the sole agency responsible for regulating clinical xenotransplantation trials in the United States. This agency has published diverse guidance documents with recommendations for sponsors of clinical xenotransplantation trials (Source Animal, Product, Preclinical, and Clinical Issues Concerning the Use of Xenotransplantation Products in Humans; Guidance for Industry 2016) [117]. The European Medical Agency (EMA) has also published key guidance documents for xenogeneic cell-based therapies, such as the "Guideline on Xenogeneic Cell-based Medicinal Products" (2009). The principles laid down in the guideline should be considered by applicants entering into clinical trials. The recommendations included in these xenotransplantation guidelines add to the respective regulations (xenogeneic CBMPs regulated by European regulatory agencies as ATMPs and in the United States as cell and gene therapy products) [117]. In South Korea, the law, "The Act on the Safety and Support of Advanced Regenerative Medicine and Advanced Biopharmaceuticals," came into force after August 2020, which includes regulation on the clinical transplantation of a xenogeneic product. Under this Act, the ethical problems, the supervision of the xenotransplantation research, the role of the governmental agencies, and the protection of the society from potential zoonotic infections should be clarified and assured [118]. In Japan, there has been no progress since the "Public Health Guidelines on Infectious Disease Issues in Xenotransplantation (original version in 2001)" was revised in 2016 and the "Act on the Safety of Regenerative Medicine" was brought into force in 2014. Xenogeneic cell therapy such as islet transplantation can be covered by the earlier mentioned law [119].

Strategies for the successful transplantation

Combination of proper techniques

Minimizing the immune response that occurs when islets from source pigs are transplanted is the most important factor in determining the success of islet xenotransplantation. The strategies are (1) development of multiple transgenic pigs to minimize the use of immunosuppressants and to overcome IBMIR, acute humoral and cellular rejection, and chronic rejection; (2) development of immunosuppressive regimen that can be used clinically with good immunosuppressive effect and few complications; and (3) development of immune tolerance such as regulatory T cells, mixed chimerism, and donor-specific tolerance. For the safety concern of PERV infection, PERV-free pigs have been produced by Niu et al. group using CRISPR-Cas9 that may be suitable as a source pig in combination with other genetic modifications [83].

Future perspectives

As a method to eliminate αGal, galactosyltransferase knock-out pigs were produced in 2002, which led to the first boom in xenotransplantation studies [120]. With the development of CRISPR Cas9, it is possible to produce a number of transgenic pigs (multiple human gene knock-in to avoid immune responses; multiple carbohydrate antigen knock-out including Gal, 5NeuGc, and B4GalNT2; PERV-free pigs, etc.). This technology has led to a second xenotransplantation boom [121]. It is unknown what theme the third xenotransplantation boom will take in the future.

The recent activity for the preparation of a clinical trial of porcine islet xenotransplantation under South Korean legislative coverage is “The Act on the Safety and Support of Advanced Regenerative Medicine and Advanced Biopharmaceuticals” which encompasses the regulation of clinical trials for xenotransplantation. This trial will be performed using wild-type adult pig cells and will be performed at a center in South Korea which would be a significant advance to this field. The clinical trial will be a single center, open, investigator-initiated trial to evaluate the safety and efficacy of xenotransplantation of naked islet in diabetic patients suffering from severe hypoglycemia. The accomplishment of the safe and transparent trial standardized and prepared based on the global consensus under the control of proper regulatory body will certainly move the islet xenotransplantation closer to the clinical realization.

Before porcine islet xenotransplantation can be considered to replace or supplement human islet allotransplantation, pig islet products produced in GMP facilities will need to be cost effective, immunosuppression protocol should match that of human islet allotransplantation in terms of patient health and safety, graft survival of porcine islet xenotransplantation should be similar to or longer than that of islet allotransplantation, and safety concerns for PERV should be solved. While experts believe the risks to be very low, the potential perception of the threat of xenozoonosis calls for additional debate. When this comprehensive situation can be achieved by clinicians with public support, porcine islet xenotransplantation can be successfully utilized in clinical practice in the future.

Acknowledgment

This study was supported partly by a grant from the Korea Healthcare Technology R&D Project through the Korea Health Industry Development Institute (KHIDI), funded by the Ministry for Health and Welfare, Republic of Korea (Grant No. HI13C0954) and partly by the National Research Foundation of Korea (NRF) grant funded by the Korea government (MSIT) (Grant No. NRF-2019R1A2C2085287). The authors thank Michael F. Knoll for his editorial help.

References

[1] Geddes J, Schopman JE, Zammitt NN, Frier BM. Prevalence of impaired awareness of hypoglycaemia in adults with type 1 diabetes. Diabet Med 2008;25(4):501–4.

[2] Ramesh A, Chhabra P, Brayman KL. Pancreatic islet transplantation in type 1 diabetes mellitus: an update on recent developments. Curr Diabetes Rev 2013;9(4):294–311.

[3] Coe TM, Markmann JF, Rickert CG. Current status of porcine islet xenotransplantation. Curr Opin Organ Transplant 2020;25(5):449–56.

[4] Kemter E, Wolf E. Recent progress in porcine islet isolation, culture and engraftment strategies for xenotransplantation. Curr Opin Organ Transplant 2018;23(6):633–41.

[5] Dufrane D, Gianello P. Pig islet for xenotransplantation in human: structural and physiological compatibility for human clinical application. Transplant Rev-Orlan 2012;26(3):183–8.

[6] Wright JR, Pohajdak B, Xu BY, Leventhal JR. Piscine islet xenotransplantation. ILAR J 2004;45(3):314–23.

[7] Wright JR, Yang H, Hyrtsenko O, Xu BY, Yu WM, Pohajdak B. A review of piscine islet xenotransplantation using wild-type tilapia donors and the production of transgenic tilapia expressing a "humanized" tilapia insulin. Xenotransplantation 2014;21(6):485–95.

[8] Hering BJ, Wijkstrom M, Graham ML, Harstedt M, Aasheim TC, Jie T, et al. Prolonged diabetes reversal after intraportal xenotransplantation of wild-type porcine islets in immunosuppressed nonhuman primates. Nat Med 2006;12(3):301–3.
[9] Cardona K, Korbutt GS, Milas Z, Lyon J, Cano J, Jiang W, et al. Long-term survival of neonatal porcine islets in nonhuman primates by targeting costimulation pathways. Nat Med 2006;12(3):304–6.
[10] van der Windt DJ, Bottino R, Casu A, Campanile N, Smetanka C, He J, et al. Long-term controlled Normoglycemia in diabetic non-human Primates after transplantation with hCD46 transgenic porcine islets. Am J Transplant 2009;9(12):2716–26.
[11] Thompson P, Badell IR, Lowe M, Cano J, Song M, Leopardi F, et al. Islet xenotransplantation using gal-deficient neonatal donors improves engraftment and function. Am J Transplant 2011;11(12):2593–602.
[12] Bottino R, Wijkstrom M, van der Windt DJ, Hara H, Ezzelarab M, Murase N, et al. Pig-to-monkey islet xenotransplantation using multi-transgenic pigs. Am J Transplant 2014;14(10):2275–87.
[13] Thompson P, Badell IR, Lowe M, Turner A, Cano J, Avila J, et al. Alternative immunomodulatory strategies for xenotransplantation: CD40/154 pathway-sparing regimens promote xenograft survival. Am J Transplant 2012;12(7):1765–75.
[14] Thompson P, Cardona K, Russell M, Badell IR, Shaffer V, Korbutt G, et al. CD40-specific costimulation blockade enhances neonatal porcine islet survival in nonhuman primates. Am J Transplant 2011;11(5):947–57.
[15] Shin JS, Kim JM, Kim JS, Min BH, Kim YH, Kim HJ, et al. Long-term control of diabetes in immunosuppressed nonhuman primates (NHP) by the transplantation of adult porcine islets. Am J Transplant 2015;15(11):2837–50.
[16] Shin JS, Min BH, Kim JM, Kim JS, Yoon IH, Kim HJ, et al. Failure of transplantation tolerance induction by autologous regulatory T cells in the pig-to-non-human primate islet xenotransplantation model. Xenotransplantation 2016;23(4):300–9.
[17] Shin JS, Kim JM, Min BH, Yoon IH, Kim HJ, Kim JS, et al. Pre-clinical results in pig-to-non-human primate islet xenotransplantation using anti-CD40 antibody (2C10R4)-based immunosuppression. Xenotransplantation 2018;25(1).
[18] Kim JM, Hong SH, Chung H, Shin JS, Min BH, Kim HJ, et al. Long-term porcine islet graft survival in diabetic non-human primates treated with clinically available immunosuppressants. Xenotransplantation 2021;28(2):e12659.
[19] Sun Y, Ma X, Zhou D, Vacek I, Sun AM. Normalization of diabetes in spontaneously diabetic cynomologus monkeys by xenografts of microencapsulated porcine islets without immunosuppression. J Clin Invest 1996;98(6):1417–22.
[20] Dufrane D, Goebbels RM, Gianello P. Alginate macroencapsulation of pig islets allows correction of Streptozotocin-induced diabetes in Primates up to 6 months without immunosuppression. Transplantation 2010;90(10):1054–62.
[21] Samy KP, Martin BM, Turgeon NA, Kirk AD. Islet cell xenotransplantation: a serious look toward the clinic. Xenotransplantation 2014;21(3):221–9.
[22] Hawthorne WJ, Cowan PJ, Buhler LH, Yi SN, Bottino R, Pierson RN, et al. Third WHO Global Consultation on Regulatory Requirements for Xenotransplantation Clinical Trials, Changsha, Hunan, China December 12-14, 2018 "The 2018 Changsha communique" the 10-year anniversary of the international consultation on xenotransplantation. Xenotransplantation 2019;26(2).
[23] Park CG, Bottino R, Hawthorne WJ. Current status of islet xenotransplantation. Int J Surg 2015;23(Pt B):261–6.
[24] Dufrane D, Gianello P. Pig islet xenotransplantation into non-human primate model. Transplantation 2008;86(6):753–60.
[25] Bottino R, Balamurugan AN, Smetanka C, Bertera S, He J, Rood PP, et al. Isolation outcome and functional characteristics of young and adult pig pancreatic islets for transplantation studies. Xenotransplantation 2007;14(1):74–82.
[26] Dufrane D, D'Hoore W, Goebbels RM, Saliez A, Guiot Y, Gianello P. Parameters favouring successful adult pig islet isolations for xenotransplantation in pig-to-primate models. Xenotransplantation 2006;13(3):204–14.
[27] Rajotte RV. Isolation and assessment of islet quality. Xenotransplantation 2008;15(2):93–5.
[28] Hecht G, Eventov-Friedman S, Rosen C, Shezen E, Tchorsh D, Aronovich A, et al. Embryonic pig pancreatic tissue for the treatment of diabetes in a nonhuman primate model. Proc Natl Acad Sci U S A 2009;106(21):8659–64.
[29] Eventov-Friedman S, Tchorsh D, Katchman H, Shezen E, Aronovich A, Hecht G, et al. Embryonic pig pancreatic tissue transplantation for the treatment of diabetes. PLoS Med 2006;3(7), e215.
[30] Matsumoto S, Tan P, Baker J, Durbin K, Tomiya M, Azuma K, et al. Clinical porcine islet xenotransplantation under comprehensive regulation. Transplant Proc 2014;46(6):1992–5.
[31] Matsumoto S, Abalovich A, Wechsler C, Wynyard S, Elliott RB. Clinical benefit of islet xenotransplantation for the treatment of type 1 diabetes. EBioMedicine 2016;12:255–62.
[32] Korbutt GS, Elliott JF, Ao Z, Smith DK, Warnock GL, Rajotte RV. Large scale isolation, growth, and function of porcine neonatal islet cells. J Clin Invest 1996;97(9):2119–29.
[33] Soderlund J, Wennberg L, Castanos-Velez E, Biberfeld P, Zhu S, Tibell A, et al. Fetal porcine islet-like cell clusters transplanted to cynomolgus monkeys: an immunohistochemical study. Transplantation 1999;67(6):784–91.
[34] Dufrane D, Gianello P. Pig islets for clinical islet xenotransplantation. Curr Opin Nephrol Hypertens 2009;18(6):495–500.
[35] Omori T, Nishida T, Komoda H, Fumimoto Y, Ito T, Sawa Y, et al. A study of the xenoantigenicity of neonatal porcine islet-like cell clusters (NPCC) and the efficiency of adenovirus-mediated DAF (CD55) expression. Xenotransplantation 2006;13(5):455–64.
[36] Bloch K, Assa S, Lazard D, Abramov N, Shalitin S, Weintrob N, et al. Neonatal pig islets induce a lower T-cell response than adult pig islets in IDDM patients. Transplantation 1999;67(5):748–52.
[37] Elliott RB, Escobar L, Tan PL, Muzina M, Zwain S, Buchanan C. Live encapsulated porcine islets from a type 1 diabetic patient 9.5 yr after xenotransplantation. Xenotransplantation 2007;14(2):157–61.
[38] Wang W, Mo Z, Ye B, Hu P, Liu S, Yi S. A clinical trial of xenotransplantation of neonatal pig islets for diabetic patients. Zhong Nan Da Xue Xue Bao Yi Xue Ban 2011;36(12):1134–40.

[39] Kim HI, Lee SY, Jin SM, Kim KS, Yu JE, Yeom SC, et al. Parameters for successful pig islet isolation as determined using 68 specific-pathogen-free miniature pigs. Xenotransplantation 2009;16(1):11–8.
[40] Rijkelijkhuizen JK, van der Burg MP, Tons A, Terpstra OT, Bouwman E. Pretransplant culture selects for high-quality porcine islets. Pancreas 2006;32(4):403–7.
[41] Bennet W, Groth CG, Larsson R, Nilsson B, Korsgren O. Isolated human islets trigger an instant blood mediated inflammatory reaction: implications for intraportal islet transplantation as a treatment for patients with type 1 diabetes. Ups J Med Sci 2000;105(2):125–33.
[42] Nilsson B, Ekdahl KN, Korsgren O. Control of instant blood-mediated inflammatory reaction to improve islets of Langerhans engraftment. Curr Opin Organ Transplant 2011;16(6):620–6.
[43] Moberg L, Johansson H, Lukinius A, Berne C, Foss A, Kallen R, et al. Production of tissue factor by pancreatic islet cells as a trigger of detrimental thrombotic reactions in clinical islet transplantation. Lancet 2002;360(9350):2039–45.
[44] O'Connell PJ. Current status of clinical islet transplantation. Diabetes Res Clin Pract 2016;120:S16.
[45] Johansson H, Lukinius A, Moberg L, Lundgren T, Berne C, Foss A, et al. Tissue factor produced by the endocrine cells of the islets of Langerhans is associated with a negative outcome of clinical islet transplantation. Diabetes 2005;54(6):1755–62.
[46] Levi M, van der Poll T, Buller HR. Bidirectional relation between inflammation and coagulation. Circulation 2004;109(22):2698–704.
[47] Bennet W, Sundberg B, Lundgren T, Tibell A, Groth CG, Richards A, et al. Damage to porcine islets of Langerhans after exposure to human blood in vitro, or after intraportal transplantation to cynomologus monkeys: protective effects of sCR1 and heparin. Transplantation 2000;69(5):711–9.
[48] Robson SC, Cooper DK, d'Apice AJ. Disordered regulation of coagulation and platelet activation in xenotransplantation. Xenotransplantation 2000;7(3):166–76.
[49] Kang HJ, Lee H, Ha JM, Lee JI, Shin JS, Kim KY, et al. The role of the alternative complement pathway in early graft loss after Intraportal porcine islet xenotransplantation. Transplantation 2014;97(10):999–1008.
[50] Tjernberg J, Ekdahl KN, Lambris JD, Korsgren O, Nilsson B. Acute antibody-mediated complement activation mediates lysis of pancreatic islets cells and may cause tissue loss in clinical islet transplantation. Transplantation 2008;85(8):1193–9.
[51] van der Windt DJ, Marigliano M, He J, Votyakova TV, Echeverri GJ, Ekser B, et al. Early islet damage after direct exposure of pig islets to blood: has humoral immunity been underestimated? Cell Transplant 2012;21(8):1791–802.
[52] Ricklin D, Hajishengallis G, Yang K, Lambris JD. Complement: a key system for immune surveillance and homeostasis. Nat Immunol 2010;11(9):785–97.
[53] Biancone L, Cantaluppi V, Romanazzi GM, Russo S, Figliolini F, Beltramo S, et al. Platelet-activating factor synthesis and response on pancreatic islet endothelial cells: relevance for islet transplantation. Transplantation 2006;81(4):511–8.
[54] Vanichakarn P, Blair P, Wu C, Freedman JE, Chakrabarti S. Neutrophil CD40 enhances platelet-mediated inflammation. Thromb Res 2008;122(3):346–58.
[55] Kanak MA, Takita M, Kunnathodi F, Lawrence MC, Levy MF, Naziruddin B. Inflammatory response in islet transplantation. Int J Endocrinol 2014;2014.
[56] Kirchhof N, Shibata S, Wijkstrom M, Kulick DM, Salerno CT, Clemmings SM, et al. Reversal of diabetes in non-immunosuppressed rhesus macaques by intraportal porcine islet xenografts precedes acute cellular rejection. Xenotransplantation 2004;11(5):396–407.
[57] Friedman T, Smith RN, Colvin RB, Iacomini J. A critical role for human CD4(+) T-cells in rejection of porcine islet cell xenografts. Diabetes 1999;48(12):2340–8.
[58] Lindeborg E, Kumagai-Braesch M, Moller E. Phenotypic and functional characterization of human T cell clones indirectly activated against adult pig islet cells. Xenotransplantation 2006;13(1):41–52.
[59] Rayat GR, Rajotte RV, Hering BJ, Binette TM, Korbutt GS. In vitro and in vivo expression of Galalpha-(1,3) gal on porcine islet cells is age dependent. J Endocrinol 2003;177(1):127–35.
[60] Komoda H, Miyagawa S, Kubo T, Kitano E, Kitamura H, Omori T, et al. A study of the xenoantigenicity of adult pig islets cells. Xenotransplantation 2004;11(3):237–46.
[61] Wolf E, Kemter E, Klymiuk N, Reichart B. Genetically modified pigs as donors of cells, tissues, and organs for xenotransplantation. Anim Front 2019;9(3):13–20.
[62] Kang HJ, Lee H, Park EM, Kim JM, Min BH, Park CG. D-dimer level, in association with humoral responses, negatively correlates with survival of porcine islet grafts in non-human primates with immunosuppression. Xenotransplantation 2017;24(3).
[63] Kang HJ, Lee H, Park EM, Kim JM, Shin JS, Kim JS, et al. Dissociation between anti-porcine albumin and anti-gal antibody responses in non-human primate recipients of intraportal porcine islet transplantation. Xenotransplantation 2015;22(2):124–34.
[64] Li S, Yan Y, Lin Y, Bullens DM, Rutgeerts O, Goebels J, et al. Rapidly induced, T-cell-independent xenoantibody production is mediated by marginal zone B cells and requires help from NK cells. Blood 2007;110(12):3926–35.
[65] Loveland BE, Milland J, Kyriakou P, Thorley BR, Christiansen D, Lanteri MB, et al. Characterization of a CD46 transgenic pig and protection of transgenic kidneys against hyperacute rejection in non-immunosuppressed baboons. Xenotransplantation 2004;11(2):171–83.
[66] White DJ, Yannoutsos N. Production of pigs transgenic for human DAF to overcome complement-mediated hyperacute xenograft rejection in man. Res Immunol 1996;147(2):88–94.
[67] Diamond LE, McCurry KR, Martin MJ, McClellan SB, Oldham ER, Platt JL, et al. Characterization of transgenic pigs expressing functionally active human CD59 on cardiac endothelium. Transplantation 1996;61(8):1241–9.
[68] Ayares D, Phelps C, Vaught T, Ball S, Mendicino M, Ramsoondar J, et al. Multi-transgenic pigs for vascularized pig organ xenografts. Xenotransplantation 2011;18(5):269.

[69] Petersen B, Ramackers W, Tiede A, Lucas-Hahn A, Herrmann D, Barg-Kues B, et al. Pigs transgenic for human thrombomodulin have elevated production of activated protein C. Xenotransplantation 2009;16(6):486–95.
[70] Le Bas-Bernardet S, Tillou X, Poirier N, Dilek N, Chatelais M, Devalliere J, et al. Xenotransplantation of galactosyl-transferase knockout, CD55, CD59, CD39, and fucosyl-transferase transgenic pig kidneys into baboons. Transplant Proc 2011;43(9):3426–30.
[71] Phelps CJ, Koike C, Vaught TD, Boone J, Wells KD, Chen SH, et al. Production of alpha 1,3-galactosyltransferase-deficient pigs. Science 2003;299(5605):411–4.
[72] Lutz AJ, Li P, Estrada JL, Sidner RA, Chihara RK, Downey SM, et al. Double knockout pigs deficient in N-glycolylneuraminic acid and galactose alpha-1,3-galactose reduce the humoral barrier to xenotransplantation. Xenotransplantation 2013;20(1):27–35.
[73] Whitworth KM, Lee K, Benne JA, Beaton BP, Spate LD, Murphy SL, et al. Use of the CRISPR/Cas9 system to produce genetically engineered pigs from in vitro-derived oocytes and embryos. Biol Reprod 2014;91(3).
[74] Hara H, Long C, Lin YJ, Tai HC, Ezzelarab M, Ayares D, et al. In vitro investigation of pig cells for resistance to human antibody-mediated rejection. Transpl Int 2008;21(12):1163–74.
[75] Miyagawa S, Murakami H, Takahagi Y, Nakai R, Yamada M, Murase A, et al. Remodeling of the major pig xenoantigen by N-acetylglucosaminyltransferase III in transgenic pig. J Biol Chem 2001;276(42):39310–9.
[76] Phelps CJ, Ball SF, Vaught TD, Vance AM, Mendicino M, Monahan JA, et al. Production and characterization of transgenic pigs expressing porcine CTLA4-Ig. Xenotransplantation 2009;16(6):477–85.
[77] Klymiuk N, van Buerck L, Bahr A, Offers M, Kessler B, Wuensch A, et al. Xenografted islet cell clusters from INSLEA29Y transgenic pigs rescue diabetes and prevent immune rejection in humanized mice. Diabetes 2012;61(6):1527–32.
[78] Ezzelarab M, Ezzelarab C, Hara H, Ayares D, Cooper D. Characterization of mesenchymal stromal cells (PMSC) from GTKO pigs transgenic for CD46 and their modulatory effect on the primate cellular response. Am J Transplant 2010;10:188.
[79] Weiss EH, Lilienfeld BG, Muller S, Muller E, Herbach N, Kessler B, et al. HLA-E/human beta2-microglobulin transgenic pigs: protection against xenogeneic human anti-pig natural killer cell cytotoxicity. Transplantation 2009;87(1):35–43.
[80] Petersen B, Lucas-Hahn A, Lemme E, Queisser AL, Oropeza M, Herrmann D, et al. Generation and characterization of pigs transgenic for human hemeoxygenase-1 (hHO-1). Xenotransplantation 2010;17(2):102–3.
[81] Oropeza M, Petersen B, Carnwath JW, Lucas-Hahn A, Lemme E, Hassel P, et al. Transgenic expression of the human A20 gene in cloned pigs provides protection against apoptotic and inflammatory stimuli. Xenotransplantation 2009;16(6):522–34.
[82] Ide K, Wang H, Tahara H, Liu J, Wang X, Asahara T, et al. Role for CD47-SIRPalpha signaling in xenograft rejection by macrophages. Proc Natl Acad Sci U S A 2007;104(12):5062–6.
[83] Niu D, Wei HJ, Lin L, George H, Wang T, Lee H, et al. Inactivation of porcine endogenous retrovirus in pigs using CRISPR-Cas9. Science 2017;357(6357):1303–7.
[84] Borg DJ, Bonifacio E. The use of biomaterials in islet transplantation. Curr Diab Rep 2011;11(5):434–44.
[85] Prochorov AV, Tretjak SI, Goranov VA, Glinnik AA, Goltsev MV. Treatment of insulin dependent diabetes mellitus with intravascular transplantation of pancreatic islet cells without immunosuppressive therapy. Adv Med Sci-Poland 2008;53(2):240–4.
[86] de Vos P, Spasojevic M, Faas MM. Treatment of diabetes with encapsulated islets. Adv Exp Med Biol 2010;670:38–53.
[87] Buder B, Alexander M, Krishnan R, Chapman DW, Lakey JR. Encapsulated islet transplantation: strategies and clinical trials. Immune Netw 2013;13(6):235–9.
[88] Veriter S, Gianello P, Igarashi Y, Beaurin G, Ghyselinck A, Aouassar N, et al. Improvement of subcutaneous bioartificial pancreas vascularization and function by coencapsulation of pig islets and mesenchymal stem cells in primates. Cell Transplant 2014;23(11):1349–64.
[89] Valdes-Gonzalez R, Rodriguez-Ventura AL, White DJ, Bracho-Blanchet E, Castillo A, Ramirez-Gonzalez B, et al. Long-term follow-up of patients with type 1 diabetes transplanted with neonatal pig islets. Clin Exp Immunol 2010;162(3):537–42.
[90] Dufrane D, Goebbels RM, Saliez A, Guiot Y, Gianello P. Six-month survival of microencapsulated pig islets and alginate biocompatibility in primates: proof of concept. Transplantation 2006;81(9):1345–53.
[91] Elliott RB, Escobar L, Tan PL, Garkavenko O, Calafiore R, Basta P, et al. Intraperitoneal alginate-encapsulated neonatal porcine islets in a placebo-controlled study with 16 diabetic cynomolgus primates. Transplant Proc 2005;37(8):3505–8.
[92] Zhu HT, Lu L, Liu XY, Yu L, Lyu Y, Wang B. Treatment of diabetes with encapsulated pig islets: an update on current developments. J Zhejiang Univ Sci B 2015;16(5):329–43.
[93] Hering BJ, Clarke WR, Bridges ND, Eggerman TL, Alejandro R, Bellin MD, et al. Phase 3 trial of transplantation of human islets in type 1 diabetes complicated by severe hypoglycemia. Diabetes Care 2016;39(7):1230–40.
[94] Teramura Y, Iwata H. Islets surface modification prevents blood-mediated inflammatory responses. Bioconjug Chem 2008;19(7):1389–95.
[95] Teramura Y, Iwata H. Surface modification of islets with PEG-lipid for improvement of graft survival in intraportal transplantation. Transplantation 2009;88(5):624–30.
[96] Kizilel S, Scavone A, Liu XA, Nothias JM, Ostrega D, Witkowski P, et al. Encapsulation of pancreatic islets within Nano-thin functional polyethylene glycol coatings for enhanced insulin secretion. Tissue Eng Pt A 2010;16(7):2217–28.
[97] Tomei AA, Manzoli V, Fraker CA, Giraldo J, Velluto D, Najjar M, et al. Device design and materials optimization of conformal coating for islets of Langerhans. Proc Natl Acad Sci U S A 2014;111(29):10514–9.
[98] Zhu HT, Yu L, He YY, Lyu Y, Wang B. Microencapsulated pig islet xenotransplantation as an alternative treatment of diabetes. Tissue Eng Part B-Re 2015;21(5):474–89.
[99] Krishnan R, Ko D, Foster III CE, Liu W, Smink AM, de Haan B, et al. Immunological challenges facing translation of alginate encapsulated porcine islet xenotransplantation to human clinical trials. Methods Mol Biol 2017;1479:305–33.

[100] Park HS, Kim JW, Lee SH, Yang HK, Ham DS, Sun CL, et al. Antifibrotic effect of rapamycin containing polyethylene glycol-coated alginate microcapsule in islet xenotransplantation. J Tissue Eng Regen Med 2017;11(4):1274–84.
[101] Vaithilingam V, Kollarikova G, Qi M, Larsson R, Lacik I, Formo K, et al. Beneficial effects of coating alginate microcapsules with macromolecular heparin conjugates-in vitro and in vivo study. Tissue Eng Part A 2014;20(1–2):324–34.
[102] Dang TT, Thai AV, Cohen J, Slosberg JE, Siniakowicz K, Doloff JC, et al. Enhanced function of immuno-isolated islets in diabetes therapy by co-encapsulation with an anti-inflammatory drug. Biomaterials 2013;34(23):5792–801.
[103] Chen T, Yuan J, Duncanson S, Hibert ML, Kodish BC, Mylavaganam G, et al. Alginate Encapsulant incorporating CXCL12 supports Long-term Allo- and Xenoislet transplantation without systemic immune suppression. Am J Transplant 2015;15(3):618–27.
[104] Amara U, Flierl MA, Rittirsch D, Klos A, Chen H, Acker B, et al. Molecular intercommunication between the complement and coagulation systems. J Immunol 2010;185(9):5628–36.
[105] Oikonomopoulou K, Ricklin D, Ward PA, Lambris JD. Interactions between coagulation and complement-their role in inflammation. Semin Immunopathol 2012;34(1):151–65.
[106] Cardona K, Milas Z, Strobert E, Cano J, Jiang W, Safley SA, et al. Engraftment of adult porcine islet xenografts in diabetic nonhuman primates through targeting of costimulation pathways. Am J Transplant 2007;7(10):2260–8.
[107] Janssen BJC, Gomes L, Koning RI, Svergun DI, Koster AJ, Fritzinger DC, et al. Insights into complement convertase formation based on the structure of the factor B-cobra venom factor complex. EMBO J 2009;28(16):2469–78.
[108] Rood PPM, Bottino R, Balamurugan AN, Smetanka C, Ayares D, Groth CG, et al. Reduction of early graft loss after intraportal porcine islet transplantation in monkeys. Transplantation 2007;83(2):202–10.
[109] Jordan SC, Toyoda M, Vo AA. Intravenous immunoglobulin a natural regulator of immunity and inflammation. Transplantation 2009;88(1):1–6.
[110] Cooper DKC, Bottino R, Gianello P, Graham M, Hawthorne WJ, Kirk AD, et al. First update of the international xenotransplantation association consensus statement on conditions for undertaking clinical trials of porcine islet products in type 1 diabetes chapter 4: pre-clinical efficacy and complication data required to justify a clinical trial. Xenotransplantation 2016;23(1):46–52.
[111] Kawai T, Andrews D, Colvin RB, Sachs DH, Cosimi AB. Thromboembolic complications after treatment with monoclonal antibody against CD40 ligand. Nat Med 2000;6(2):114.
[112] Schuurman HJ. The international xenotransplantation association consensus statement on conditions for undertaking clinical trials of porcine islet products in type 1 diabetes—chapter 2: source pigs. Xenotransplantation 2009;16(4):215–22.
[113] Spizzo T, Denner J, Gazda L, Martin M, Nathu D, Scobie L, et al. First update of the international xenotransplantation association consensus statement on conditions for undertaking clinical trials of porcine islet products in type 1 diabetes chapter 2a: source pigspreventing xenozoonoses. Xenotransplantation 2016;23(1):25–31.
[114] Denner J. How active are porcine endogenous retroviruses (PERVs)? Viruses-Basel 2016;8(8).
[115] Wynyard S, Nathu D, Garkavenko O, Denner J, Elliott R. Microbiological safety of the first clinical pig islet xenotransplantation trial in New Zealand. Xenotransplantation 2014;21(4):309–23.
[116] Noordergraaf J, Schucker A, Martin M, Schuurman HJ, Ordway B, Cooley K, et al. Pathogen elimination and prevention within a regulated, designated pathogen free, closed pig herd for long-term breeding and production of xenotransplantation materials. Xenotransplantation 2018;25(4).
[117] Jorqui-Azofra M. Regulation of clinical xenotransplantation: a reappraisal of the legal, ethical, and social aspects involved. Methods Mol Biol 2020;2110:315–58.
[118] Park CG, Shin JS, Min BH, Kim H, Yeom SC, Ahn C. Current status of xenotransplantation in South Korea. Xenotransplantation 2019;26(1).
[119] Kobayashi T, Miyagawa S. Current activity of xenotransplantation in Japan. Xenotransplantation 2019;26(1).
[120] Lai L, Kolber-Simonds D, Park KW, Cheong HT, Greenstein JL, Im GS, et al. Production of alpha-1,3-galactosyltransferase knockout pigs by nuclear transfer cloning. Science 2002;295(5557):1089–92.
[121] Anon. Xenotransplantation 2.0. Nat Biotechnol 2016;34(1):1.

Chapter 11

The development of stem cell therapies to treat diabetes utilizing the latest science and medicine have to offer

Giuseppe Pettinato[a], Lev T. Perelman[a], and Robert A. Fisher[b]

[a]*Center for Advanced Biomedical Imaging and Photonics, Division of Gastroenterology, Department of Medicine, Beth Israel Deaconess Medical Center, Harvard Medical School, Boston, MA, United States,* [b]*SP Global, Chantilly, VA, United States*

Introduction to stem cells—What they are and how they can help treat disease

Stem cells possess a number of characteristics that distinguish them from any other type of cells: (i) self-renewal, which is the ability to divide indefinitely making daughter cells that will be identical to the mother cell, and with the ability not to age, and (ii) pluripotency, which is the capacity of differentiating into any kind of cell in the body. These are fundamental features that confer to the stem cells a high potential for clinical applicability, as they potentially can be used as an inexhaustible source of cells in cell replacement therapy.

There are other additional properties associated with stem cells that also make them attractive to use for cellular therapies and these include: (i) their ability to divide asymmetrically giving rise to two different types of cells, one identical daughter cell that will keep the pluripotent abilities, and another daughter cell that will undergo terminal differentiation, and (ii) they also possess the ability to be quiescent (dormant) to be maintained in culture without dividing. However, these properties cannot be considered valid for all kinds of stem cells [1].

To this end stem cells can be classified according to their pluripotency or differentiation ability, which depends upon their origins. Nevertheless, both classifications are somehow related, depending upon their origin they will have a higher or lower ability to differentiate.

The first type of classification is based upon their differentiation potential. We can distinguish mainly four types of stem cells: (i) Totipotent cells: these are the ones that possess the ability to differentiate into any type of cells, including the trophoblast from which the placenta arises, extraembryonic tissue, and umbilical cord. These types of stem cells can be found only in the early stages of embryonic cellular division (when the embryo possesses approximately 8–16 cells). (ii) Pluripotent cells: these are the cells that can give rise to two different types of cells belonging to the three germ layers (endoderm, mesoderm, and ectoderm), including the cells from the germline (oocyte and spermatozoa). (iii) Multipotent cells: these are the cells that can give rise to only one type of cell line. An example of this type of multipotent cell is the hematopoietic stem cells, from which all the blood cells arise (erythrocytes and white cells). (iv) Unipotent cells, which are cell lines that can give rise to only one cell type.

The second type of classification is based upon their origins. We can distinguish: (i) embryonic stem cells (ESCs), (ii) germline stem cells (EGCs), (iii) embryonic carcinoma cells (ECCs), (iv) stem cells from the trophoblast (TSCs), and (v) adult stem cells or fetal stem cells (ASCs and FSCs).

Embryonic stem cells are derived from embryos at the blastocyst stage, before their implantation in the uterus, which in the case of the human is a week after zygote formation. The blastocyst is composed of approximately 200 cells that can be divided into two types: (i) trophoblast cells, which are those that compose the outer layer of the embryo and (ii) cells from the inner cell mass (ICM) from which all the cells in the body will originate. The last ones, when isolated and cultured in a specific growth media, are the ESCs. They are pluripotent cells, able to give rise to any type of cells that originate from the three germ layers, but not from the placental cells. The first ESCs line was derived from mice (mESCs) in the early 1980s [2,3]. Between 1998 and 2000 the first human embryonic stem cells (hESCs) were isolated [4,5]. Nowadays ESCs can be readily derived from different species.

Pancreas and Beta Cell Replacement. https://doi.org/10.1016/B978-0-12-824011-3.00002-3

Germline stem cells are derived from the primordial germ cells (PGCs), the gametes precursors. In the mouse these develop in the first week of embryonic development. The first EGCs line was obtained from mouse in the 1990s [6,7]. Mouse EGCs are very similar to mESCs and possess the same markers, characteristics, and properties. In order to obtain EGCs it is only possible by generating a spherical structure called the embryoid body (EB) starting from the PGCs. These EBs are constituted of a variety of cells at various differentiation stages [8]. EGCs of human origin (hEGCs) were isolated for the first time from Shamblott in 1998. They are derived from the future gonads that are developing in the embryo at the fifth week after implantation. EGCs differ from ESC for: (i) morphology, (ii) presence of different surface markers, (iii) different culture protocols to growth and differentiate them, (iv) they cannot form teratomas or teratocarcinomas as ESCs when they are transplanted, and (v) they possess a limited capacity to proliferate and divide [9].

Stem cells from carcinoma or embryonic carcinoma cells (ECCs) are undifferentiated cells that are located inside teratocarcinomas (germline tumors). The first ECCs line was derived from mouse, and from studies carried out on these cells it was possible to find that these cells were able to differentiate into various types of cells and inexhaustible growth [10]. Seven years later the first human ECCs line was derived from human teratocarcinomas [11]. In general, ECCs possess the same characteristics of the ESCs but with some differences: (i) their differentiation potential is limited, therefore they cannot differentiate into all types of tissues; (ii) they possess an unstable karyotype, therefore they can potentially accumulate mutations during their divisions (mitosis).

Stem cells from the trophoblast (TSCs) are derived from the outer layer of the murine blastocyst at day three of differentiation. In addition, it can also be obtained from extraembryonic ectodermal tissue coming from 7-day-old embryos. All the cell lines from trophoblasts are generally obtained from mice, and they possess similar characteristics and gene expression. They also possess the same differentiation potential, and they can be cultured with the same culture media [12].

Adult stem cells (ASCs) possess a limited capacity to grow and differentiate as compared to ESCs. Currently we know the existence of different types of tissues that possess adult stem cells with multipotent abilities [12]. Nevertheless, multiple studies have highlighted the presence of ASCs with pluripotent abilities in different tissues: (i) mesangioblast [13]; (ii) dermis [14]; (iii) muscle tissue; (iv) bone marrow; and (v) blood [15]. There are several explanations for these pluripotent adult stem cells: (i) these pluripotent stem cells are located in the bone marrow and they travel toward other tissues, (ii) they are pluripotent cells that are localized in different tissues, (iii) they are multipotent stem cells that can lose their differentiated state and acquire pluripotent capacities, and (iv) these cells might derive from phenomena of cellular fusion. There are several studies that demonstrate that cell fusion is very important for the formation of these cell types [16–18] and yet other studies indicate this is not the case [19].

Embryonic stem cells (ESCs)

The first ESCs were derived from the mouse approximately 40 years ago when they were isolated from the inner cell mass (ICM) of mouse embryos at the blastocyst stage and cultivated in vitro [2,3]. Cells from the ICM are the ones that give rise to the embryo from which all the tissues in the body arise. The isolation of ESCs occurs from preimplantation embryos and it is carried out by an immunosurgery technique, which consists in the incubation of the blastocyst with antibodies that are specific for the trophoectoderm (eliminating the layer of glycoproteins also known as zona pellucida), and then transferred to an enriched complement protein solution that is extracted from the guinea pig serum. In this way the trophoectodermal cells are lysate by the complement, obtaining the intact ICM. The cell mass is then kept on a mitotically inactivated mouse embryonic fibroblasts (MEFs) layer, with the addition of LIF to keep the cells undifferentiated in order to culture and expand them [20].

Since the isolation of the very first stem cells, many stem cells have been derived from different species, such as mouse ESC (mESCs) [2,3,21,22], rabbit [23], chicken [24], hamster [21], mink [25], pig [26], chimpanzee (Rhesus monkey) [27], marmot [28], and human [4,5,8,29–31].

The ESCs that have been mostly used since their first development have been the mouse and human embryonic stem cells (hESCs). Usually, mESCs are used as the initial step to develop and optimize studies that are ultimately aimed at developing new cell types from these cells. Subsequently, the same differentiation protocols have been modified and tailored to be applied to hESCs.

hESCs have very substantial differences when compared to mESCs in regard to culture in vitro. hESCs grow slowly and tend to form flat round shaped colonies, and they are easy to dissociate into single cells compared to the mESCs [32]. hESCs do not need LIF to maintain their undifferentiated state, but they need to be cultivated on feeder layers of mitotically inactivated MEF with the addition of basic fibroblast growth factor (bFGF) [4,5,32,33], or using Matrigel or laminin plus the addition of MEF's conditioned medium [34].

Uniquely hESCs are also characterized by the expression of surface-specific antigen markers, such as SSEA-3 and SSEA-4, as well as the glycoproteins TRA-1-60, TRA-1-81, and GCTM-2. None of these markers are expressed on the cell surface of mESCs, which express only the embryonic antigen SSEA-1, which is not present on hESCs [4,33].

To test their pluripotency, mESCs are normally employed to generate chimeric mice. However, for obvious reasons, this test cannot be used on hESCs, therefore, in order to test their pluripotency, the elective test for these cells is teratoma formation in SCID mice [4,5,33]. The number of hESC lines has increased over the years and nowadays many cell lines are available and sold commercially.

For therapeutic use, hESCs cannot be cultivated on MEF feeder layers or by using media that contain animal-derived products, as there would be the potential of creating a zoonosis by transferring animal pathogens into humans (xeno-contamination). For this purpose, over the past several years many commercial maintenance medias have been developed, which contain formulas to keep the hESCs in an undifferentiated state and using matrices that are also suitable for GMP purposes.

In summary, the properties that define an ESC are as follows:

- They are cells derived from the ICM or from the epiblast of the blastocyst.
- They can replicate themselves an unlimited number of times without differentiating.
- They possess a stable karyotype, diploid, and with normal chromosomes.
- They are pluripotent and can generate all the cells belonging to the three germ layers.
- They are able to integrate themselves into fetal tissues during the development. mESCs kept in culture for a long period of time can generate any type of tissue when reintroduced into a developing embryo [35].
- They can colonize the germline and form oocyte and spermatozoa [36].
- They are clonogenic, therefore, starting from one stem cell it is possible to obtain an entire colony genetically identical to the starting cell (clone), which possess the same properties of the mother cell [5].
- They are able to form teratoma when injected into a mouse, under the skin or into the renal capsule [3,4].
- They can be induced to continuously proliferate or to differentiate [33].
- They express the transcription factor Oct3/4, which activate or inhibit a large number of master genes to keep the ESCs in a proliferative and undifferentiated state.
- They must express undifferentiation markers, such as SSEA, TRA-1-60, and TRA-1-81 [4,33].
- They are most of the time in S phase of the cell cycle. Differently from somatic cells, ESCs do not require any external stimulus to initiate DNA replication [37].

In vitro differentiation of hESCs

The most important aspect of ESCs is their ability to differentiate into all the cell types in our body. The method currently used to initiate differentiation is through the formation of a three-dimensional structure called an embryoid body (EB). Due to its unique characteristics, they are similar to the preimplantation embryonic tissue, as such they can be considered as a prerequisite for the generation of most of the cells of the three germ layers [38–40]. Studies have shown that by the formation of EBs it is possible to recreate properties of early embryogenesis, including the complex three-dimensional architecture [41,42].

Several techniques have been developed to obtain EBs from mESCs and then for hESC [43], and these methods differ according to the origin of the ESCs. There are specific techniques that are used for mESCs such as: (i) withdrawing of LIF and/or feeder layers to let the ESC differentiate spontaneously [44]; (ii) culture in nonadherent plates at a high cell density; (iii) culture in media containing methylcellulose [45]; or (iv) hanging drop culture. This last technique allows generation of mEBs starting with a defined number of cells being placed into each individual drop of culture medium [46,47]. Timing for the EB formation heavily depends on the type of ESCs that have been used.

Over the past 10 years other technologies for the generation of hEBs have been developed and they are mainly placed into three major categories: (i) low adherence multiwell plates with or without centrifugation steps [48]; (ii) bioreactors for mass cell culture and scalable formation of hEBs [49]; and (iii) micromold technologies that use different cell repellent materials for their fabrication [41,42,50].

The differentiation of these EBs is similar to the postimplantation of embryos in vivo [44]. Mesoderm and ectoderm precursors form within a few days, while endodermal precursors need more time, taking up to 10 days. This will lead to a state where the majority of the EBs will cavitate and form a cystic structure [38,39].

After EBs formation there are several ways to induce the differentiation of ESCs toward a specific lineage and these can be: (i) by adding a combination of growth factors into the media, (ii) coculture ESCs with inductor cells or tissues, (iii) implant ESC into specific areas or organs (mainly for experimental purposes in animal studies), (iv) overexpression of transcription factors associated with the development of a specific tissue, (v) selection of cells by programmed gene expression, and (vi) isolation of pure cell populations by fluorescence-activated cell sorting (FACS) using specific surface markers.

There have been several different tissues that have been derived from hESCs. These can be subdivided into three lineages. Firstly, from the mesodermal germ layer, which include the cardiomyocytes [51–53], chondrocytes [54,55], adipocytes [56–58], smooth muscle cells [59,60], hematopoietic cells [61,62], endothelial cells [63,64], and osteoblasts [65,66].

Those hESCs derived from the ectodermal germ layer are keratinocytes [67,68], hair cells [69,70], and sensory epithelium of the eye [71,72].

Also, the endodermal cells derived from hESCs are the alveolar epithelial cells [73,74], hepatocytes [75–77], and importantly to this chapter and book the pancreatic beta cells [78–81]. As can be seen, there are a number of very specific ways in which we can direct the development of the specific stem cell lineages that we may wish to develop. As such, the next sections of this chapter will focus on the more specific ways we can target treatments for diabetes.

Human-induced pluripotent stem cells

Since their discovery in 1998 [4], human embryonic stem cells (hESCs) have held great promise for regenerative medicine, as these cells possess the transformative features of pluripotency, the ability to become any cell in our body; and self-renewal, the capacity of maintaining their pluripotent state indefinitely. However, despite such promise, hESCs have encountered several hurdles mainly due to ethical controversies and the potential of teratoma formation after transplantation [82,83]. Teratomas are groups of cells that can come from different lineages and form a tumor; hESCs have great capacity to form them. To eliminate these challenges, scientists have worked on developing technologies to generate synthetic embryonic stem cells from adult cells. In 2007 for the first time ever, Thompson's group and independently Yamanaka's group developed a new strategy to obtain what they called "human-induced pluripotent stem cells" (hiPSCs) [84,85]. These cells were initially derived from human adult foreskin fibroblast by forcing the expression of four genes (Oct3/4, Sox2, Klf4, c-Myc) that were deemed to be essential for the induction of pluripotency in these adult cells. These factors when overexpressed lead to a stem cell like phenotype, independently to the cell type found in the body that undergo this treatment.

Subsequently, these hiPSCs have been used to produce different tissues in vitro for regenerative medicine purposes, by using different cocktails of growth factors, small molecules and various chemicals in different concentrations and timings, leading to what are called "differentiation protocols." By the application of such differentiation protocols, researchers from all over the world have tried to obtain mature tissues that can be used for transplantation for the treatment of degenerative diseases including type 1 diabetes.

Pancreatic embryological structure and cell signaling

Knowing the basic mechanisms of pancreatic organogenesis is of utmost importance in order to be able to acquire a deep understanding of pancreatic development and homeostasis. This in turn will lead to a better understanding of the mechanisms causing diabetes. By acquiring this knowledge, it will be possible to generate functional insulin-producing cells that can be made available for therapeutic purposes.

The pancreas arises starting from the middle portion of the endoderm during the initial formation of the primitive gut. Epithelial cells from the dorsal and ventral area start to evaginate to form the ventral and dorsal pancreatic buds. Subsequently, these pancreatic buds will branch out occupying the surrounding mesenchyme. At day 16 of embryogenesis, ventral and dorsal pancreatic buds fuse together, leading to the formation of the exocrine pancreatic ducts and acinar cells. Precursors of endocrine cells will arise from these structures forming interstitial clusters adjacent to the ductal epithelium. At this stage, islets of Langerhans do not show the typical organization and do not present all the different cell types that can be found in the mature pancreas. Between day 14 and day 18 of embryogenesis, endocrine cells split from the exocrine portion of the arising pancreas, and they start to proliferate and form mature islets of Langerhans. The typical distribution of the beta cells in the central area of the islet and the outer layer of alpha cells will start appearing by the end of the gestational period (day 18.5) [86]. After birth the pancreas develops the ability to sense glucose and secrete insulin accordingly [87].

Over the past decades there have been several studies looking at the molecular pathways that lead to the development of insulin-secreting cells in vivo during embryogenesis. Such studies have highlighted how inductive signaling coming from neighboring tissue territories in the embryo can be of utmost importance for the development of an organ. In the specific case of the pancreas organogenesis, this arises from the dorsal and ventral pancreatic buds. Each of these pancreatic buds is induced through signaling coming from the notochord, for the dorsal pancreatic bud, and from the cardiac mesenchyme for the ventral pancreatic bud [88]. The inductive signaling coming from the notochord is based on the inhibition of the sonic hedgehog (SSH) pathway, which is a negative regulator of PDX-1, the key factor for pancreatic development [89]. Therefore when the SSH pathway is active, it will repress pancreatic formation, by inhibiting PDX-1 expression, while when it is inhibited it will lead to an increased PDX-1 gene expression with consequential pancreatic development.

The factors that can inhibit the SSH pathway include: (i) TGF-β family factor Fibroblast Growth Factor-2 (FGF-2) and Activin βb, (ii) the bone morphogenetic protein (BMP), and (iii) Activin A [90].

PDX-1 plays an important role during the pancreas ontogenesis: (i) during the early stage it is essential for pancreas development, and later its expression becomes limited to the cells that's fate is to become mature β-cells. PDX-1 is involved in the gene expression of insulin and somatostatin [91,92], and other important genes of β-cells, such as: (i) GLUT-2; (ii) glucokinase, and (iii) amyloid polypeptide. Embryos that are deficient in PDX-1 expression do not develop a pancreas, indicating a key role of this factor for pancreatic organogenesis [93,94].

The deletion of genes such as Isl-1 and Hlxb9 has documented that the development of the dorsal and ventral pancreatic buds is independent and requires different groups of transcription factors. Mutations in these genes lead to the blockage of development of the dorsal bud and not the ventral bud. These experiments provide great insights in identifying specific markers to follow the development of the pancreatic dorsal bud [95–97].

Neurogenin 3 (Ngn-3) expression is induced by Hnf6 and is considered a marker for endocrine cells. Knockout mice for Ngn-3 do not show the presence of any pancreatic endocrine structure, and they die from diabetes after birth [98]. Ngn-3 is not expressed in mature islets although it is coexpressed with other endocrine markers such as Nkx2.2 or Nkx6.1, suggesting that its expression is limited to the earlier stages of development of immature pancreatic endocrine cells [99,100].

The transcription factor Pax-6 is normally expressed in a subgroup of endocrine cell precursors, as well as in the mature endocrine pancreas. Knockout mice for Pax-6 do not develop α-cells, indicating that this factor is important for this cell type. On the other hand, Pax-4 is expressed exclusively during islet development, and it is not detected in adult pancreatic islets [101]. Pax-4 deficient mice develop severe diabetes at birth and the islets show a β- and δ-cell deficiency, together with an increased α-cell population [102].

Genes belonging to the homeodomain protein NK, such as Nkx-6.1 and Nkx-2.2, play an important role during endocrine differentiation. During E9.5, Nkx-2.2 expression becomes limited to α-, β-, and PP-cells in mature islets. Mice with homozygote deletions for both genes show a reduced number of β-cells. Nkx-2.2 mutant mice display a reduction in the number of α- and PP-cells, with the consequent cancellation of the expression of Nkx-6.1 gene, suggesting that Nkx-2.2 acts as a transcriptional regulator of this gene [103,104].

These genetic studies that have been performed with the transcription factors that are expressed during specific points in the development of pancreatic endocrine differentiation have highlighted important genetic markers that are useful in defining the differentiation status of pancreatic precursors (Fig. 1). Importantly, these markers are not uniquely linked to pancreatic differentiation, as they are also shared with development of the central nervous system. In Table 1, there is a list of the main genetic markers that are linked with a specific stage of the pancreatic differentiation in vivo.

Pancreatic beta-cells physiology in relation to stem cells

β-cells represent one of the most important cell types within the islets of Langerhans. These cell clusters are composed of a heterogeneous population of cells that secrete different hormones, such as: (i) insulin from β-cells, which constitute the majority of the cell mass within the islets of Langerhans (60%–70%); (ii) α-cells, which produce glucagon and they are located on the outer layer of the islets (15%); (iii) δ-cells, which produce somatostatin (5%) and are localized near the (iv) PP-cells, which produce pancreatic polypeptide (10%). These last two types of cells are located around the β-cells and α-cells.

The main function of β-cells is to secrete measured amounts of insulin in response to blood glucose levels in a biofeedback loop. The blood glucose levels are produced in response to nutrients, hormones, and nerve stimulus, to maintain plasma glucose levels within a tightly maintained range, in order to allow the physiologic function of all the body tissues [119–121].

Insulin secretion is strictly correlated with body metabolism, and it is coupled with electrical activity that initiates synthesis and secretion from the β-cells [122,123]. This chain of events is called "coupling of stimulus-secretion." Increased levels of glucose in the blood, and other nutrients, such as amino acids and fats, trigger the secretion of insulin from the β-cells [120,124,125]. Hormones and neurotransmitters act on the β-cells by inducing secretion of insulin in response to nutrients.

The efficiency of the β-cells to respond to glucose concentrations is due to the glucose transporter type 2 (GLUT-2) and to an enzyme, glucokinase (GK). Through its cellular membrane, β-cells can transport high amounts of glucose inside the cytosol by GLUT-2. An important characteristic that GLUT-2 has is the high K_m for glucose, which means that this transporter will work only when the glucose levels are high, allowing for a quick balance between the extracellular and intracellular concentration.

The enzyme GK (M_m 8 mM) phosphorylates glucose, allowing it to enter in the glycolysis cycle [126]. Following phosphorylation, the enzymatic machinery within the β-cells metabolizes the glucose through the oxidative pathway, leading to

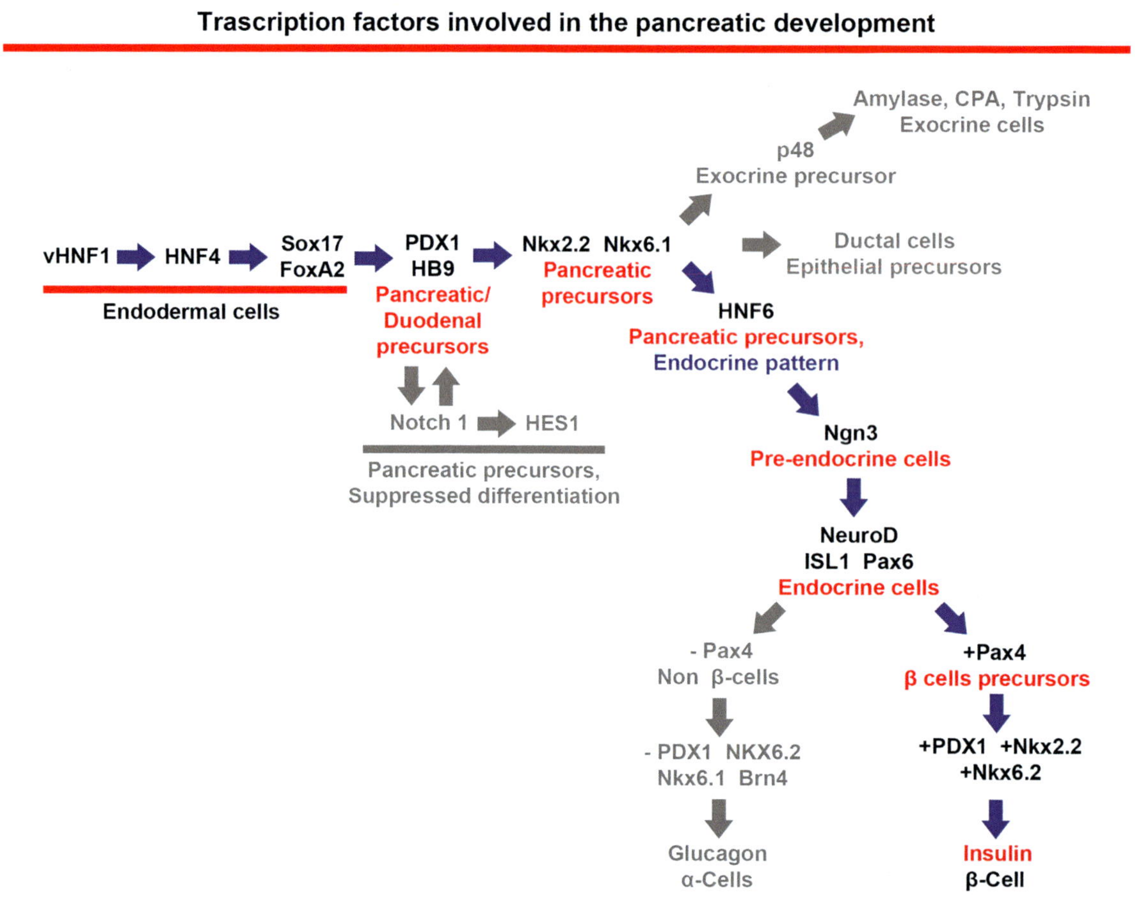

FIG. 1 Transcription factor progression from endodermal cells toward insulin-producing cells during embryological development.

an increase of mitochondrial ATP [127]. An increase between the ratio of ATP/ADP and of the diadenosine polyphosphate within the β-cells triggers the blockage of the potassium ATP-dependent channels (K_{ATP}) [119]. This K_{ATP} channel, which is situated in the β-cells plasmatic membrane, consists of a protein that forms a pore (Kir-6.2), and one regulatory subunit, the sulfonylurea acceptor protein (Sur-1).

This channel complex plays an important role for the insulin secretion response. The closing of the channel inhibits the flux of potassium, leading to the β-cells plasmatic membrane depolarization and the Ca^{2+} voltage-dependent channels activation, allowing for the intake of calcium from the extracellular side to the intracellular compartments, thereby triggering the exocytosis of insulin-containing vesicles [120,121,128,129].

Stem cell differentiation strategies—An overview of the past two decades

In in vivo embryogenesis, stem cells undergo a natural differentiation process. Since the discovery of hESCs scientists have worked to mimic the complex process in an in vitro setting, to create artificial pancreatic organoids that can be used in trials for drug discovery and as potential cell replacement therapy in regenerative medicine.

The first attempt to create pancreatic beta cells was by Soria et al. in 2000 and 2001, using mouse embryonic stem cells (mESCs) in a gene trapping technology [128,130]. This work led to the formation of cells that were able to restore normoglycemia in streptozotocin-treated mice.

Following this first attempt, other groups tried to differentiate mESCs into insulin-producing cells, obtaining some successful results [131–137]. The major step for clinical research was to apply this knowledge to create human embryonic stem cells.

In 2001 Assady et al. were one of the first groups to develop a protocol obtaining insulin-positive cells. This was a major step forward which demonstrated that we could generate insulin-positive cells, in spite of the success, these cells represented only 1% of the differentiated population and they were not able to respond to glucose stimulation assays [138].

TABLE 1 Transcription factors important during the in vivo embryological pancreatic organogenesis.

Gene name	Stages of pancreatic embryonic expression	Expression in pancreatic adult cells	Genetic evidences of gene mutated functions	References
Hnf1a	Expressed in the developing pancreas starting at E13.5	β-cells, perhaps also in other cell types	Severely reduced glucose-induced insulin secretion	[105,106]
Hnf3b/Foxa2	Expressed in the early endoderm, progenitors, and all pancreatic cell types during the whole development	All pancreatic cell types	Endoderm formation dysfunction; β-cell mutants show altered insulin secretion	[107]
Ipf1/Pdx1	Highly expressed in pancreatic progenitor cells at E8-E10. Expressed at high level in differentiating β-cells	β-cells	Absence of pancreas in null mutants; reduced insulin gene expression and β-cell dysfunction in β-cell mutants leading to diabetes	[93,108–110]
HB9/Hlxb9	Expressed in pancreatic progenitor cells at E8-E9 and in differentiating β-cells	β-cells	Dorsal and ventral pancreas absence has mild disturbance on endocrine cell differentiation	[96,111]
Nkx2.2	Expressed in pancreatic progenitors at late E8-E9, and it becomes limited to pancreatic endocrine cells at late stages	α-, β-, and PP-cells	α-, β-, and PP-cells are reduced β-cells cannot express insulin	[103]
Nkx6.1	Expressed in early progenitor cells starting at ~ E9.0-E9.5 and becomes limited to β-cells	β-cells	Decreased amount of mature β-cells	[104]
Hnf6	Expressed in the pancreatic progenitor cells and all the pancreatic cell types starting at ~ E9 and during the whole development	All pancreatic cell types	Decreased amount of endocrine cells and late endocrine cell differentiation	[98,112,113]
Ngn3	Expressed in endocrine progenitor cells starting at E9	Not expressed	Absence of endocrine cells and disturbance of acinar polarity	[114]
Neurod	Expressed in endocrine cells starting at E9.5	All four differentiated endocrine cell types	Decreased amount of endocrine cells; disrupted apical-basal cell polarity of acinar cells	[115]
Isl1	Expressed in endocrine cells starting at E9. *Isl1* is likewise expressed in dorsal mesenchyme starting at E8-E9	All four differentiated endocrine cell types	Reduced endocrine cell differentiation; absence of dorsal pancreatic mesenchyme	[97]
Pax6	Expressed in endocrine cells starting at E9.0-E9.5	All four differentiated endocrine cell types	Reduced amount of endocrine cells and decreased hormone gene expression	[116,117]
Pax4	Expressed in pancreatic progenitor cells at E9.0-E9.5	Not expressed or expressed at very low levels in pancreas	Absence of β- and δ-cells	[102]
p48	Expressed in pancreatic progenitor cells from E9.5. After it will be limited to exocrine cells	Exocrine cells	Absence of exocrine pancreas; endocrine cells form but stays in the spleen	[118]

E, embryonic day; *Foxa*, forkhead box A; *Hlxb9*, homeobox gene HB9; *Hnf*, hepatocyte nuclear factor; *Ipf1*, insulin promoter factor 1; *Isl1*, Insulin Gene Enhancer Protein ISL-1; *NeuroD*, neurogenic differentiation; *Ngn3*, neurogenin 3; *Nkx*, NK-related homeobox; *Pax*, paired box gene; *Pdx1*, pancreatic and duodenal homeobox gene 1.

In 2006 several groups published new protocols using hESCs with improved results. Xu et al. published research describing a spontaneous differentiation process using hESCs, obtaining endodermal cells, and subsequently, insulin-producing cells, using human embryoid bodies (hEBs) [139]. D'Amour et al. developed a differentiation protocol using different growth factors at precise concentrations that initiated the era of mimicking natural embryogenesis [140]. They first described how Activin A was able to generate endodermal cells from hESCs. Following this their team demonstrated that by applying a sequence of growth factors at specific doses, they were able to differentiate cells that expressed pancreatic markers, such as C-peptide, together with other specific markers for endodermal precursors (NKX-6.1 and PDX-1) and adult pancreatic beta cells (Insulin) [78]. This work was the first one of its type, and again showed proof of principle despite not performing any in vivo experiments to determine if these differentiated cells were able to reverse hyperglycemia in diabetic mouse models.

As a result of this work, several other groups developed and published different protocols that were based on the pancreatic in vivo organogenesis cues [79–81].

Two of these protocols developed by Shim et al. and Wei Jiang shared the use of two common factors, the Activin A and retinoic acid (RA), although with different doses and times. Shim et al. started with the formation of hEB, followed by a partial dissociation to allow the maturation into β-cells, conducted totally in suspension culture. Wei Jiang et al. used the same factors while the cells were cultured in adherence, and then switched to suspension culture, scrapping the hESC colonies and adding another two factors, nicotinamide and the basic fibroblast growth factor (bFGF). This suspension culture led to the formation of aggregates. In both works the differentiated cells expressed insulin and were employed to rescue hyperglycemia in animal models, confirming the efficacy of these cells.

Strategies behind developing a pancreatic differentiation protocol

Organogenesis is a very complex and fascinating process in vivo, and it is only by mastering it that it will be possible to obtain mature and develop functional organoids in vitro. Most of the organs in our body arise from an intricate interaction between several neighboring embryonic territories that secrete inducing factors that trigger the development of a specific organ. It is by carefully studying the activation and repression of the main organogenetic pathways that it was possible to discover critical key points, controlling the differentiation of an organ in vitro by using stem cells [141–144].

In the specific case of pancreatic development, this organ arises from the endoderm and from the intestinal epithelium. Corentin et al. showed how the cells coming from this epithelium first proliferate and then differentiate into endocrine cells to form islets of Langerhans and the exocrine pancreas [145]. According to this observation, from one cell type we can obtain either gut cells or pancreatic cells. The system that rules the fate that these precursor cells will take is the Sonic hedgehog (Shh) pathway. Shh activation promotes epithelial intestinal differentiation, while Shh inhibition induces pancreatic endocrine cell formation [146].

Activin A is a soluble factor, belonging to the TGF-β superfamily, and as other members of this group, interact with two types of transmembrane receptors on the cell surface (type I and II), which possess intrinsic serine/threonine kinase activity in their cytoplasmic domains. When Activin A binds to the type II receptor, it will start a series of reactions that will ultimately lead to the recruitment, phosphorylation, and activation of the type I receptor. Type I receptor will interact with the type II receptor and together they will phosphorylate Smad2 and Smad3, two of the cytoplasmic Smad proteins. Smad3 will enter into the nucleus where it will bind with Smad4 acting as transcription factors, which will activate a variety of genes [147].

The next step of pancreatic differentiation is to obtain pancreatic endocrine cells. The endocrine pancreatic differentiation is ruled by the Notch pathway. This active pathway will repress the differentiation of the pancreatic progenitor cells, by keeping them in a standby state, and when blocked, will lead to the development of endocrine pancreatic cells [148]. Neurogenin 3 (Ngn3) belongs to the transcription factor basic helix-loop-helix family, and it participates in the endocrine pancreatic development. It has been seen that transgenic mice that overexpressed Ngn3 in the early developmental phases have shown a strong increase in pancreatic endocrine cells' generation, indicating that Ngn3 is involved in the differentiation of islets of Langerhans' precursors [149]. Ngn3 acts through the inhibition of the Notch pathway [149,150].

After obtaining pancreatic endocrine progenitor cells, the next step is to allow these cells to mature and increase their number, to obtain an enriched population of hormone-secreting cells. Several factors can be used for this purpose which could stimulate the expression of the β-cells master gene PDX-1, insulin, and Glut-2. For example, by stimulating the protein-tyrosine kinase activity of certain receptors, it can activate a signal transduction cascade which will lead to several biochemical changes within the cells, including an increase of intracellular calcium levels, enhanced by protein synthesis and glycolysis activity, which ultimately lead to DNA synthesis and cell proliferation [145].

State of the art for the pancreatic differentiation

Over the two past decades many research groups have devoted their attention to developing differentiation protocols that could give rise to a product that would be suitable for clinical applications, as stem cell therapies. Although there are many strategies for the generation of pancreatic organoids, two major approaches have succeeded in reaching the clinical trial phase of development.

The first approach to obtain pancreatic organoids was the derivation of definitive endodermal cells from human embryonic stem cells (hESCs) developed by Kevin D'Amour in 2005. This author showed how Activin A was able to induce endoderm differentiation in hESCs [140]. Following this first article, the author published in 2006, a more detailed protocol for the obtainment of pancreatic progenitor cells (NKX-6.1 and PDX-1 positive cells), which once transplanted into a diabetic mouse model were able to maturate into insulin-secreting cells in vivo [78]. Researches published to improve on this first protocol include one from Kroon et al. in 2008 and more recently another one from Rezania et al. in 2013 [151,152]. The culmination of the combined protocols has become one of the most used methodologies for pancreatic differentiation moving toward the clinical setting. This has become possible after critical success in restoring euglycemia in animal models. For clinical application, the use of hESCs dictated the need to isolate such cells from the host immune system. One model has developed pancreatic islet progenitor cells encapsulated into a retrievable cell delivery device in order to implant this combination of cells/device under the skin of diabetic animal models [153]. The cells and two delivery devices have undergone preclinical testing and have moved to clinical trials. One of these devices delivers the pancreatic islet progenitor cells in a non-immune-protective device and is being developed for patients with type 1 diabetes who have hypoglycemia unawareness, extreme glycemic lability, and/or recurrent severe hypoglycemic episodes. The second device delivers the same pancreatic islet progenitor cells in an immune-protective device and is being developed for all patients with type 1 and type 2 diabetes [153].

Data from the first clinical trials have shown the presence of C-peptide and insulin production through histological and biochemical measurements at several time points in multiple patients [154]. The detection of C-peptide has been associated with the engraftment efficiency of this product. (C-peptide is a short sequence of amino acids that is cogenerated during the formation of insulin by the pancreas and is coreleased into the blood stream in the same proportion with insulin. C-peptide is used as a standard biomarker for assessing the amount of functional insulin-producing pancreatic beta cells, because its measurement is not mistaken by the injected insulin.) Histological data have shown that, after the pancreatic islet progenitor cells were implanted under the skin, and they properly engrafted, they can mature into functional beta cells and other cells of the islet that are responsible for controlling blood glucose levels [155].

Another strategy for obtaining insulin-secreting cells is based upon the application of a stepwise differentiation protocol that ultimately leads to mature pancreatic organoids that are able to secrete insulin after multiple glucose stimulations in vitro. This protocol was first published in 2014 [49] and used hESCs as the primary cell type for the differentiation process. Over the past 5 years several other articles have been published to improve on their protocol and to apply it to human-induced pluripotent stem cells derived from patients [156]. To prevent immune rejection researchers have created a coating for their stem cell-derived (SC)-β cells and encapsulate them with alginate derivatives capable of reducing foreign body responses in vivo [157]. Vegas's group implanted these encapsulated SC-β cells into the intraperitoneal (IP) space of streptozotocin-treated (STZ) C57BL/6J mice showing effective glucose control over a period of 174 days posttransplantation.

Further adjustments of the differentiation protocols have improved the specificity to obtain β-like cells in vitro with a high rate of mono-hormonal, insulin-responsive β-like cells [158–165]. One of the most important improvements consisted in culturing a large amount of human pluripotent stem cells into 3D clusters, called human embryoid bodies (hEBs), as they are meant to recapitulate what happens in embryos in vivo. These hEBs are then cultured in a stepwise manner, by the administration of specific growth factors and small molecules that are able to activate or repress different pathways important for the pancreatic differentiation in vivo, such as Nodal, WNT, RA, FGF, bone morphogenetic protein (BMP), and Notch. Many of these protocols are composed of 5–7 differentiation stages and take from 20 to 35 days to obtain β-like cells from human pluripotent stem cells. These cells keep their identity for a long time in vitro, making them a valuable tool for drug discovery and cell transplantation.

In most of these protocols, the first two to three stages of differentiation produce a uniform population of pancreatic progenitor cells, characterized by the expression of the master gene for pancreatic differentiation PDX-1 (also known as Ipf-1 in human). Following these steps, C-peptide positive cells will appear together with other pancreatic markers such as NKX-6.1. The overall pancreatic organoids express beta cell markers (insulin and NKX-6.1), alpha cells that express glucagon and nonpancreatic cells, that although have endocrine characteristics, they resemble intestinal endocrine cells. A percentage of cells constituting the pancreatic organoids can be referred to as nonendocrine cells, and these cells can be produced by the dissociation and reaggregation of these pancreatic organoids [161,165].

The advances that have been made in the formation of countless quantities of functional insulin-secreting cells from human ESCs and iPSCs represent the most promising strategy for stem cell therapy to treat T1D. Furthermore, the possibility to produce functional β-cells in large amounts constitutes an innovative means for understanding the mechanisms of human β-cell generation and maturation. This work creates an opportunity to study diseases in vitro, including for drug and molecule screening that could affect proliferation and/or function of β-cells in identifying new targets for diabetes therapy [166–168].

Many questions remain, such as the importance of having a pancreatic organoid that possesses all the cells that are present in a natural islet of Langerhans. Beta cell physiology depends heavily on the interaction with its neighboring cells, such as alpha, delta, PP, and ghrelin cells, to finely regulate insulin secretion in response to physiological glucose stimulation [169]. *β*-cells are connected with each other via Cx36, Pnx2, multiple cadherins, N-CAM, occludin, and several claudin isoforms. *β*-cells also adhere to their basal lamina and to those of the endothelial cells of islet capillaries by a variety of integrins. *β*-cells further interact with nearby *α*-cells via Pnx1 and N-CAM. Whether a similar heterocellular interaction takes place with the somatostatin-producing *δ*-cells is not well known. There is no direct evidence for an interaction mediated by integral membrane proteins between *β*-cells and either the islet ghrelin-producing *ε*-cells or the pancreatic polypeptide-producing PP-cells [170]. One very important question remains: must all these interactions be reproduced in artificial organoids to maintain lifelong function once transplanted in TD1 patients or are there other potential ways of doing this?

A next step that will surely allow the advancement of this field and bring the stem cell therapy to the bedside will be the use of hypoimmunogenic hiPSCs that can be obtained by the depletion of HLA epitopes to make these cells stealth to the host immune system. CRISPR technology and DNA recombination methods allow the production of such cells that are currently used by several groups in order to obtain patient-friendly organoids [171–176]. The generation of pancreatic organoids using these immune elusive cells will allow us to establish an organoid bank which will serve to supply the high demand of human organ transplantation.

Another approach to obtain pancreatic organoids that could be suitable for low-immunogenic cell therapy transplantation would be the patient-derived induced pluripotent stem cells, which allow to obtain cells that possess the same HLA system of the patient from which they are derived [156,177].

Final considerations

Amelioration in the ability of generating functional differentiated β-cells rely based upon the knowledge of the extracellular signaling that rule the cell fate during embryonic and postnatal development. We still have a lack of understanding of the cues necessary to completely master ESC differentiation to push it toward comprehensive induction of all the endocrine cells that comprise an islet cell. Typically a differentiated pancreatic organoid could contain around 40%–60% endocrine cells that are normally present in a naturally developed islet, while the remaining cells in the organoid consist of immature cells that may not reach terminal differentiation plus additional unwanted nonpancreatic endocrine cells. To obviate to this undesired off target effect to the cells within the pancreatic organoids, some strategies have been used, such as β- and α-cell enrichment using FACS and magnetic bead sorting. However, the overall ability to produce β-like cell is remarkably impaired during the sorting process as a high cell loss occurs [165]. To obtain pancreatic organoids that are functionally as close as possible to human islets, it would be essential to control the ratio between β and non-β endocrine cells in the organoids. Differentiation protocols that are currently used are focused on obtaining a higher number of β-like cells respect α-and δ-like cells. It is still unclear if having a mono-hormonal organoid that produce only insulin will lead to an adequate amount of insulin secretion in response of a glucose stimulation, or if the other endocrine cells are needed to reach such of physiological response from differentiated pancreatic organoids. Although in vitro differentiated β-cells resemble those in vivo, however, they miss the expression of some maturation markers, such as UCN3, MAFA, and SIX3 that are observed in mature human β-cell [178–180]. This incapacity to reach a similar result in vitro is probably due to the missing information about the postnatal maturation factors that occurs in vivo. One example is given by the extreme nutritional and metabolic modifications that happen after birth to which the β-cell are subject and that we cannot recapitulate in vitro. Of note is that, when some progenitor pancreatic cells created in vitro are transplanted, these cells will acquire mature features only after transplantation, validating this hypothesis. While mature β-cells secrete insulin in response to glucose intake and other factors in a glucose load-dependent manner, in contrast, fetal β-cells, as well as in vitro differentiated β-like cells, are not responsive to glucose, therefore their insulin release is not adequate to the amount of glucose stimulation. One reason of such behavior of these in vitro differentiated β-cells could be the absence of glucose sensors, such as GLUT2, which impair their ability to secrete insulin appropriately; instead, as it happens during the embryonic development, these cells release a continuous amount of insulin constantly. A study that was recently published showed that the difference between fetal β-cells and adult β-cells resides in the way these cells react to glucose metabolism, mitochondrial activity, and the way they sense glucose stimulus [181].

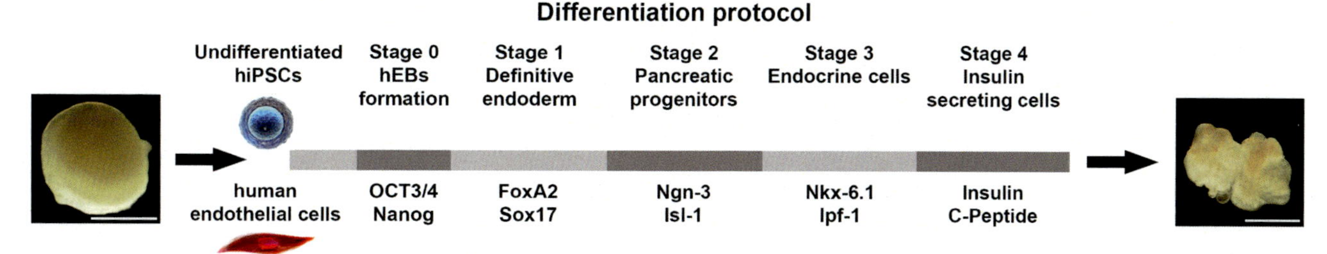

FIG. 2 Schematic representation of our pancreatic differentiation protocol showing the various differentiation steps. Embryoid body formation using hiPSCs interlaced with human endothelial cells in a specific ratio, is a key start point of our differentiation protocol, which ultimately lead to a pancreatic organoid composed of various functional cells. Scale bars 500 μm.

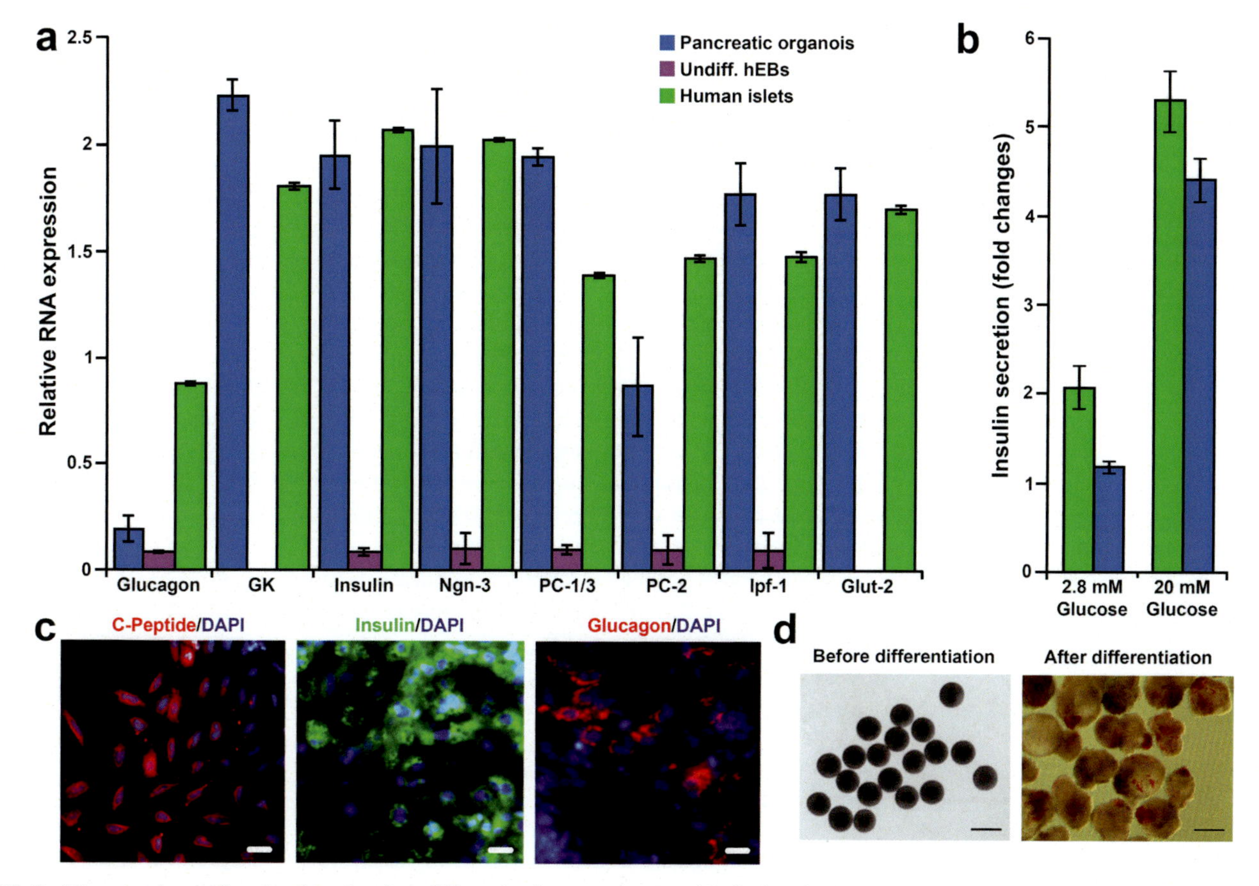

FIG. 3 Biomolecular and functional results of our differentiated pancreatic organoids displayed (A) gene expression of the main pancreatic markers; (B) insulin release in response to glucose stimulation; (C) staining positive for three important pancreatic hormones such as C-peptide, insulin, and glucagon (scale bar 10 μm); and (D) staining positive for the insulin chelating agent dithizone (scale bar 500 μm).

In this regard, our group has been working on developing a pancreatic differentiation protocol, following embryogenetic cues that occur in in vivo development, therefore using key factors that are able to lead to the generation of appropriate progenitor cells that ultimately will lead to mature insulin-secreting cells (Fig. 2). Our work is centered on obtaining pancreatic organoids that possess all the features of a mini-pancreas, so that once transplanted would be sufficient to subsidize the lack of function that the host pancreas is displaying.

Functional analysis of our pancreatic organoids demonstrated, together with other results, that they can express pancreatic gene markers (Fig. 3A), respond to glucose stimulation in vitro (Fig. 3B), produce key pancreatic proteins (Fig. 3C), and display a positivity for the zinc insulin chelating agent Dithizone (Fig. 3D).

Another important obstacle that needs to be addressed to be able to use these stem cell-derived pancreatic organoids in stem cell therapy is protecting them from immune attack. One way to do so is by encapsulating them in alginate microspheres or macro devices as described before [182]. Although these methods proven to be effective in controlling blood

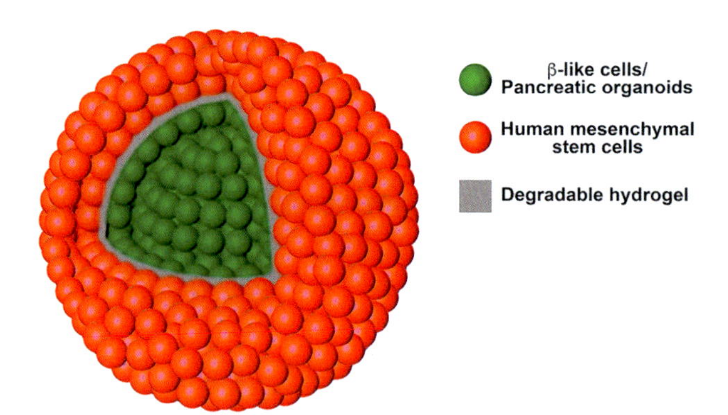

FIG. 4 hMSC biocoating of pancreatic organoids after differentiation. Schematic representation of how the biocoating with hMSCs would be obtained. The pancreatic organoid would be first coated with a biodegradable hydrogel and then coated with different thicknesses of hMSCs.

glucose in diabetic animal models for a long period of time [153,155,157,183–187], however, an hurdle that all of them have to overcome is the ability to provide acceptable access to nutrients and oxygen yet avoiding being in direct contact with the blood stream. An alternative to mechanical encapsulation methods could derive from using a coating of naturally immune evasive cells, such as human mesenchymal stem cells (hMSC), which could immune modulate the host immune system by masking the foreign organoids (Fig. 2). These cells can be derived from the same patient; therefore, they could be the ideal candidate for this purpose. Our group is currently working in this direction to create a viable biocoating that would mitigate the immune response after organoids transplantation, much the same way as the studies into biocoating of various forms of islet cells. Fig. 4 shows a schematic representation of the biocoating strategy to obtain different thickness of MSC layer and test them in vivo for their efficiency in modulating the host immune system.

Another tactic to prevent immune rejection of transplanted pancreatic organoids is the use of engineered cells, either hiPSC or cells involved in regulating the immune response toward foreigners bodies, such as T-reg cells [188]. By modulating the HLA system of these cells, one can elude the host immune system and create a universal donor cell that can be received by any patient potentially without the need of immune suppression. To further protect the patient to unwanted events, such as the nidation or mutation of these cells that could become harmful, an inducible suicide genes system may provide safety against this unfortunate event [189].

By using the strategies that have been shown here, it will be possible to ameliorate the outcome of pancreatic organoids transplantation, allowing for the possibility to have an unlimited source of "organoids on demand" to be able to treat all the diabetic patients on our current transplant waitlists without the need for donor organs.

References

[1] Watt FM, Hogan BL. Out of Eden: stem cells and their niches. Science 2000;287(5457):1427–30.

[2] Martin GR. Isolation of a pluripotent cell line from early mouse embryos cultured in medium conditioned by teratocarcinoma stem cells. Proc Natl Acad Sci U S A 1981;78(12):7634–8.

[3] Evans MJ, Kaufman MH. Establishment in culture of pluripotential cells from mouse embryos. Nature 1981;292(5819):154–6.

[4] Thomson JA, Itskovitz-Eldor J, Shapiro SS, Waknitz MA, Swiergiel JJ, Marshall VS, Jones JM. Embryonic stem cell lines derived from human blastocysts. Science 1998;282(5391):1145–7.

[5] Amit M, Carpenter MK, Inokuma MS, Chiu CP, Harris CP, Waknitz MA, Itskovitz-Eldor J, Thomson JA. Clonally derived human embryonic stem cell lines maintain pluripotency and proliferative potential for prolonged periods of culture. Dev Biol 2000;227(2):271–8.

[6] Matsui Y, Zsebo K, Hogan BL. Derivation of pluripotential embryonic stem cells from murine primordial germ cells in culture. Cell 1992;70(5):841–7.

[7] Resnick JL, Bixler LS, Cheng L, Donovan PJ. Long-term proliferation of mouse primordial germ cells in culture. Nature 1992;359(6395):550–1.

[8] Shamblott MJ, Axelman J, Wang S, Bugg EM, Littlefield JW, Donovan PJ, Blumenthal PD, Huggins GR, Gearhart JD. Derivation of pluripotent stem cells from cultured human primordial germ cells. Proc Natl Acad Sci U S A 1998;95(23):13726–31.

[9] Shamblott MJ, Axelman J, Littlefield JW, Blumenthal PD, Huggins GR, Cui Y, Cheng L, Gearhart JD. Human embryonic germ cell derivatives express a broad range of developmentally distinct markers and proliferate extensively in vitro. Proc Natl Acad Sci U S A 2001;98(1):113–8.

[10] Kahan BW, Ephrussi B. Developmental potentialities of clonal in vitro cultures of mouse testicular teratoma. J Natl Cancer Inst 1970;44(5):1015–36.

[11] Hogan B, Fellous M, Avner P, Jacob F. Isolation of a human teratoma cell line which expresses F9 antigen. Nature 1977;270(5637):515–8.

[12] Soria B, Skoudy A, Martín F. From stem cells to beta cells: new strategies in cell therapy of diabetes mellitus. Diabetologia 2001;44(4):407–15.

[13] Cossu G, Bianco P. Mesoangioblasts—vascular progenitors for extravascular mesodermal tissues. Curr Opin Genet Dev 2003;13(5):537–42.

[14] Toma JG, Akhavan M, Fernandes KJ, Barnabé-Heider F, Sadikot A, Kaplan DR, Miller FD. Isolation of multipotent adult stem cells from the dermis of mammalian skin. Nat Cell Biol 2001;3(9):778–84.

[15] Ruhnke M, Ungefroren H, Nussler A, Martin F, Brulport M, Schormann W, Hengstler JG, Klapper W, Ulrichs K, Hutchinson JA, Soria B, Parwaresch RM, Heeckt P, Kremer B, Fändrich F. Differentiation of in vitro-modified human peripheral blood monocytes into hepatocyte-like and pancreatic islet-like cells. Gastroenterology 2005;128(7):1774–86.

[16] Ying QL, Nichols J, Evans EP, Smith AG. Changing potency by spontaneous fusion. Nature 2002;416(6880):545–8.

[17] Terada N, Hamazaki T, Oka M, Hoki M, Mastalerz DM, Nakano Y, Meyer EM, Morel L, Petersen BE, Scott EW. Bone marrow cells adopt the phenotype of other cells by spontaneous cell fusion. Nature 2002;416(6880):542–5.

[18] Alvarez-Dolado M, Pardal R, Garcia-Verdugo JM, Fike JR, Lee HO, Pfeffer K, Lois C, Morrison SJ, Alvarez-Buylla A. Fusion of bone-marrow-derived cells with Purkinje neurons, cardiomyocytes and hepatocytes. Nature 2003;425(6961):968–73.

[19] Ianus A, Holz GG, Theise ND, Hussain MA. In vivo derivation of glucose-competent pancreatic endocrine cells from bone marrow without evidence of cell fusion. J Clin Invest 2003;111(6):843–50.

[20] Smith AG, Heath JK, Donaldson DD, Wong GG, Moreau J, Stahl M, Rogers D. Inhibition of pluripotential embryonic stem cell differentiation by purified polypeptides. Nature 1988;336(6200):688–90.

[21] Doetschman T, Williams P, Maeda N. Establishment of hamster blastocyst-derived embryonic stem (ES) cells. Dev Biol 1988;127(1):224–7.

[22] Iannaccone PM, Taborn GU, Garton RL, Caplice MD, Brenin DR. Pluripotent embryonic stem cells from the rat are capable of producing chimeras. Dev Biol 1994;163(1):288–92.

[23] Graves KH, Moreadith RW. Derivation and characterization of putative pluripotential embryonic stem cells from preimplantation rabbit embryos. Mol Reprod Dev 1993;36(4):424–33.

[24] Pain B, Clark ME, Shen M, Nakazawa H, Sakurai M, Samarut J, Etches RJ. Long-term in vitro culture and characterisation of avian embryonic stem cells with multiple morphogenetic potentialities. Development 1996;122(8):2339–48.

[25] Sukoyan MA, Vatolin SY, Golubitsa AN, Zhelezova AI, Semenova LA, Serov OL. Embryonic stem cells derived from morulae, inner cell mass, and blastocysts of mink: comparisons of their pluripotencies. Mol Reprod Dev 1993;36(2):148–58.

[26] Wheeler MB. Development and validation of swine embryonic stem cells: a review. Reprod Fertil Dev 1994;6(5):563–8.

[27] Thomson JA, Marshall VS. Primate embryonic stem cells. Curr Top Dev Biol 1998;38:133–65.

[28] Thomson JA, Kalishman J, Golos TG, Durning M, Harris CP, Hearn JP. Pluripotent cell lines derived from common marmoset (Callithrix jacchus) blastocysts. Biol Reprod 1996;55(2):254–9.

[29] Richards M, Fong CY, Chan WK, Wong PC, Bongso A. Human feeders support prolonged undifferentiated growth of human inner cell masses and embryonic stem cells. Nat Biotechnol 2002;20(9):933–6.

[30] Hovatta O, Mikkola M, Gertow K, Strömberg AM, Inzunza J, Hreinsson J, Rozell B, Blennow E, Andäng M, Ahrlund-Richter L. A culture system using human foreskin fibroblasts as feeder cells allows production of human embryonic stem cells. Hum Reprod 2003;18(7):1404–9.

[31] Mitalipova M, Calhoun J, Shin S, Wininger D, Schulz T, Noggle S, Venable A, Lyons I, Robins A, Stice S. Human embryonic stem cell lines derived from discarded embryos. Stem Cells 2003;21(5):521–6.

[32] Laslett AL, Filipczyk AA, Pera MF. Characterization and culture of human embryonic stem cells. Trends Cardiovasc Med 2003;13(7):295–301.

[33] Reubinoff BE, Pera MF, Fong CY, Trounson A, Bongso A. Embryonic stem cell lines from human blastocysts: somatic differentiation in vitro. Nat Biotechnol 2000;18(4):399–404.

[34] Xu C, Inokuma MS, Denham J, Golds K, Kundu P, Gold JD, Carpenter MK. Feeder-free growth of undifferentiated human embryonic stem cells. Nat Biotechnol 2001;19(10):971–4.

[35] Bradley A, Evans M, Kaufman MH, Robertson E. Formation of germ-line chimaeras from embryo-derived teratocarcinoma cell lines. Nature 1984;309(5965):255–6.

[36] Schwartzberg PL, Goff SP, Robertson EJ. Germ-line transmission of a c-abl mutation produced by targeted gene disruption in ES cells. Science 1989;246(4931):799–803.

[37] Ivanova NB, Dimos JT, Schaniel C, Hackney JA, Moore KA, Lemischka IR. A stem cell molecular signature. Science 2002;298(5593):601–4.

[38] Abe K, Niwa H, Iwase K, Takiguchi M, Mori M, Abé SI, Abe K, Yamamura KI. Endoderm-specific gene expression in embryonic stem cells differentiated to embryoid bodies. Exp Cell Res 1996;229(1):27–34.

[39] Leahy A, Xiong JW, Kuhnert F, Stuhlmann H. Use of developmental marker genes to define temporal and spatial patterns of differentiation during embryoid body formation. J Exp Zool 1999;284(1):67–81.

[40] Itskovitz-Eldor J, Schuldiner M, Karsenti D, Eden A, Yanuka O, Amit M, Soreq H, Benvenisty N. Differentiation of human embryonic stem cells into embryoid bodies compromising the three embryonic germ layers. Mol Med 2000;6(2):88–95.

[41] Pettinato G, Vanden Berg-Foels WS, Zhang N, Wen X. ROCK inhibitor is not required for embryoid body formation from singularized human embryonic stem cells. PLoS One 2014;9(11), e100742.

[42] Pettinato G, Wen X, Zhang N. Formation of well-defined embryoid bodies from dissociated human induced pluripotent stem cells using microfabricated cell-repellent microwell arrays. Sci Rep 2014;4:7402.

[43] Pettinato G, Wen X, Zhang N. Engineering strategies for the formation of embryoid bodies from human pluripotent stem cells. Stem Cells Dev 2015;24(14):1595–609.

[44] Keller GM. In vitro differentiation of embryonic stem cells. Curr Opin Cell Biol 1995;7(6):862–9.

[45] Wiles MV, Keller G. Multiple hematopoietic lineages develop from embryonic stem (ES) cells in culture. Development 1991;111(2):259–67.

[46] Wobus AM, Wallukat G, Hescheler J. Pluripotent mouse embryonic stem cells are able to differentiate into cardiomyocytes expressing chronotropic responses to adrenergic and cholinergic agents and Ca2+ channel blockers. Differentiation 1991;48(3):173–82.

[47] Boheler KR, Czyz J, Tweedie D, Yang HT, Anisimov SV, Wobus AM. Differentiation of pluripotent embryonic stem cells into cardiomyocytes. Circ Res 2002;91(3):189–201.

[48] Ng ES, Davis R, Stanley EG, Elefanty AG. A protocol describing the use of a recombinant protein-based, animal product-free medium (APEL) for human embryonic stem cell differentiation as spin embryoid bodies. Nat Protoc 2008;3(5):768–76.

[49] Pagliuca FW, Millman JR, Gürtler M, Segel M, Van Dervort A, Ryu JH, Peterson QP, Greiner D, Melton DA. Generation of functional human pancreatic β cells in vitro. Cell 2014;159(2):428–39.

[50] Mansour AA, Gonçalves JT, Bloyd CW, Li H, Fernandes S, Quang D, Johnston S, Parylak SL, Jin X, Gage FH. An in vivo model of functional and vascularized human brain organoids. Nat Biotechnol 2018;36(5):432–41.

[51] Mummery C, Ward-van Oostwaard D, Doevendans P, Spijker R, van den Brink S, Hassink R, van der Heyden M, Opthof T, Pera M, de la Riviere AB, Passier R, Tertoolen L. Differentiation of human embryonic stem cells to cardiomyocytes: role of coculture with visceral endoderm-like cells. Circulation 2003;107(21):2733–40.

[52] Huber I, Itzhaki I, Caspi O, Arbel G, Tzukerman M, Gepstein A, Habib M, Yankelson L, Kehat I, Gepstein L. Identification and selection of cardiomyocytes during human embryonic stem cell differentiation. FASEB J 2007;21(10):2551–63.

[53] Niebruegge S, Bauwens CL, Peerani R, Thavandiran N, Masse S, Sevaptisidis E, Nanthakumar K, Woodhouse K, Husain M, Kumacheva E, Zandstra PW. Generation of human embryonic stem cell-derived mesoderm and cardiac cells using size-specified aggregates in an oxygen-controlled bioreactor. Biotechnol Bioeng 2009;102(2):493–507.

[54] Oldershaw RA, Baxter MA, Lowe ET, Bates N, Grady LM, Soncin F, Brison DR, Hardingham TE, Kimber SJ. Directed differentiation of human embryonic stem cells toward chondrocytes. Nat Biotechnol 2010;28(11):1187–94.

[55] Vats A, Bielby RC, Tolley N, Dickinson SC, Boccaccini AR, Hollander AP, Bishop AE, Polak JM. Chondrogenic differentiation of human embryonic stem cells: the effect of the micro-environment. Tissue Eng 2006;12(6):1687–97.

[56] van Harmelen V, Aström G, Strömberg A, Sjölin E, Dicker A, Hovatta O, Rydén M. Differential lipolytic regulation in human embryonic stem cell-derived adipocytes. Obesity (Silver Spring) 2007;15(4):846–52.

[57] Xiong C, Xie CQ, Zhang L, Zhang J, Xu K, Fu M, Thompson WE, Yang LJ, Chen YE. Derivation of adipocytes from human embryonic stem cells. Stem Cells Dev 2005;14(6):671–5.

[58] Hannan NR, Wolvetang EJ. Adipocyte differentiation in human embryonic stem cells transduced with Oct4 shRNA lentivirus. Stem Cells Dev 2009;18(4):653–60.

[59] Huang H, Zhao X, Chen L, Xu C, Yao X, Lu Y, Dai L, Zhang M. Differentiation of human embryonic stem cells into smooth muscle cells in adherent monolayer culture. Biochem Biophys Res Commun 2006;351(2):321–7.

[60] Xie CQ, Zhang J, Villacorta L, Cui T, Huang H, Chen YE. A highly efficient method to differentiate smooth muscle cells from human embryonic stem cells. Arterioscler Thromb Vasc Biol 2007;27(12):e311–2.

[61] Ledran MH, Krassowska A, Armstrong L, Dimmick I, Renström J, Lang R, Yung S, Santibanez-Coref M, Dzierzak E, Stojkovic M, Oostendorp RA, Forrester L, Lako M. Efficient hematopoietic differentiation of human embryonic stem cells on stromal cells derived from hematopoietic niches. Cell Stem Cell 2008;3(1):85–98.

[62] Narayan AD, Chase JL, Lewis RL, Tian X, Kaufman DS, Thomson JA, Zanjani ED. Human embryonic stem cell-derived hematopoietic cells are capable of engrafting primary as well as secondary fetal sheep recipients. Blood 2006;107(5):2180–3.

[63] Levenberg S, Golub JS, Amit M, Itskovitz-Eldor J, Langer R. Endothelial cells derived from human embryonic stem cells. Proc Natl Acad Sci U S A 2002;99(7):4391–6.

[64] Sriram G, Tan JY, Islam I, Rufaihah AJ, Cao T. Efficient differentiation of human embryonic stem cells to arterial and venous endothelial cells under feeder- and serum-free conditions. Stem Cell Res Ther 2015;6:261.

[65] Lee KW, Yook JY, Son MY, Kim MJ, Koo DB, Han YM, Cho YS. Rapamycin promotes the osteoblastic differentiation of human embryonic stem cells by blocking the mTOR pathway and stimulating the BMP/Smad pathway. Stem Cells Dev 2010;4:557–68.

[66] Bielby RC, Boccaccini AR, Polak JM, Buttery LD. In vitro differentiation and in vivo mineralization of osteogenic cells derived from human embryonic stem cells. Tissue Eng 2004;10(9-10):1518–25.

[67] Kidwai FK, Liu H, Toh WS, Fu X, Jokhun DS, Movahednia MM, Li M, Zou Y, Squier CA, Phan TT, Cao T. Differentiation of human embryonic stem cells into clinically amenable keratinocytes in an autogenic environment. J Invest Dermatol 2013;133(3):618–28.

[68] Movahednia MM, Kidwai FK, Zou Y, Tong HJ, Liu X, Islam I, Toh WS, Raghunath M, Cao T. Differential effects of the extracellular microenvironment on human embryonic stem cell differentiation into keratinocytes and their subsequent replicative life span. Tissue Eng Part A 2015;21(7-8):1432–43.

[69] Ronaghi M, Nasr M, Ealy M, Durruthy-Durruthy R, Waldhaus J, Diaz GH, Joubert LM, Oshima K, Heller S. Inner ear hair cell-like cells from human embryonic stem cells. Stem Cells Dev 2014;23(11):1275–84.

[70] Ding J, Tang Z, Chen J, Shi H, Chen J, Wang C, Zhang C, Li L, Chen P, Wang J. Induction of differentiation of human embryonic stem cells into functional hair-cell-like cells in the absence of stromal cells. Int J Biochem Cell Biol 2016;81(Pt A):208–22.

[71] Klimanskaya I, Hipp J, Rezai KA, West M, Atala A, Lanza R. Derivation and comparative assessment of retinal pigment epithelium from human embryonic stem cells using transcriptomics. Cloning Stem Cells 2004;6(3):217–45.

[72] Shi F, Corrales CE, Liberman MC, Edge AS. BMP4 induction of sensory neurons from human embryonic stem cells and reinnervation of sensory epithelium. Eur J Neurosci 2007;26(11):3016–23.

[73] Samadikuchaksaraei A, Cohen S, Isaac K, Rippon HJ, Polak JM, Bielby RC, Bishop AE. Derivation of distal airway epithelium from human embryonic stem cells. Tissue Eng 2006;12(4):867–75.

[74] Wang D, Haviland DL, Burns AR, Zsigmond E, Wetsel RA. A pure population of lung alveolar epithelial type II cells derived from human embryonic stem cells. Proc Natl Acad Sci U S A 2007;104(11):4449–54.

[75] Hay DC, Zhao D, Ross A, Mandalam R, Lebkowski J, Cui W. Direct differentiation of human embryonic stem cells to hepatocyte-like cells exhibiting functional activities. Cloning Stem Cells 2007;9(1):51–62.

[76] Basma H, Soto-Gutiérrez A, Yannam GR, Liu L, Ito R, Yamamoto T, Ellis E, Carson SD, Sato S, Chen Y, Muirhead D, Navarro-Alvarez N, Wong RJ, Roy-Chowdhury J, Platt JL, Mercer DF, Miller JD, Strom SC, Kobayashi N, Fox IJ. Differentiation and transplantation of human embryonic stem cell-derived hepatocytes. Gastroenterology 2009;136(3):990–9.
[77] Li Q, Hutchins AP, Chen Y, Li S, Shan Y, Liao B, Zheng D, Shi X, Li Y, Chan WY, Pan G, Wei S, Shu X, Pei D. A sequential EMT-MET mechanism drives the differentiation of human embryonic stem cells towards hepatocytes. Nat Commun 2017;8:15166.
[78] D'Amour KA, Bang AG, Eliazer S, Kelly OG, Agulnick AD, Smart NG, Moorman MA, Kroon E, Carpenter MK, Baetge EE. Production of pancreatic hormone-expressing endocrine cells from human embryonic stem cells. Nat Biotechnol 2006;24(11):1392–401.
[79] Jiang W, Shi Y, Zhao D, Chen S, Yong J, Zhang J, Qing T, Sun X, Zhang P, Ding M, Li D, Deng H. In vitro derivation of functional insulin-producing cells from human embryonic stem cells. Cell Res 2007;17(4):333–44.
[80] Shim JH, Kim SE, Woo DH, Kim SK, Oh CH, McKay R, Kim JH. Directed differentiation of human embryonic stem cells towards a pancreatic cell fate. Diabetologia 2007;50(6):1228–38.
[81] Jiang J, Au M, Lu K, Eshpeter A, Korbutt G, Fisk G, Majumdar AS. Generation of insulin-producing islet-like clusters from human embryonic stem cells. Stem Cells 2007;25(8):1940–53.
[82] Lee AS, Tang C, Cao F, Xie X, van der Bogt K, Hwang A, Connolly AJ, Robbins RC, Wu JC. Effects of cell number on teratoma formation by human embryonic stem cells. Cell Cycle 2009;8(16):2608–12.
[83] Rong Z, Fu X, Wang M, Xu Y. A scalable approach to prevent teratoma formation of human embryonic stem cells. J Biol Chem 2012;287(39):32338–45.
[84] Takahashi K, Tanabe K, Ohnuki M, Narita M, Ichisaka T, Tomoda K, Yamanaka S. Induction of pluripotent stem cells from adult human fibroblasts by defined factors. Cell 2007;131(5):861–72.
[85] Yu J, Vodyanik MA, Smuga-Otto K, Antosiewicz-Bourget J, Frane JL, Tian S, Nie J, Jonsdottir GA, Ruotti V, Stewart R, Slukvin II, Thomson JA. Induced pluripotent stem cell lines derived from human somatic cells. Science 2007;318(5858):1917–20.
[86] Herrera PL, Huarte J, Sanvito F, Meda P, Orci L, Vassalli JD. Embryogenesis of the murine endocrine pancreas; early expression of pancreatic polypeptide gene. Development 1991;113(4):1257–65.
[87] Boschero AC, Crepaldi SC, Carneiro EM, Delattre E, Atwater I. Prolactin induces maturation of glucose sensing mechanisms in cultured neonatal rat islets. Endocrinology 1993;133(2):515–20.
[88] Wessells NK, Cohen JH. The influence of collagen and embryo extract on the development of pancreatic epithelium. Exp Cell Res 1966;43(3):680–4.
[89] Hebrok M, Kim SK, St Jacques B, McMahon AP, Melton DA. Regulation of pancreas development by hedgehog signaling. Development 2000;127(22):4905–13.
[90] Kumar M, Jordan N, Melton D, Grapin-Botton A. Signals from lateral plate mesoderm instruct endoderm toward a pancreatic fate. Dev Biol 2003;259(1):109–22.
[91] Hui H, Perfetti R. Pancreas duodenum homeobox-1 regulates pancreas development during embryogenesis and islet cell function in adulthood. Eur J Endocrinol 2002;146(2):129–41.
[92] Itkin-Ansari P, Demeterco C, Bossie S, de la Tour DD, Beattie GM, Movassat J, Mally MI, Hayek A, Levine F. PDX-1 and cell-cell contact act in synergy to promote delta-cell development in a human pancreatic endocrine precursor cell line. Mol Endocrinol 2000;14(6):814–22.
[93] Offield MF, Jetton TL, Labosky PA, Ray M, Stein RW, Magnuson MA, Hogan BL, Wright CV. PDX-1 is required for pancreatic outgrowth and differentiation of the rostral duodenum. Development 1996;122(3):983–95.
[94] Shih DQ, Heimesaat M, Kuwajima S, Stein R, Wright CV, Stoffel M. Profound defects in pancreatic beta-cell function in mice with combined heterozygous mutations in Pdx-1, Hnf-1alpha, and Hnf-3beta. Proc Natl Acad Sci U S A 2002;99(6):3818–23.
[95] Li H, Edlund H. Persistent expression of Hlxb9 in the pancreatic epithelium impairs pancreatic development. Dev Biol 2001;240(1):247–53.
[96] Li H, Arber S, Jessell TM, Edlund H. Selective agenesis of the dorsal pancreas in mice lacking homeobox gene Hlxb9. Nat Genet 1999;23(1):67–70.
[97] Ahlgren U, Pfaff SL, Jessell TM, Edlund T, Edlund H. Independent requirement for ISL1 in formation of pancreatic mesenchyme and islet cells. Nature 1997;385(6613):257–60.
[98] Jacquemin P, Durviaux SM, Jensen J, Godfraind C, Gradwohl G, Guillemot F, Madsen OD, Carmeliet P, Dewerchin M, Collen D, Rousseau GG, Lemaigre FP. Transcription factor hepatocyte nuclear factor 6 regulates pancreatic endocrine cell differentiation and controls expression of the proendocrine gene ngn3. Mol Cell Biol 2000;20(12):4445–54.
[99] Servitja JM, Ferrer J. Transcriptional networks controlling pancreatic development and beta cell function. Diabetologia 2004;47(4):597–613.
[100] Schwitzgebel VM, Scheel DW, Conners JR, Kalamaras J, Lee JE, Anderson DJ, Sussel L, Johnson JD, German MS. Expression of neurogenin3 reveals an islet cell precursor population in the pancreas. Development 2000;127(16):3533–42.
[101] Dohrmann C, Gruss P, Lemaire L. Pax genes and the differentiation of hormone-producing endocrine cells in the pancreas. Mech Dev 2000;92(1):47–54.
[102] Sosa-Pineda B, Chowdhury K, Torres M, Oliver G, Gruss P. The Pax4 gene is essential for differentiation of insulin-producing beta cells in the mammalian pancreas. Nature 1997;386(6623):399–402.
[103] Sussel L, Kalamaras J, Hartigan-O'Connor DJ, Meneses JJ, Pedersen RA, Rubenstein JL, German MS. Mice lacking the homeodomain transcription factor Nkx2.2 have diabetes due to arrested differentiation of pancreatic beta cells. Development 1998;125(12):2213–21.
[104] Sander M, Sussel L, Conners J, Scheel D, Kalamaras J, Dela-Cruz F, Schwitzgebel V, Hayes-Jordan A, German M. Homeobox gene Nkx6.1 lies downstream of Nkx2.2 in the major pathway of beta-cell formation in the pancreas. Development 2000;127(24):5533–40.
[105] Boj SF, Parrizas M, Maestro MA, Ferrer J. A transcription factor regulatory circuit in differentiated pancreatic cells. Proc Natl Acad Sci U S A 2001;98(25):14481–6.
[106] Dukes ID, Sreenan S, Roe MW, Levisetti M, Zhou YP, Ostrega D, Bell GI, Pontoglio M, Yaniv M, Philipson L, Polonsky KS. Defective pancreatic beta-cell glycolytic signaling in hepatocyte nuclear factor-1alpha-deficient mice. J Biol Chem 1998;273(38):24457–64.

[107] Sund NJ, Vatamaniuk MZ, Casey M, Ang SL, Magnuson MA, Stoffers DA, Matschinsky FM, Kaestner KH. Tissue-specific deletion of Foxa2 in pancreatic beta cells results in hyperinsulinemic hypoglycemia. Genes Dev 2001;15(13):1706–15.
[108] Jonsson J, Carlsson L, Edlund T, Edlund H. Insulin-promoter-factor 1 is required for pancreas development in mice. Nature 1994;371(6498):606–9.
[109] Ahlgren U, Jonsson J, Edlund H. The morphogenesis of the pancreatic mesenchyme is uncoupled from that of the pancreatic epithelium in IPF1/PDX1-deficient mice. Development 1996;122(5):1409–16.
[110] Ahlgren U, Jonsson J, Jonsson L, Simu K, Edlund H. Beta-cell-specific inactivation of the mouse Ipf1/Pdx1 gene results in loss of the beta-cell phenotype and maturity onset diabetes. Genes Dev 1998;12(12):1763–8.
[111] Harrison KA, Thaler J, Pfaff SL, Gu H, Kehrl JH. Pancreas dorsal lobe agenesis and abnormal islets of Langerhans in Hlxb9-deficient mice. Nat Genet 1999;23(1):71–5.
[112] Rausa F, Samadani U, Ye H, Lim L, Fletcher CF, Jenkins NA, Copeland NG, Costa RH. The cut-homeodomain transcriptional activator HNF-6 is coexpressed with its target gene HNF-3 beta in the developing murine liver and pancreas. Dev Biol 1997;192(2):228–46.
[113] Landry C, Clotman F, Hioki T, Oda H, Picard JJ, Lemaigre FP, Rousseau GG. HNF-6 is expressed in endoderm derivatives and nervous system of the mouse embryo and participates to the cross-regulatory network of liver-enriched transcription factors. Dev Biol 1997;192(2):247–57.
[114] Gradwohl G, Dierich A, LeMeur M, Guillemot F. Neurogenin3 is required for the development of the four endocrine cell lineages of the pancreas. Proc Natl Acad Sci U S A 2000;97(4):1607–11.
[115] Naya FJ, Huang HP, Qiu Y, Mutoh H, DeMayo FJ, Leiter AB, Tsai MJ. Diabetes, defective pancreatic morphogenesis, and abnormal enteroendocrine differentiation in BETA2/neuroD-deficient mice. Genes Dev 1997;11(18):2323–34.
[116] Sander M, Neubüser A, Kalamaras J, Ee HC, Martin GR, German MS. Genetic analysis reveals that PAX6 is required for normal transcription of pancreatic hormone genes and islet development. Genes Dev 1997;11(13):1662–73.
[117] St-Onge L, Sosa-Pineda B, Chowdhury K, Mansouri A, Gruss P. Pax6 is required for differentiation of glucagon-producing alpha-cells in mouse pancreas. Nature 1997;387(6631):406–9.
[118] Krapp A, Knöfler M, Ledermann B, Bürki K, Berney C, Zoerkler N, Hagenbüchle O, Wellauer PK. The bHLH protein PTF1-p48 is essential for the formation of the exocrine and the correct spatial organization of the endocrine pancreas. Genes Dev 1998;12(23):3752–63.
[119] Martín F, Pintor J, Rovira JM, Ripoll C, Miras-Portugal MT, Soria B. Intracellular diadenosine polyphosphates: a novel second messenger in stimulus-secretion coupling. FASEB J 1998;12(14):1499–506.
[120] Pertusa JA, Sanchez-Andres JV, Martín F, Soria B. Effects of calcium buffering on glucose-induced insulin release in mouse pancreatic islets: an approximation to the calcium sensor. J Physiol 1999;520(Pt 2(Pt 2)):473–83.
[121] Quesada I, Martin F, Roche E, Soria B. Nutrients induce different Ca(2+) signals in cytosol and nucleus in pancreatic beta-cells. Diabetes 2004;53(Suppl 1):S92–5.
[122] Rorsman P. The pancreatic beta-cell as a fuel sensor: an electrophysiologist's viewpoint. Diabetologia 1997;40(5):487–95.
[123] Martín F, Andreu E, Rovira JM, Pertusa JA, Raurell M, Ripoll C, Sanchez-Andrés JV, Montanya E, Soria B. Mechanisms of glucose hypersensitivity in beta-cells from normoglycemic, partially pancreatectomized mice. Diabetes 1999;48(10):1954–61.
[124] Khalid Q, Rahman MA. Studies on synergism between glucose and amino acids with respect to insulin release in vitro. Z Naturforsch, C: Biosci 1980;35(1-2):72–5.
[125] Opara EC, Garfinkel M, Hubbard VS, Burch WM, Akwari OE. Effect of fatty acids on insulin release: role of chain length and degree of unsaturation. Am J Phys 1994;266(4 Pt 1):E635–9.
[126] Matschinsky FM, Meglasson M, Ghosh A, Appel M, Bedoya F, Prentki M, Corkey B, Shimizu T, Berner D, Najafi H, Manning C. Biochemical design features of the pancreatic islet cell glucose-sensory system. Adv Exp Med Biol 1986;211:459–69.
[127] Malaisse WJ, Rasschaert J, Conget I, Sener A. Hexose metabolism in pancreatic islets. Regulation of aerobic glycolysis and pyruvate decarboxylation. Int J Biochem 1991;23(9):955–9.
[128] Soria B, Roche E, Berná G, León-Quinto T, Reig JA, Martín F. Insulin-secreting cells derived from embryonic stem cells normalize glycemia in streptozotocin-induced diabetic mice. Diabetes 2000;49(2):157–62.
[129] Soria B, Quesada I, Ropero AB, Pertusa JA, Martín F, Nadal A. Novel players in pancreatic islet signaling: from membrane receptors to nuclear channels. Diabetes 2004;53(Suppl 1):S86–91.
[130] Soria B. In-vitro differentiation of pancreatic beta-cells. Differentiation 2001;68(4–5):205–19.
[131] Lumelsky N, Blondel O, Laeng P, Velasco I, Ravin R, McKay R. Differentiation of embryonic stem cells to insulin-secreting structures similar to pancreatic islets. Science 2001;292(5520):1389–94.
[132] Rajagopal J, Anderson WJ, Kume S, Martinez OI, Melton DA. Insulin staining of ES cell progeny from insulin uptake. Science 2003;299(5605):363.
[133] Sipione S, Eshpeter A, Lyon JG, Korbutt GS, Bleackley RC. Insulin expressing cells from differentiated embryonic stem cells are not beta cells. Diabetologia 2004;47(3):499–508.
[134] Hansson M, Tonning A, Frandsen U, Petri A, Rajagopal J, Englund MC, Heller RS, Håkansson J, Fleckner J, Sköld HN, Melton D, Semb H, Serup P. Artifactual insulin release from differentiated embryonic stem cells. Diabetes 2004;53(10):2603–9.
[135] Hori Y, Rulifson IC, Tsai BC, Heit JJ, Cahoy JD, Kim SK. Growth inhibitors promote differentiation of insulin-producing tissue from embryonic stem cells. Proc Natl Acad Sci U S A 2002;99(25):16105–10.
[136] Blyszczuk P, Czyz J, Kania G, Wagner M, Roll U, St-Onge L, Wobus AM. Expression of Pax4 in embryonic stem cells promotes differentiation of nestin-positive progenitor and insulin-producing cells. Proc Natl Acad Sci U S A 2003;100(3):998–1003.
[137] Miyazaki S, Yamato E, Miyazaki J. Regulated expression of pdx-1 promotes in vitro differentiation of insulin-producing cells from embryonic stem cells. Diabetes 2004;53(4):1030–7.

[138] Assady S, Maor G, Amit M, Itskovitz-Eldor J, Skorecki KL, Tzukerman M. Insulin production by human embryonic stem cells. Diabetes 2001;50(8):1691–7.
[139] Xu X, Kahan B, Forgianni A, Jing P, Jacobson L, Browning V, Treff N, Odorico J. Endoderm and pancreatic islet lineage differentiation from human embryonic stem cells. Cloning Stem Cells 2006;8(2):96–107.
[140] D'Amour KA, Agulnick AD, Eliazer S, Kelly OG, Kroon E, Baetge EE. Efficient differentiation of human embryonic stem cells to definitive endoderm. Nat Biotechnol 2005;23(12):1534–41.
[141] Edlund H. Pancreas: how to get there from the gut? Curr Opin Cell Biol 1999;11(6):663–8.
[142] Edlund H. Developmental biology of the pancreas. Diabetes 2001;50(Suppl 1):S5–9.
[143] Edlund H. Factors controlling pancreatic cell differentiation and function. Diabetologia 2001;44(9):1071–9.
[144] Edlund H. Pancreatic organogenesis-developmental mechanisms and implications for therapy. Nat Rev Genet 2002;3(7):524–32.
[145] Cras-Méneur C, Elghazi L, Czernichow P, Scharfmann R. Epidermal growth factor increases undifferentiated pancreatic embryonic cells in vitro: a balance between proliferation and differentiation. Diabetes 2001;50(7):1571–9.
[146] Lau J, Kawahira H, Hebrok M. Hedgehog signaling in pancreas development and disease. Cell Mol Life Sci 2006;63(6):642–52.
[147] Mehta S, Gittes GK. Pancreatic differentiation. J Hepato-Biliary-Pancreat Surg 2005;12(3):208–17.
[148] Murtaugh LC. Pancreas and beta-cell development: from the actual to the possible. Development 2007;134(3):427–38.
[149] Kaneto H, Nakatani Y, Miyatsuka T, Matsuoka TA, Matsuhisa M, Hori M, Yamasaki Y. PDX-1/VP16 fusion protein, together with NeuroD or Ngn3, markedly induces insulin gene transcription and ameliorates glucose tolerance. Diabetes 2005;54(4):1009–22.
[150] Apelqvist A, Li H, Sommer L, Beatus P, Anderson DJ, Honjo T, Hrabe de Angelis M, Lendahl U, Edlund H. Notch signalling controls pancreatic cell differentiation. Nature 1999;400(6747):877–81.
[151] Kroon E, Martinson LA, Kadoya K, Bang AG, Kelly OG, Eliazer S, Young H, Richardson M, Smart NG, Cunningham J, Agulnick AD, D'Amour KA, Carpenter MK, Baetge EE. Pancreatic endoderm derived from human embryonic stem cells generates glucose-responsive insulin-secreting cells in vivo. Nat Biotechnol 2008;26(4):443–52.
[152] Rezania A, Bruin JE, Xu J, Narayan K, Fox JK, O'Neil JJ, Kieffer TJ. Enrichment of human embryonic stem cell-derived NKX6.1-expressing pancreatic progenitor cells accelerates the maturation of insulin-secreting cells in vivo. Stem Cells 2013;31(11):2432–42.
[153] Agulnick AD, Ambruzs DM, Moorman MA, Bhoumik A, Cesario RM, Payne JK, Kelly JR, Haakmeester C, Srijemac R, Wilson AZ, Kerr J, Frazier MA, Kroon EJ, D'Amour KA. Insulin-producing endocrine cells differentiated in vitro from human embryonic stem cells function in macroencapsulation devices in vivo. Stem Cells Transl Med 2015;4(10):1214–22.
[154] Henry RR, Pettus J, Wilensky J, Shapiro JAM, Senior PA, Roep B, Wang R, Kroon EJ, Scott M, D'amour K, Howard L. Initial clinical evaluation of VC-01TM combination product -a stem cell-derived islet replacement for type 1 diabetes (T1D). Diabetes 2018;67(Suppl 1).
[155] Motté E, Szepessy E, Suenens K, Stangé G, Bomans M, Jacobs-Tulleneers-Thevissen D, Ling Z, Kroon E, Pipeleers D. Beta cell therapy consortium EU-FP7. Composition and function of macroencapsulated human embryonic stem cell-derived implants: comparison with clinical human islet cell grafts. Am J Physiol Endocrinol Metab 2014;307(9):E838–46.
[156] Millman JR, Xie C, Van Dervort A, Gürtler M, Pagliuca FW, Melton DA. Generation of stem cell-derived β-cells from patients with type 1 diabetes. Nat Commun 2016;7:12379.
[157] Vegas AJ, Veiseh O, Gürtler M, Millman JR, Pagliuca FW, Bader AR, Doloff JC, Li J, Chen M, Olejnik K, Tam HH, Jhunjhunwala S, Langan E, Aresta-Dasilva S, Gandham S, McGarrigle JJ, Bochenek MA, Hollister-Lock J, Oberholzer J, Greiner DL, Weir GC, Melton DA, Langer R, Anderson DG. Long-term glycemic control using polymer-encapsulated human stem cell-derived beta cells in immune-competent mice. Nat Med 2016;22(3):306–11.
[158] Rezania A, Bruin JE, Arora P, Rubin A, Batushansky I, Asadi A, O'Dwyer S, Quiskamp N, Mojibian M, Albrecht T, Yang YH, Johnson JD, Kieffer TJ. Reversal of diabetes with insulin-producing cells derived in vitro from human pluripotent stem cells. Nat Biotechnol 2014;32(11):1121–33.
[159] Nostro MC, Sarangi F, Yang C, Holland A, Elefanty AG, Stanley EG, Greiner DL, Keller G. Efficient generation of NKX6-1+ pancreatic progenitors from multiple human pluripotent stem cell lines. Stem Cell Rep 2015;4(4):591–604.
[160] Russ HA, Parent AV, Ringler JJ, Hennings TG, Nair GG, Shveygert M, Guo T, Puri S, Haataja L, Cirulli V, Blelloch R, Szot GL, Arvan P, Hebrok M. Controlled induction of human pancreatic progenitors produces functional beta-like cells in vitro. EMBO J 2015;34(13):1759–72.
[161] Nair GG, Liu JS, Russ HA, Tran S, Saxton MS, Chen R, Juang C, Li ML, Nguyen VQ, Giacometti S, Puri S, Xing Y, Wang Y, Szot GL, Oberholzer J, Bhushan A, Hebrok M. Recapitulating endocrine cell clustering in culture promotes maturation of human stem-cell-derived β cells. Nat Cell Biol 2019;21(2):263–74.
[162] Rosado-Olivieri EA, Anderson K, Kenty JH, Melton DA. YAP inhibition enhances the differentiation of functional stem cell-derived insulin-producing β cells. Nat Commun 2019;10(1):1464.
[163] Sharon N, Vanderhooft J, Straubhaar J, Mueller J, Chawla R, Zhou Q, Engquist EN, Trapnell C, Gifford DK, Melton DA. Wnt signaling separates the progenitor and endocrine compartments during pancreas development. Cell Rep 2019;27(8):2281–2291.e5.
[164] Velazco-Cruz L, Song J, Maxwell KG, Goedegebuure MM, Augsornworawat P, Hogrebe NJ, Millman JR. Acquisition of dynamic function in human stem cell-derived β cells. Stem Cell Rep 2019;12(2):351–65.
[165] Veres A, Faust AL, Bushnell HL, Engquist EN, Kenty JH, Harb G, Poh YC, Sintov E, Gürtler M, Pagliuca FW, Peterson QP, Melton DA. Charting cellular identity during human in vitro β-cell differentiation. Nature 2019;569(7756):368–73.
[166] Huch M, Koo BK. Modeling mouse and human development using organoid cultures. Development 2015;142(18):3113–25.
[167] Dutta D, Heo I, Clevers H. Disease modeling in stem cell-derived 3D organoid systems. Trends Mol Med 2017;23(5):393–410.
[168] Bakhti M, Böttcher A, Lickert H. Modelling the endocrine pancreas in health and disease. Nat Rev Endocrinol 2019;15(3):155–71.

[169] Aamodt KI, Powers AC. Signals in the pancreatic islet microenvironment influence β-cell proliferation. Diabetes Obes Metab 2017;19(Suppl 1):124–36.
[170] Meda P. Protein-mediated interactions of pancreatic islet cells. Scientifica (Cairo) 2013;2013, 621249.
[171] Zhao W, Lei A, Tian L, Wang X, Correia C, Weiskittel T, Li H, Trounson A, Fu Q, Yao K, Zhang J. Strategies for genetically engineering hypoimmunogenic universal pluripotent stem cells. iScience 2020;23(6):101162.
[172] Jang Y, Choi J, Park N, Kang J, Kim M, Kim Y, Ju JH. Development of immunocompatible pluripotent stem cells via CRISPR-based human leukocyte antigen engineering. Exp Mol Med 2019;51(1):1–11.
[173] Han X, Wang M, Duan S, Franco PJ, Kenty JH, Hedrick P, Xia Y, Allen A, Ferreira LMR, Strominger JL, Melton DA, Meissner TB, Cowan CA. Generation of hypoimmunogenic human pluripotent stem cells. Proc Natl Acad Sci U S A 2019;116(21):10441–6.
[174] Shi L, Li W, Liu Y, Chen Z, Hui Y, Hao P, Xu X, Zhang S, Feng H, Zhang B, Zhou S, Li N, Xiao L, Liu L, Ma L, Zhang X. Generation of hypoimmunogenic human pluripotent stem cells via expression of membrane-bound and secreted β2m-HLA-G fusion proteins. Stem Cells 2020;38(11):1423–37.
[175] Ye Q, Sung TC, Yang JM, Ling QD, He Y, Higuchi A. Generation of universal and hypoimmunogenic human pluripotent stem cells. Cell Prolif 2020;53(12), e12946.
[176] Xu H, Wang B, Ono M, Kagita A, Fujii K, Sasakawa N, Ueda T, Gee P, Nishikawa M, Nomura M, Kitaoka F, Takahashi T, Okita K, Yoshida Y, Kaneko S, Hotta A. Targeted disruption of HLA genes via CRISPR-Cas9 generates iPSCs with enhanced immune compatibility. Cell Stem Cell 2019;24(4):566–578.e7.
[177] Millman JR, Pagliuca FW. Autologous pluripotent stem cell-derived β-like cells for diabetes cellular therapy. Diabetes 2017;66(5):1111–20.
[178] Blum B, Hrvatin S, Schuetz C, Bonal C, Rezania A, Melton DA. Functional beta-cell maturation is marked by an increased glucose threshold and by expression of urocortin 3. Nat Biotechnol 2012;30(3):261–4.
[179] Hrvatin S, O'Donnell CW, Deng F, Millman JR, Pagliuca FW, DiIorio P, Rezania A, Gifford DK, Melton DA. Differentiated human stem cells resemble fetal, not adult, β cells. Proc Natl Acad Sci U S A 2014;111(8):3038–43.
[180] Arda HE, Li L, Tsai J, Torre EA, Rosli Y, Peiris H, Spitale RC, Dai C, Gu X, Qu K, Wang P, Wang J, Grompe M, Scharfmann R, Snyder MS, Bottino R, Powers AC, Chang HY, Kim SK. Age-dependent pancreatic gene regulation reveals mechanisms governing human β cell function. Cell Metab 2016;23(5):909–20.
[181] Yoshihara E, Wei Z, Lin CS, Fang S, Ahmadian M, Kida Y, Tseng T, Dai Y, Yu RT, Liddle C, Atkins AR, Downes M, Evans RM. ERRγ is required for the metabolic maturation of therapeutically functional glucose-responsive β cells. Cell Metab 2016;23(4):622–34.
[182] Sneddon JB, Tang Q, Stock P, Bluestone JA, Roy S, Desai T, Hebrok M. Stem cell therapies for treating diabetes: progress and remaining challenges. Cell Stem Cell 2018;22(6):810–23.
[183] Robert T, De Mesmaeker I, Stangé GM, Suenens KG, Ling Z, Kroon EJ, Pipeleers DG. Functional beta cell mass from device-encapsulated hesc-derived pancreatic endoderm achieving metabolic control. Stem Cell Rep 2018;10(3):739–50.
[184] Bruin JE, Rezania A, Xu J, Narayan K, Fox JK, O'Neil JJ, Kieffer TJ. Maturation and function of human embryonic stem cell-derived pancreatic progenitors in macroencapsulation devices following transplant into mice. Diabetologia 2013;56(9):1987–98.
[185] An D, Chiu A, Flanders JA, Song W, Shou D, Lu YC, Grunnet LG, Winkel L, Ingvorsen C, Christophersen NS, Fels JJ, Sand FW, Ji Y, Qi L, Pardo Y, Luo D, Silberstein M, Fan J, Ma M. Designing a retrievable and scalable cell encapsulation device for potential treatment of type 1 diabetes. Proc Natl Acad Sci U S A 2018;115(2):E263–72.
[186] Alagpulinsa DA, Cao JJL, Driscoll RK, Sîrbulescu RF, Penson MFE, Sremac M, Engquist EN, Brauns TA, Markmann JF, Melton DA, Poznansky MC. Alginate-microencapsulation of human stem cell-derived β cells with CXCL12 prolongs their survival and function in immunocompetent mice without systemic immunosuppression. Am J Transplant 2019;19(7):1930–40.
[187] Chang R, Faleo G, Russ HA, Parent AV, Elledge SK, Bernards DA, Allen JL, Villanueva K, Hebrok M, Tang Q, Desai TA. Nanoporous immunoprotective device for stem-cell-derived β-cell replacement therapy. ACS Nano 2017;11(8):7747–57.
[188] Bluestone JA, Tang Q. Treg cells-the next frontier of cell therapy. Science 2018;362(6411):154–5.
[189] Liang Q, Monetti C, Shutova MV, Neely EJ, Hacibekiroglu S, Yang H, Kim C, Zhang P, Li C, Nagy K, Mileikovsky M, Gyongy I, Sung HK, Nagy A. Linking a cell-division gene and a suicide gene to define and improve cell therapy safety. Nature 2018;563(7733):701–4.

Chapter 12

Regeneratively speaking: Reflections on organ transplantation and beta cell replacement in the regenerative medicine era

Justine M. Aziz[a], Paul A. Grisales[a], John R. Savino[a], Lori Nicole Byers[a], Antonio Citro[b], Andrea Peloso[c], Amish Asthana[a], and Giuseppe Orlando[a]

[a]*Wake Forest University School of Medicine, Winston Salem, NC, United States,* [b]*San Raffaele Diabetes Research Institute, IRCCS San Raffaele Scientific Institute, Milan, Italy,* [c]*Department of General and Transplantation Surgery, University of Geneva Hospitals, University of Geneva, Geneva, Switzerland*

Current available technologies have not adequately addressed the two most urgent needs of modern organ transplantations, namely the identification of an inexhaustible source of organs and immunosuppression-free transplantation. Regenerative medicine (RM) provides a potential solution to these critical challenges and promises to revolutionize the way we think and implement transplant medicine. In this context, beta cell replacement represents a formidable platform where regenerative and transplant medicine meet. This chapter concludes the present book, the first of a series with the overarching goal to bring the two fields together, in order to join forces and shape our common future. The present chapter will explain how we got here and will propose a path forward in the quest for the new Holy Grail of modern transplantation, namely the bioengineering, regeneration and repair of functionally impaired organs [1].

A historical perspective

The history of organ transplantation can be split into three phases or "eras" (Fig. 1) [2]. The first phase spans from the early days of transplantation as an accepted therapy to the advent of cyclosporine and can be referred to as the surgery era. In those days, the idea that diseased organs could be replaced with viable organs procured from healthy donors opened a new era. Technical feasibility was proven but outcomes were not good enough due to the lack of effective therapies to prevent rejection. However, during that era, the foundations for organ preservation, tissue histocompatibility, and transplant immunology were laid, and the field of xenotransplantation was born. The advent of cyclosporine revolutionized the field by allowing transplantation to become the treatment modality of choice for a myriad of clinical settings characterized by irreversible, terminal organ dysfunction; this projected the field into its second phase, the immunology era. During this time, transplantation became a victim of its own success because, while results and outcomes improved spectacularly, more and more patients started being referred for transplantation, in spite of a disproportionate increase of available donors. Consequently, waiting times and mortality on the waiting list have increased dramatically. At the same time, we have realized that antirejection therapy is a double-edged sword, a friend at the beginning that—as the time from the transplant goes by—turns into a foe. In fact, while on one hand the immunosuppression has represented the "magic stick" that allowed transplantation to become the standard of care for many diseases and one of the major milestones in the history of medicine in the 20th century, on the other hand it does so at a price. The price is the myriad of side effects deriving from its long-term administration that is detrimental to the transplant outcome itself and patients' quality of life; not to mention that some patients may not tolerate the immunosuppressive regimen at all. For this reason, research has focused on strategies to achieve an immunosuppression-free status (IFS, otherwise said, tolerance) whereby the recipient accepts the allograft without mounting any immune response, while not being administered any immunosuppressant, or immunosuppression is weaned off after tolerance-inducing strategies are applied.

Pancreas and Beta Cell Replacement. https://doi.org/10.1016/B978-0-12-824011-3.00003-5

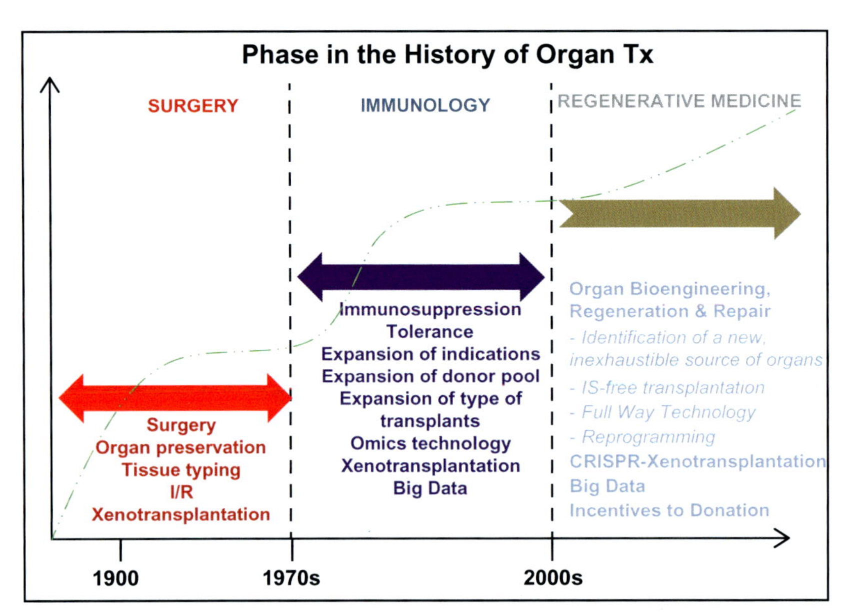

FIG. 1 In the history of organ transplantation, we can identify three eras. The first can be referred to as the surgery phase and spans from the early days to the advent of cyclosporine. The introduction of this potent immunosuppressant allowed transplantation to become a lifesaving procedure for a myriad of clinical scenarios characterized by irreversible organ failure. The second phase (immunology) spans from the advent of cyclosporine to nowadays. During that phase, we have learned how to manage antirejection medications and their impact on patient's quality of life. Importantly, given the burden of side effects that comes with lifelong immunosuppression, we have realized that we should devise strategies to minimize the immunosuppression if not withdrawing it completely sometime after the transplant. Unfortunately, tolerance remains unrealistic and cannot be proposed on a large scale, despite intense research and multiple attempts to translate promising laboratory findings into the clinic. The third phase is referred to as the RM phase. RM bears promise to revolutionize the way we think and do transplantation. *(From Orlando G, Murphy SV, Bussolati B, Clancy M, Cravedi P, Migliaccio G, Murray P. Rethinking regenerative medicine from a transplant perspective (and vice versa). Transplantation 2019;103(2):237–249, with permission.)*

In more recent years, the field has been transitioning into its third phase, the RM era. At a time that seems dominated by new immunosuppressants, big data, exchanged pair donation chains, transplants across blood groups or among incompatible donors, this transition may not be perceived by a distracted observer but it's actually occurring. Recent advances in organ bioengineering and regeneration have shown that the application of organ bioengineering and regeneration technologies to manufacture organs for transplant purposes from patients' own cells may offer the quickest route to the implementation of *organ-on-demand* and to the achievement of an IFS.

Organ transplantation: A halfway technology

Organ transplantation is a last-line, lifesaving therapy for patients with end-stage organ disease and some forms of hyperacute organ failure. However, successful transplantation requires a toll to be paid which is represented by the potential burden of side effects deriving from lifelong antirejection medications. Moreover, organ transplantation does not address the risk of recurrence of the baseline disease. The long-term management of transplant recipients is focused on reduction in morbidity and mortality, improved quality of life, while balancing side effects of immunosuppressive drugs with risk of graft failure. Halfway technologies save patients' lives but do not cure the baseline disease. Going back to the case of KT, function wise, the majority of our patients are chronic kidney disease stage 3, based on their glomerular filtration rate. In other words, for a patient to be well and a KT to be considered successful, a glomerular filtration rate (GFR) >90 mL/min/1.73 m^2 (otherwise said, normal renal function or chronic kidney disease stage 1) is not required, as long as patient is asymptomatic and conducts a normal life with no restrictions (as a matter of fact, the majority of KT recipients have a GFR ranging between 30 and 60, and therefore may be categorized as stage 3). Chronic disease management is classified as a halfway technology [3], because it does not cure the clinical condition. Organ transplantation should be considered a "halfway technology" for two reasons. First, because it does not directly target or treat the underlying disease (otherwise said, it does not eliminate the etiology of the disease, which may recur after transplantation). Second, because the lifelong immunosuppression needed to prevent rejection causes the onset of a new clinical syndrome consisting in the side effects of the different drugs patients will be on. Moreover, organ allografts have an intrinsic "expiration" time, which is estimated at 10 years for deceased donor organs and potentially 20 years for living donors.

RM may offer a solution to not only the unmet need for transplantable organs but may also render organ transplantation a "whole technology" instead of a halfway technology. This may happen in patients developing end-stage organ failure due to a disease that would not recur after the transplant, like polycystic kidney treated by KT or alcoholic liver cirrhosis treated by liver transplantation, provided—in this latter case—that the patient will remain abstinent from alcohol. If organs manufactured from patient's own cells could be used, then the patient will never need any form of antirejection medication posttransplant and therefore will never experience the burden of complications that may derive from it. Feasibility of this approach has been proven with the implantation of segments of the urinary tract or upper airways. An increasing body of evidence, however, suggests that it is only a matter of time before complex organs are bioengineered [4–9]. The length of this time will be determined by which field of health sciences will invest in the field of whole organ bioengineering. Common sense suggests that transplantation should take the lead in RM research, because—as explained in a recent state of the art manuscript—"*no field in health sciences has more interest than organ transplantation in fostering progress in RM because the future of no other field more than the future of organ transplantation will be forged by progress occurring in RM*" [2].

Regenerative medicine technologies

The most promising RM-inspired technologies currently under development to manufacture transplantable organs are succinctly presented as follows.

Decellularization

General concepts

The ECM is a three-dimensional structural framework of proteins in a state of dynamic reciprocity with the cells of any given organ. Once thought to be a simple scaffold, the ECM regulates virtually all aspects of cell biology, including development, morphology and differentiation, intracellular signaling, gene expression, adhesion and migration, proliferation, secretion, and survival. The ECM mediates these functions via physical and special cues, signaling molecules, and the secretion and storage of growth factors and cytokines, in a way that the life of multicellular organisms would not be possible without it. Therefore if we plan to build organs, we cannot do that without including a template that mimics the ECM. This is why decellularization has become of age in RM. Decellularization is a process in which an acellularized extracellular matrix scaffold (aECMs) can be obtained by using chemical or physical means to remove cellular components of living tissues [3]. The product of decellularization is a 3D ultrastructure of ECM that may be used as a natural scaffold for application in tissue engineering and RM. Ideally, the ECM of the native tissue retains both structural integrity and existing biochemical properties of the native tissue nanostructure. Studies have shown that aECMs exhibit bioinductive properties for cellular chemotaxis, including attachment, migration, proliferation, and function comparable to native tissue. These constructs have the advantage of maintaining tissue-specific cell functions and phenotypes to induce host tissue remodeling. Previous studies have had success in obtaining ECM from virtually all mammal organs, including the liver, respiratory tract, nerve, adipose, mammary gland, heart, artery, cardiac valve, kidney, small intestine submucosa, dental pulp, bone, tendon, complex structures like vascularized composite tissues (Fig. 2 [10]), urinary tract, and cochlear tissues [11]. By utilizing host tissue remodeling processes, creation of new site-specific tissue is possible. Furthermore, ECMs preserve native vasculature, allowing for adequate perfusion of constructs while being able to withstand physiologic blood pressures [12]. It has been suggested that RM technologies in the past decade have been largely unsuccessful due to inefficient vascularization techniques, resulting in inadequate perfusion of oxygen and nutrients to maintain viability of constructs. Thus native ECMs may offer better feasibility compared to other RM methods relying on embryogenesis alone from a single progenitor cell. The morbidity and mortality, however, associated with bioengineered tissues and organs remain high and the implantation in vivo of viable and functioning organoids has never been reported. Challenges of decellularization revolve around minimizing damage to innate ECM during tissue processing techniques and inadequate recellularization of structurally complex organs, including the kidney. Moving forward, optimal techniques of decellularization and recellularization warrant further study.

Beta cell replacement

Current protocols for isolating islets from the human pancreas rely on a process that involves enzymatic digestion of the organ (the so-called Ricordi's method [13]), whereby the pancreas is broken down, its connections to the body are lost, and a significant number of islets are irreversibly damaged while those that are isolated are then transplanted into a hostile environment where they can undergo further impacts such as the instant blood-mediated inflammatory reaction (IBMIR). Islet damage is determined by critical factors such as the destruction of the ECM, which represents the 3D framework of the islet niche and whose loss has been associated with the progression of type 1 diabetes [14]. Moreover, as ECM signaling to islets is critical for islet function [15–19], destruction of the islet niche contributes to the limited graft survival observed after islet transplantation when compared to whole pancreas transplantation. As strong data have shown that the incorporation

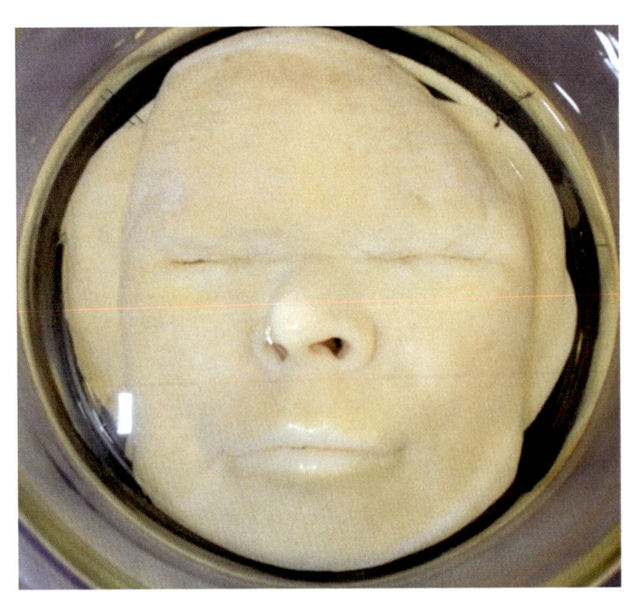

FIG. 2 A scaffold of the human face, after decellularization. *(From Duisit J, Maistriaux L, Taddeo A, Orlando G, Joris V, Coche E, Behets C, Lerut J, Dessy C, Cossu G, Vogelin E, Rieben R, Gianello P, Lengele B. Bioengineering a human face graft: the matrix of identity. Ann Surg 2017;266(5):754–764, with permission.)*

of ECM within capsules dramatically enhances islet function and allows to obtain euglycemia with a significantly lower than usual number of islets [20], recapitulating the islet niche has become a critical goal in the field of beta cell replacement bioengineering.

Our group at Wake Forest has significantly contributed to the field. Preliminary results obtained with the successful decellularization and partial recellularization of the porcine pancreas [21] led to scaling up to the human pancreas. The initial idea was to build an implantable mini pancreas obtained through the decellularization of the human pancreas. After decellularization, the organ was sectioned at the level of emergence of the splenic artery and vein which would represent the pedicle of the mini pancreas (see Fig. 3). However, in front of the evidence that current technology does not allow the regeneration of the cellular compartment of a clinically relevant organ, we momentarily put aside this project and opted for an alternative plan. In fact, acellular ECMs can be further processed to obtain gels, soluble powder, or crystal clear liquid for a myriad of possible applications (Fig. 4). Gels, for example, can either be incorporated within alginate capsules in order to recreate the islet niche and ultimately improve their lifespan and performance or can be used as islet delivery tool [22–26]. Moreover, despite the pancreas may intuitively be seen as the ideal source of ECM, other organs are also being investigated [27].

3D printing

General concepts

The advent of printable, cytocompatible materials and appropriate hardware provides an unprecedented ability to design and create 3D structures throughout which living cells and bioactive components are strategically distributed [28]. This technology offers a promising new approach for tissue engineering with the ability to fabricate specific constructs with desirable structural and mechanical properties, and to directly deposit living cells with demanding biological functions for the regenerative building of scaffolds, tissues, and organs [29]. Advantages of 3D printing include standardization and automation of construct fabrication, reliable reproducibility, increased resolution, and potential for mass production. Possibly, the most anticipated and exponential applications of 3D printing are those in tissue bioengineering, which may collectively be referred to as 3D bioprinting (3DBP). Biomaterials used for bioprinting constructs compatible with human physiology include both naturally derived polymers (e.g., alginate, gelatin, collagen, fibrin, hyaluronic acid) and synthetic polymers (e.g., polyethylene glycol [PEG]) [2]. A combination of naturally derived and synthetic polymers produces a functionally superior material that retains both the structural integrity of the synthetic polymers and the innate physiological interactions at a cellular level of the naturally derived polymers. Currently, bioprinter technology is able to fabricate multicomponent biological constructs with resolution of 2 μm for constructs with only biomaterials, and 50 μm for constructs including

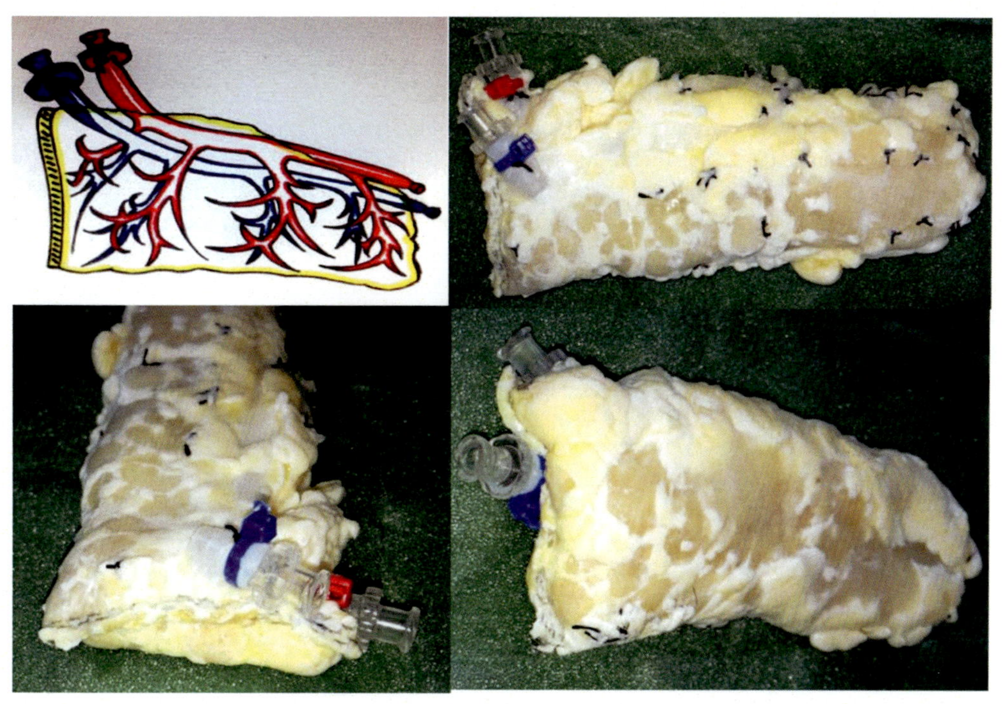

FIG. 3 ECM scaffold of a human pancreas after perfusion decellularization. The acellular scaffold was sectioned with a stapler at the level of the splenic vessels.

FIG. 4 The human pancreas can be processed to produce powder, gel, and crystal clear solution for tissue engineering applications.

encapsulated cells [30]. Successful preclinical studies have implemented 3D bioprinted cartilage and bone tissues in animal models, with examples including direct human cartilage repair and cartilage regeneration in dorsal subcutaneous spaces in mice [31–33]. Despite exponential advances in 3D bioprinting, the lack of ability for creation of viable vasculature, innervation, and lymphatic networks limits the viability of biologically printed constructs. Modern bioprinting techniques fail to produce viable tissue that exceeds one centimeters thickness, and techniques have thus far been unsuccessful to reproduce tissue types with complex vasculature. Additionally, slow printing speeds and printer resolution restrictions may also limit the size of 3D bioprinted constructs. It is theorized that current bioprinting techniques lack geometrical complexity due to insufficient usage of multipolymer constructs and multiple cell lines, which consequently extends to inadequate replication of organ vasculature leading to inadequate tissue perfusion. Overall, while bioprinting has become a large and growing subfield of RM with significant potential applications—such as in the development of skin, heart, liver, bone, and cartilage tissue [34]—challenges remain in improving printing resolution, speed and biomaterial compatibilities, as well as in identifying a broader range of suitable 3D printed materials [29].

Beta cell replacement

Despite rapid growth of the field, clinical translation has been slow and has focused on avascular, acellular scaffolds to promote bone regeneration in orthopedics, dentistry, and maxillofacial surgery; instead, the implantation or transplantation in patients of more complex soft, cellular and vascularized tissues has not been reported yet [34]. However, beta cell replacement may offer an ideal platform for the application of 3DBP technologies because, instead of a whole pancreas, what is required is a rather smaller construct that, once implanted, may warrant the same outcome of whole pancreas or (at least) islet transplantation, in terms of half-life, duration of the insulin-free status, and blood sugar control. Ideally, a 3D bioprinted endocrine pancreas to treat diabetes in humans should meet the following criteria: (1) should embed a renewable insulin-producing cell source, (2) should feature a vascular pedicle to be anastomosed to the recipient's bloodstream, (3) should have a sufficient vascularization to provide adequate nutrient and oxygen support to the cells, and (4) should feature beta cell –/islet-compatible and print-compatible bioinks that mimic the innate extracellular matrix of the native organ.

Since the first report in 2011 [35], the literature on 3DBP applied to beta cell replacement has shown feasibility in incorporating different bioink and cells (beta cells for blood sugar control, vascular cells for neoangiogenesis, neuronal cells for innervation, and mesenchymal stromal cells) in an appropriate spatial arrangement, as well as strong potential for automation, manufacturing scale up, and reproducibility. As in the case of other applications of 3DBP, major challenges remain in the ability to recapitulate an adequate vascularization of the 3D bioconstruct, in the printing resolution and fabrication time, not to mention the lack of adequately supportive bioink. Nevertheless, the immense potential of 3DBP to meet the unmet needs of beta cell replacement is well recognized, while the limited studies on bioprinting of endocrine tissues motivate the need for more research to determine the components necessary and the optimal methods for creating a printed endocrine pancreas. This is the reason why, for example, JDRF hosted in 2019 the first 3D Bioprinting Workshop for Type 1 Diabetes (https://www.jdrf.org/blog/2019/08/12/jdrf-hosts-3d-bioprinting-workshop-type-1-diabetes/, accessed on March 22, 2021) and is actively supporting projects that utilize bioprinting as an alternative RM strategy for beta cell replacement [34].

Stem cell technologies

General concepts

Utilization of embryonic stem cells (ESCs) remains one of the earliest developed forms of RM technologies. ESCs are cells found within the inner cell mass of blastocyst stage embryos that retain their ability to differentiate into all adult cell types. ESC-based modalities, however, remain limited due to potential tumorigenic risk, risk of immune rejection, and ethical dilemmas surrounding use of human embryos. Another well-described method is harvesting multipotent mesenchymal stromal cells (MSCs) from the patient's bone marrow, periosteum, periodontal ligament, adipose tissue, and umbilical cord, circumventing the ethical question of using human blastocysts. Unfortunately, this method is intrinsically limited by the evidence that these progenitor cells are no longer pluripotent [36]. Instead, MSC-based therapies are useful for both direct and indirect stimulation of endogenous repair, relying on paracrine factors to mediate this process [2]. Alternatively, data from Yamanaka's groundbreaking discovery of induced pluripotent stem cells (iPSCs) navigates these challenges by reprogramming somatic stem cells to have the same potential plasticity as ESCs [37]. Since then, iPSCs have also emerged as the cell of preference in RM because of the ethical implications which have limited ESC technology.

Beta cell replacement

Beta cell replacement offers a formidable platform for the application of stem cell-based technologies to transplant medicine. In fact, due to the fact that islets represent only a minimal part (less than 2% of the volume) of the organ (the pancreas) in which they reside, bioengineers and RM scientists are not required to recapitulate the whole organ in all its complexity. Groundbreaking work led by the groups at ViaCyte, Inc. and Harvard has shown that a new, potentially inexhaustible source of transplantable insulin-producing cells, beta cells, or islets to treat type 1 diabetes may be available soon [38–40]. In the last two decades, evidence has been provided that islet derivatives may be obtained from different progenitors, including iPS obtained from diabetic patients and pancreatic progenitors. The various protocols reported so far have reached a high differentiation efficiency, yielding organoids that show the same phenotypical features and functional behavior of native human islets, both in vitro and in vivo. Moreover, these organoids can be produced in large quantities to satisfy the demand. Despite clinical trials have not shown yet the desired efficacy [41] and many hurdles remain to overcome, the potential of this approach is immense and it is anticipated that in the near future it may replace other forms of beta cell replacement [42].

An alternative stem cell-based approach for the production of autologous islets contemplates the harnessing of iPSCs technology with blastocyst manipulation and is referred to as interspecies blastocyst complementation (IBC). Targeted removal of cells destined for development of a specific organ allows for manipulation of a host blastocyst. This permits insertion of donor iPSCs into the host blastocyst to produce a new patient-autologous organ as compensation. Previous work has shown implementation of IBC in both allogenic species and interspecies blastocysts for development of missing tissues and organs [43,44]. Overall, these concepts demonstrate the possibility of using a compatible host blastocyst, such as porcine hosts, to create organs for patients that will retain size, function, and adequate vascularization comparable to host organs. Comparatively, the IBC method holds great potential because of its ease of application and increased viability of engineered tissue in host animals as opposed to other RM-derived organ models. More work, however, is needed because IBC has yet to produce certain organ systems in animal models despite efforts to do so [45,46]. Additional concerns include potential host contamination of the autologous tissue which could cause organ rejection in patients or a reverse contamination of the host by the donor iPSCs, manifesting unknown health and ethical concerns for host animals. Low yields of chimerism and evidence of extensive human chimerism in a large livestock animal embryo, which have not yet been achieved, are also hurdles to overcome. While this approach bears potential for solid organs, it cannot be contemplated for vascularized composite tissues for obvious reasons.

Organoids

General concepts

Organoids are 3D in vitro constructs created from multiple cell types that recapitulate the intricate pattern and functionality of the original tissue and organ. Introduced by Clevers in 2009 [47], organoid technology has been developed to investigate organ development, pathogenicity, disease models, and drug discovery by recreating the tissue and organ of interest on a dish [47,48]. By creating microenvironments of the simulated organ tissue, organoids can model many medical conditions such as cancer, developmental disease, degenerative conditions, and genetic disorders. Organoids serving as disease models may also function as preclinical drug testing models to reduce animal studies required in pharmacologic research. As a corollary, organoids have the potential to bring personalized medicine into the forefront of pharmacologic research, particularly in treatment-resistant oncologic diseases. Issues with the creation of proper vascularization, however, often limit the size of organoids, in which organoids incorporating higher levels of complexity often suffer from poor nutritional diffusion resulting in poor development and cellular death. Further optimization is required to address these concerns and work to increase complexity of existing models, especially in organs consisting of functional units such as the liver and kidneys.

Beta cell replacement

The building blocks required to orchestrate a controlled and functional engraftment are (i) ECM, (ii) vascularization, and (iii) an improved beta cell mass [49]. Insulin-secreting organoids are 3D complex aggregates that can produce and secrete insulin in a regular manner [50]. It has been demonstrated that differences in islet diameter may differently affect the islet function both in vitro and in vivo [51]. In this scenario, small islets may survive and perform better compared to the bigger ones. This may be explained by the activation of the necrosis of the islet core in the bigger ones. To generate an improved beta cell mass, recently it has been shown that islet could be disassociated and reaggregated to control their size and composition [52]. The final product (referred to as pseudo-islets, PI) showed not only the possibility to have a more homogenous cell preparation but also an improved function compared to the native islet preparation [53]. Despite the option to control PI size, insulin-secreting organoids allow to improve their cell composition to offer additional function by incorporating new cell types. In this context, human amniotic epithelial cells [54], endothelial cells [55], and cholinergic neurons [56] have been used to shield organoids and improve their vascularization and innervation. Thus the possibility to add new function to the original islet preparation by assembly organoid as chimeric new cell aggregates may open to the possibility to damp the immunological reaction against the beta cell after implantation. Indeed by adding or stratifying, in the organoids structure, cells with immunomodulatory and antiinflammatory properties, it would be possible to define an immune-protected environment able to improve long-term beta cell engraftment and function. This has been successfully demonstrated in preclinical models of T1D by coculturing insulin-secreting cells with mesenchymal stem cells [57] and amniotic epithelial cells [27,54], demonstrating improved function and viability in vitro and in vivo. In the beta cell replacement evolution, most of the approaches have been focused on reshaping the scaffold to house the beta cell mass. As of now, organoids are offering the opportunity to reshape the beta cell mass. In this scenario both strategies need to be complementary to overcome the present bottlenecks in clinical beta cell replacement field.

Chip technology

General concepts

Organoids Organs-on-chip (OoC) has recently sparked a heightened amount of interest for its underlying potential to be contributory at multiple stages of organ development research and drug discovery pipelines [58]. Indeed, OoC technology holds promise for delivering a testing and interrogation environment that could potentially replace the complexities of existing animal models. The OoC is a microfluidic-based platform for culturing and growing living cells and organoid substructures in a controlled and coordinated microenvironment. Commonly, OoC platforms recapitulate one or more aspects of organ dynamics, function, and in vivo physiological responses under a real-time monitoring testing condition [59].

Beta cell replacement

Pancreas-on-a-chip (PoC) is a tool that investigates the endocrine part of the pancreas in a microfluidic chip. It is envisioned that it will (a) perform quality control of islets in the postisolation phase; (b) explore functionality of stem cell-derived beta-like cells and compare them with primary islets in a standardized platform; and (c) investigate beta cells' potency and interactions in other tissue types, in a multiorgan chip system.

Leclerc et al. compared PoC to the traditional Petri culture system [60]. The genc expression profile of rats' islets seeded in either a dynamic PoC platform or Petri dishes, and cultivated up to 5 days, was studied. Interestingly, mRNA levels of important pancreatic islet genes such as Ins1, App, Insr, Gcgr, Reg3a, and Neurod were higher in the PoC system. Furthermore, also insulin and glucagon secretions were more pronounced in the PoC.

PoC offers several opportunities to beta cell medicine. Microfluidic devices have been engineered with see-through materials which provide "live" in vitro images of cultured islets during all major culture phases, including glucose response, stress stimuli, and death pathways. Zbinden et al. recently explored the use of the Raman imaging technology applied to PoC [61]. The authors, through a single-channel microfluidic system with passively pumped flow, analyzed human pseudoislets response to glucose stimulation. Raman microspectroscopy allowed to physically trace glucose responsiveness and to visualize different molecular structures such as lipids, mitochondria, or nuclei. Body-on-a-chip (BoC) represents a consolidation of microfluidic technologies, moving from a system designed to investigate a single organ to a more complex solution intended to explore the response of multiple organs to specific conditions. Thus PoC has been integrated in BoC. In 2017 Bauer et al. and colleagues engineered a liver/pancreas-on-chip device connecting hepatic organoids to human islets [62]. This has resulted into a deeper understanding of islet–hepatocyte dynamics and confirmed a feedback loop between human liver organoids and human pancreatic islets. In particular, the authors detailed that as insulin discharged from pancreatic islets increases, a concomitant increase of its uptake into the liver spheroids is observed. However, a reduction in glucose concentration in the chip system led to an immediate shutdown of insulin delivery from the pancreatic islets. Importantly this system was able to coculture the two different cellular populations for up to 15 days, allowing quantifications and essays to be repeated at different time points.

Final remarks

The beta cell replacement community is investing in RM and has undertaken many initiatives to join forces with that field. For example, in 2016, the International Pancreas and Islet Transplant Association (IPITA) launched, in collaboration with the Juvenile Diabetes Research Foundation (JDRF) and the Harvard Stem Cell Institute, a one-of-a-kind conference fully dedicated to the application of stem cell technologies to beta cell replacement; in 2020 this conference series celebrated its third edition despite the COVID pandemic. More broadly and on a larger scale, in January 2021, the American Society of Transplantation (AST) signed a letter of collaboration with the Tissue Engineering and Regenerative Medicine International Society (TERMIS) with the intent to—as explained in the AST website—bringing "*together experts from both fields on the same stage for the first time, in order to share knowledge and ultimately foster progress in organ bioengineering, regeneration and repair which will shape and define the future of both worlds*" (https://www.myast.org/meetings/ast-termis-webinar-joining-forces-shape-our-mutual-future). On the editorial front, transplant journals are publishing more and more RM-oriented manuscripts and transplant societies are establishing committees focusing on RM-related topics like cell therapy or organ bioengineering. Some examples are as follows: AST launched, in 2014, the Transplant Regenerative Medicine Community of Practice. The Cell Transplantation Society rebranded its name into the Cell Transplant and Regenerative Medicine Society, while the European Society of Organ Transplantation (ESOT) established the European Cell Therapy and Organ Regeneration Section (ECTORS) in 2019.

It is clear that RM will change the face of transplant medicine and will revolutionize the way we think and perform beta cell replacement. This book is the inaugural volume of a new book series that aims at providing state-of-the-art information to anyone who has interest in the newest and most exciting concepts that are emerging in the field and will impact soon transplant and beta cell medicine. This volume focuses on the most recent advances in cutting-edge technologies that are impacting beta cell replacement, spanning from biomaterial sciences to xenotransplant and stem cell engineering. What lies behind the horizon cannot be seen in its entirety but can certainly be imagined and will be extraordinary.

References

[1] Orlando G, Soker S, Stratta RJ. Organ bioengineering and regeneration as the new Holy Grail for organ transplantation. Ann Surg 2013;258(2):221–32.

[2] Orlando G, Murphy SV, Bussolati B, Clancy M, Cravedi P, Migliaccio G, Murray P. Rethinking regenerative medicine from a transplant perspective (and vice versa). Transplantation 2019;103(2):237–49.

[3] Badylak SF, Taylor D, Uygun K. Whole-organ tissue engineering: decellularization and recellularization of three-dimensional matrix scaffolds. Annu Rev Biomed Eng 2011;13:27–53.

[4] Orlando G, Dominguez-Bendala J, Shupe T, Bergman C, Bitar KN, Booth C, Carbone M, Koch KL, Lerut JP, Neuberger JM, Petersen B, Ricordi C, Atala A, Stratta RJ, Soker S. Cell and organ bioengineering technology as applied to gastrointestinal diseases. Gut 2013;62(5):774–86.

[5] Orlando G, Wood KJ, De Coppi P, Baptista PM, Binder KW, Bitar KN, Breuer C, Burnett L, Christ G, Farney A, Figliuzzi M, Holmes JHT, Koch K, Macchiarini P, Sani SHM, Opara E, Remuzzi A, Rogers J, Saul JM, Seliktar D, Shapira-Schweitzer K, Smith T, Solomon D, Van Dyke M, Yoo JJ, Zhang Y, Atala A, Stratta RJ, Soker S. Regenerative medicine as applied to general surgery. Ann Surg 2012;255(5):867–80.

[6] Orlando G, Baptista P, Birchall M, De Coppi P, Farney A, Guimaraes-Souza NK, Opara E, Rogers J, Seliktar D, Shapira-Schweitzer K, Stratta RJ, Atala A, Wood KJ, Soker S. Regenerative medicine as applied to solid organ transplantation: current status and future challenges. Transpl Int 2011;24(3):223–32.

[7] Orlando G, Wood KJ, Stratta RJ, Yoo JJ, Atala A, Soker S. Regenerative medicine and organ transplantation: past, present, and future. Transplantation 2011;91(12):1310–7.

[8] Rogers J, Katari R, Gifford S, Tamburrini R, Edgar L, Voigt MR, Murphy SV, Igel D, Mancone S, Callese T, Colucci N, Mirzazadeh M, Peloso A, Zambon JP, Farney AC, Stratta RJ, Orlando G. Kidney transplantation, bioengineering and regeneration: an originally immunology-based discipline destined to transition towards ad hoc organ manufacturing and repair. Expert Rev Clin Immunol 2016;12(2):169–82.

[9] Orlando G, Wood KJ, Soker S, Stratta RJ. How regenerative medicine may contribute to the achievement of an immunosuppression-free state. Transplantation 2011;92(8):e36–8. author reply e39.

[10] Duisit J, Maistriaux L, Taddeo A, Orlando G, Joris V, Coche E, Behets C, Lerut J, Dessy C, Cossu G, Vogelin E, Rieben R, Gianello P, Lengele B. Bioengineering a human face graft: the matrix of identity. Ann Surg 2017;266(5):754–64.

[11] Edgar L, Pu T, Porter B, Aziz JM, La Pointe C, Asthana A, Orlando G. Regenerative medicine, organ bioengineering and transplantation. Br J Surg 2020;107(7):793–800.

[12] Orlando G, Farney AC, Iskandar SS, Mirmalek-Sani SH, Sullivan DC, Moran E, AbouShwareb T, De Coppi P, Wood KJ, Stratta RJ, Atala A, Yoo JJ, Soker S. Production and implantation of renal extracellular matrix scaffolds from porcine kidneys as a platform for renal bioengineering investigations. Ann Surg 2012;256(2):363–70.

[13] Piemonti L, Pileggi A. 25 years of the Ricordi automated method for islet isolation. CellR4 Repair Replace Regen Reprogram 2013;1(1):e128.

[14] Korpos E, Kadri N, Kappelhoff R, Wegner J, Overall CM, Weber E, Holmberg D, Cardell S, Sorokin L. The peri-islet basement membrane, a barrier to infiltrating leukocytes in type 1 diabetes in mouse and human. Diabetes 2013;62(2):531–42.

[15] Daoud J, Petropavlovskaia M, Rosenberg L, Tabrizian M. The effect of extracellular matrix components on the preservation of human islet function in vitro. Biomaterials 2010;31(7):1676–82.

[16] Kuehn C, Vermette P, Fulop T. Cross talk between the extracellular matrix and the immune system in the context of endocrine pancreatic islet transplantation. A review article. Pathol Biol (Paris) 2014;62(2):67–78.

[17] Tomei AA, Manzoli V, Fraker CA, Giraldo J, Velluto D, Najjar M, Pileggi A, Molano RD, Ricordi C, Stabler CL, Hubbell JA. Device design and materials optimization of conformal coating for islets of Langerhans. Proc Natl Acad Sci U S A 2014;111(29):10514–9.

[18] Irving-Rodgers HF, Choong FJ, Hummitzsch K, Parish CR, Rodgers RJ, Simeonovic CJ. Pancreatic islet basement membrane loss and remodeling after mouse islet isolation and transplantation: impact for allograft rejection. Cell Transplant 2014;23(1):59–72.

[19] Miao G, Zhao Y, Li Y, Xu J, Gong H, Qi R, Li J, Wei J. Basement membrane extract preserves islet viability and activity in vitro by up-regulating alpha3 integrin and its signal. Pancreas 2013;42(6):971–6.

[20] Chaimov D, Baruch L, Krishtul S, Meivar-Levy I, Ferber S, Machluf M. Innovative encapsulation platform based on pancreatic extracellular matrix achieve substantial insulin delivery. J Control Release 2016;257:91–101.

[21] Mirmalek-Sani SH, Orlando G, McQuilling JP, Pareta R, Mack DL, Salvatori M, Farney AC, Stratta RJ, Atala A, Opara EC, Soker S. Porcine pancreas extracellular matrix as a platform for endocrine pancreas bioengineering. Biomaterials 2013;34(22):5488–95.

[22] Enck K, Tamburrini R, Deborah C, Gazia C, Jost A, Khalil F, et al. Effect of alginate matrix engineered to mimic the pancreatic microenvironment on encapsulated islet function. Biotechnol Bioeng 2021;118(3):1177–85.

[23] Chaimov D, Baruch L, Krishtul S, Meivar-Levy I, Ferber S, Machluf M. Innovative encapsulation platform based on pancreatic extracellular matrix achieve substantial insulin delivery. J Control Release 2017;257:91–101.

[24] Asthana A, Tamburrini R, Chaimov D, Gazia C, Walker SJ, Van Dyke M, Tomei A, Lablanche S, Robertson J, Opara EC, Soker S, Orlando G. Comprehensive characterization of the human pancreatic proteome for bioengineering applications. Biomaterials 2020;270:120613.

[25] Jiang K, Chaimov D, Patel SN, Liang JP, Wiggins SC, Samojlik MM, Rubiano A, Simmons CS, Stabler CL. 3-D physiomimetic extracellular matrix hydrogels provide a supportive microenvironment for rodent and human islet culture. Biomaterials 2019;198:37–48.

[26] Baidal DA, Ricordi C, Berman DM, Alvarez A, Padilla N, Ciancio G, Linetsky E, Pileggi A, Alejandro R. Bioengineering of an intraabdominal endocrine pancreas. N Engl J Med 2017;376(19):1887–9.

[27] Lebreton F, Lavallard V, Bellofatto K, Bonnet R, Wassmer CH, Perez L, Kalandadze V, Follenzi A, Boulvain M, Kerr-Conte J, Goodman DJ, Bosco D, Berney T, Berishvili E. Insulin-producing organoids engineered from islet and amniotic epithelial cells to treat diabetes. Nat Commun 2019;10(1):4491.

[28] Yue Z, Liu X, Coates PT, Wallace GG. Advances in printing biomaterials and living cells: implications for islet cell transplantation. Curr Opin Organ Transplant 2016;21(5):467–75.

[29] Xie Z, Gao M, Lobo AO, Webster TJ. 3D bioprinting in tissue engineering for medical applications: the classic and the hybrid. Polymers 2020;12(8):1717.

[30] Kang HW, Lee SJ, Ko IK, Kengla C, Yoo JJ, Atala A. A 3D bioprinting system to produce human-scale tissue constructs with structural integrity. Nat Biotechnol 2016;34(3):312–9.

[31] Cui X, Breitenkamp K, Finn MG, Lotz M, D'Lima DD. Direct human cartilage repair using three-dimensional bioprinting technology. Tissue Eng Part A 2012;18(11–12):1304–12.

[32] Markstedt K, Mantas A, Tournier I, Martinez Avila H, Hagg D, Gatenholm P. 3D bioprinting human chondrocytes with nanocellulose-alginate bioink for cartilage tissue engineering applications. Biomacromolecules 2015;16(5):1489–96.

[33] Kundu J, Shim JH, Jang J, Kim SW, Cho DW. An additive manufacturing-based PCL-alginate-chondrocyte bioprinted scaffold for cartilage tissue engineering. J Tissue Eng Regen Med 2015;9(11):1286–97.

[34] Gurlin RE, Giraldo JA, Latres E. 3D bioprinting and translation of beta cell replacement therapies for type 1 diabetes. Tissue Eng Part B Rev 2021;27(3):238–52.

[35] Daoud JT, Petropavlovskaia MS, Patapas JM, Degrandpre CE, Diraddo RW, Rosenberg L, Tabrizian M. Long-term in vitro human pancreatic islet culture using three-dimensional microfabricated scaffolds. Biomaterials 2011;32(6):1536–42.

[36] Caplan AI. Mesenchymal stem cells: time to change the name! Stem Cells Transl Med 2017;6(6):1445–51.

[37] Okita K, Ichisaka T, Yamanaka S. Generation of germline-competent induced pluripotent stem cells. Nature 2007;448(7151):313–7.

[38] Veres A, Faust AL, Bushnell HL, Engquist EN, Kenty JH, Harb G, Poh YC, Sintov E, Gurtler M, Pagliuca FW, Peterson QP, Melton DA. Charting cellular identity during human in vitro beta-cell differentiation. Nature 2019;569(7756):368–73.

[39] Kelly OG, Chan MY, Martinson LA, Kadoya K, Ostertag TM, Ross KG, Richardson M, Carpenter MK, D'Amour KA, Kroon E, Moorman M, Baetge EE, Bang AG. Cell-surface markers for the isolation of pancreatic cell types derived from human embryonic stem cells. Nat Biotechnol 2011;29(8):750–6.

[40] Kroon E, Martinson LA, Kadoya K, Bang AG, Kelly OG, Eliazer S, Young H, Richardson M, Smart NG, Cunningham J, Agulnick AD, D'Amour KA, Carpenter MK, Baetge EE. Pancreatic endoderm derived from human embryonic stem cells generates glucose-responsive insulin-secreting cells in vivo. Nat Biotechnol 2008;26(4):443–52.

[41] Deinsberger J, Reisinger D, Weber B. Global trends in clinical trials involving pluripotent stem cells: a systematic multi-database analysis. NPJ Regen Med 2020;5:15.

[42] Bartlett ST, Markmann JF, Johnson P, Korsgren O, Hering BJ, Scharp D, Kay TW, Bromberg J, Odorico JS, Weir GC, Bridges N, Kandaswamy R, Stock P, Friend P, Gotoh M, Cooper DK, Park CG, O'Connell P, Stabler C, Matsumoto S, Ludwig B, Choudhary P, Kovatchev B, Rickels MR, Sykes M, Wood K, Kraemer K, Hwa A, Stanley E, Ricordi C, Zimmerman M, Greenstein J, Montanya E, Otonkoski T. Report from IPITA-TTS opinion leaders meeting on the future of beta-cell replacement. Transplantation 2016;100(Suppl 2):S1–44.

[43] Matsunari H, Nagashima H, Watanabe M, Umeyama K, Nakano K, Nagaya M, Kobayashi T, Yamaguchi T, Sumazaki R, Herzenberg LA, Nakauchi H. Blastocyst complementation generates exogenic pancreas in vivo in apancreatic cloned pigs. Proc Natl Acad Sci U S A 2013;110(12):4557–62.

[44] Chang AN, Liang Z, Dai HQ, Chapdelaine-Williams AM, Andrews N, Bronson RT, Schwer B, Alt FW. Neural blastocyst complementation enables mouse forebrain organogenesis. Nature 2018;563(7729):126–30.

[45] Wu J, Greely HT, Jaenisch R, Nakauchi H, Rossant J, Belmonte JC. Stem cells and interspecies chimaeras. Nature 2016;540(7631):51–9.

[46] Yamaguchi T, Sato H, Kato-Itoh M, Goto T, Hara H, Sanbo M, Mizuno N, Kobayashi T, Yanagida A, Umino A, Ota Y, Hamanaka S, Masaki H, Rashid ST, Hirabayashi M, Nakauchi H. Interspecies organogenesis generates autologous functional islets. Nature 2017;542(7640):191–6.

[47] Sato T, Vries RG, Snippert HJ, van de Wetering M, Barker N, Stange DE, van Es JH, Abo A, Kujala P, Peters PJ, Clevers H. Single Lgr5 stem cells build crypt-villus structures in vitro without a mesenchymal niche. Nature 2009;459(7244):262–5.

[48] Dayem AA, Lee SB, Kim K, Lim KM, Jeon TI, Cho SG. Recent advances in organoid culture for insulin production and diabetes therapy: methods and challenges. BMB Rep 2019;52(5):295–303.

[49] Citro A, Ott HC. Can we re-engineer the endocrine pancreas? Curr Diab Rep 2018;18(11):122.

[50] Wassmer CH, Lebreton F, Bellofatto K, Bosco D, Berney T, Berishvili E. Generation of insulin-secreting organoids: a step toward engineering and transplanting the bioartificial pancreas. Transpl Int 2020;33(12):1577–88.

[51] Lehmann R, Zuellig RA, Kugelmeier P, Baenninger PB, Moritz W, Perren A, Clavien PA, Weber M, Spinas GA. Superiority of small islets in human islet transplantation. Diabetes 2007;56(3):594–603.

[52] Wassmer CH, Bellofatto K, Perez L, Lavallard V, Cottet-Dumoulin D, Ljubicic S, Parnaud G, Bosco D, Berishvili E, Lebreton F. Engineering of primary pancreatic islet cell spheroids for three-dimensional culture or transplantation: a methodological comparative study. Cell Transplant 2020;29, 963689720937292.

[53] Yu Y, Gamble A, Pawlick R, Pepper AR, Salama B, Toms D, Razian G, Ellis C, Bruni A, Gala-Lopez B, Lu JL, Vovko H, Chiu C, Abdo S, Kin T, Korbutt G, Shapiro AMJ, Ungrin M. Bioengineered human pseudoislets form efficiently from donated tissue, compare favourably with native islets in vitro and restore normoglycaemia in mice. Diabetologia 2018;61(9):2016–29.

[54] Lebreton F, Bellofatto K, Wassmer CH, Perez L, Lavallard V, Parnaud G, Cottet-Dumoulin D, Kerr-Conte J, Pattou F, Bosco D, Othenin-Girard V, Martinez de Tejada B, Berishvili E. Shielding islets with human amniotic epithelial cells enhances islet engraftment and revascularization in a murine diabetes model. Am J Transplant 2020;20(6):1551–61.

[55] Urbanczyk M, Zbinden A, Layland SL, Duffy G, Schenke-Layland K. Controlled heterotypic pseudo-islet assembly of human beta-cells and human umbilical vein endothelial cells using magnetic levitation. Tissue Eng Part A 2020;26(7–8):387–99.

[56] Arzouni AA, Vargas-Seymour A, Dhadda PK, Rackham CL, Huang GC, Choudhary P, King AJF, Jones PM. Characterization of the effects of mesenchymal stromal cells on mouse and human islet function. Stem Cells Transl Med 2019;8(9):935–44.
[57] Ito T, Itakura S, Todorov I, Rawson J, Asari S, Shintaku J, Nair I, Ferreri K, Kandeel F, Mullen Y. Mesenchymal stem cell and islet co-transplantation promotes graft revascularization and function. Transplantation 2010;89(12):1438–45.
[58] Low LA, Mummery C, Berridge BR, Austin CP, Tagle DA. Organs-on-chips: into the next decade. Nat Rev Drug Discov 2021;20(5):345–61.
[59] Marx U, Akabane T, Andersson TB, Baker E, Beilmann M, Beken S, Brendler-Schwaab S, Cirit M, David R, Dehne EM, Durieux I, Ewart L, Fitzpatrick SC, Frey O, Fuchs F, Griffith LG, Hamilton GA, Hartung T, Hoeng J, Hogberg H, Hughes DJ, Ingber DE, Iskandar A, Kanamori T, Kojima H, Kuehnl J, Leist M, Li B, Loskill P, Mendrick DL, Neumann T, Pallocca G, Rusyn I, Smirnova L, Steger-Hartmann T, Tagle DA, Tonevitsky A, Tsyb S, Trapecar M, Van de Water B, Van den Eijnden-van Raaij J, Vulto P, Watanabe K, Wolf A, Zhou X, Roth A. Biology-inspired microphysiological systems to advance patient benefit and animal welfare in drug development. ALTEX 2020;37(3):365–94.
[60] Essaouiba A, Okitsu T, Jellali R, Shinohara M, Danoy M, Tauran Y, Legallais C, Sakai Y, Leclerc E. Microwell-based pancreas-on-chip model enhances genes expression and functionality of rat islets of Langerhans. Mol Cell Endocrinol 2020;514:110892.
[61] Zbinden A, Marzi J, Schlunder K, Probst C, Urbanczyk M, Black S, Brauchle EM, Layland SL, Kraushaar U, Duffy G, Schenke-Layland K, Loskill P. Non-invasive marker-independent high content analysis of a microphysiological human pancreas-on-a-chip model. Matrix Biol 2020;85–86:205–20.
[62] Bauer S, Wennberg Huldt C, Kanebratt KP, Durieux I, Gunne D, Andersson S, Ewart L, Haynes WG, Maschmeyer I, Winter A, Ammala C, Marx U, Andersson TB. Functional coupling of human pancreatic islets and liver spheroids on-a-chip: towards a novel human ex vivo type 2 diabetes model. Sci Rep 2017;7(1):14620.

Index

Note: Page numbers followed by *f* indicate figures and *t* indicate tables.

Printed in the United States
by Baker & Taylor Publisher Services